Foundations of Molecular Pharmacology

Volume 2

The Chemical Basis of Drug Action

This text has emerged from thirty years of teaching undergraduate courses and conducting research in medicinal and pharmaceutical chemistry. It is conceived essentially as a foundation course in the basic principles of organic chemistry applied to the study of medicinal agents and the formulations in which they are used. The basic philosophy underlying the text is that those concerned with the design and use of drugs and medicines are interested fundamentally in properties rather than in methods of manufacture and Professor Stenlake draws widely on drugs in common daily use for his examples.

Volume 1 provides a systematic coverage of basic medicinal and pharmaceutical chemistry and is primarily intended as a foundation course for undergraduates studying for a degree in Medicinal Chemistry and Pharmacy.

Volume 2 adopts a more broadly based approach and develops the study of some of the more general chemical factors which determine drug action. It is intended for specialists in Medicinal Chemistry and Pharmacy and should appeal also to Clinical Pharmacologists.

Foundations of Molecular Pharmacology

Volume 2

The Chemical Basis of Drug Action

Foundations of Molecular Pharmacology

Volume 2

The Chemical Basis of Drug Action

J. B. STENLAKE

PH.D., D.SC., F.P.S., C. CHEM., F.R.I.C., F.R.S.E.

Professor of Pharmacy and Pharmaceutical Chemistry
University of Strathclyde, Glasgow

THE ATHLONE PRESS *of the University of London* 1979

Published by
THE ATHLONE PRESS
UNIVERSITY OF LONDON
4 *Gower Street, London* WC1

Distributed by Tiptree Book Services Ltd
Tiptree, Essex

USA and Canada
Humanities Press Inc
New Jersey

British Library Cataloguing in Publication Data
Stenlake, John Bedford
Foundations of molecular pharmacology
Vol. 2: The chemical basis of drug action
1. Chemistry, Medical and pharmaceutical
2. Drugs
I. Title
615'.19 RS421

ISBN 0 485 11172 1

Set in Monophoto Times by
COMPOSITION HOUSE LTD, SALISBURY

Printed by photolithography in Great Britain by
WHITSTABLE LITHO LTD, WHITSTABLE, KENT

Preface

This text has emerged from some thirty years of teaching undergraduate courses and conducting research in medicinal and pharmaceutical chemistry. It is conceived essentially as a foundation course in the basic principles of organic chemistry applied to the study of medicinal agents and the formulations in which they are used. It is intended primarily to cater for the needs of undergraduate students of pharmacy and medicinal chemistry up to Honours level. References to original papers, however, should extend its use to postgraduate students and others engaged in the search for new drugs.

My intention was to contain the text within the covers of a single volume, concentrating essentially on the fundamental groundwork chemistry which must of necessity be taught in any undergraduate course. Experience, however, has shown the value of more general discussion of certain selected topics of wide general applicability in the study of drug action, and it was always my objective to conclude the book in this way. In the event, I have been defeated, partly by the ramifications of the subject, but mainly by my enthusiasm and attempts to achieve a realistic degree of coverage. Publishing costs, too, have risen enormously in the ten years of writing. Attempts to overcome the twin difficulties of coverage and cost, therefore, left no alternative other than to divide the book between two volumes.

It is just possible that some readers may find virtue in the necessity which has forced this publication of the Foundations of Molecular Pharmacology in two separate volumes. I hope, nonetheless, that serious students will not be deterred by this somewhat artificial division from pursuing the broader approach to the subject contained in Volume 2. In order, therefore, to reinforce the continuity of the subject, I have provided a system of cross-referencing between chapters, both within and between the two volumes. Such cross-references are denoted by two numbers, the first indicating volume, and the second chapter; thus, for example, (**1**, 13) indicates Volume 1, Chapter 13, and (**2**, 5) Volume 2, Chapter 5.

The basic philosophy underlying the text is that those concerned with the design and use of drugs and medicines are interested fundamentally in properties rather than in methods of manufacture. Accordingly, the chemistry in this book almost entirely ignores the synthesis of medicinal agents. Instead, attention is focused, in Volume 1, on the physical and chemical properties of medicinal agents, pharmaceutical additives and cellular components, that determine the way in which they interact with each other. To achieve this end, substantial accounts of relevant intermediary tissue metabolism, drug transport and metabolism, and other factors affecting both stability and availability of drugs from dosage forms have been brought together in the general body of the text. This approach emphasises the close similarity between chemical and biochemical transformations, and should help to give students and others engaged in the design of new drugs a better understanding of the fundamental mechanisms which control interactions between drugs and body chemistry.

The more general, but essentially similar approach to the Chemical Basis of Drug Action adopted in Volume 2, which reinforces the basic principles for the specialist, should also appeal in its own right to clinical pharmacologists and others whose interests lie rather more in the action and use of drugs than in their design.

Since this book is designed to assist in the education of students, many of whom will be engaged in later life in the handling and use of drugs in practice, I have deliberately chosen to draw my examples from drugs in current use in western medicine. My text, however, is essentially British, and British Approved Names, denoted by italics, are used throughout, notwithstanding the difficulties that this may make for North American readers. Fortunately, British and American drug nomenclature is convergent, but where important and confusing differences still exist, I have endeavoured to overcome them by also giving the United States Adopted Name.

It is an unfortunate fact of life that the vast majority of modern drugs have chemical structures which are infinitely more complex than those of the simple examples commonly used in most textbooks of organic chemistry. Indeed, their very complexity frequently presents an educational hurdle, so that students of medicinal and pharmaceutical chemistry often fail to grasp the essential simplicity of drug action mechanisms and transformations. I am, therefore, most grateful to the publishers for their help and co-operation in the use of printing devices involving bold type and colour to focus attention on the simple stepwise transformations of otherwise complex compounds.

I am very much indebted to my colleagues, past and present, and friends, who between them provided the stimulus to write this book, and all those who, once I was embarked upon it, so patiently answered my questions, and helped to resolve the many problems I inevitably encountered. I am especially grateful to Dr G. A. Smail and Dr R. E. Bowman, both of whom read the entire original draft and commented so helpfully upon it. I am sure others will still find errors, oversights and misconceptions, but there would have been many more without the help of these two colleagues. For similar reasons, I am also grateful for all the many valuable comments and criticisms I received from the Athlone Press's own anonymous referees. My most grateful thanks are also due to Tom Moody for help with the preparation of diagrams, to Dr N. C. Dhar for assistance in locating and checking references, and especially to my ever willing Secretary, Mrs Sylvia Cohen, for her invaluable help in typing the manuscript, for countless hours devoted to the dull routine of checking text and references at every stage right through to the final proofs, and for her help in compiling the index.

The time I have taken to write this book has been taken away from many things I might otherwise have done, and most of all, taken from my wife, Anne, and our family. Their tolerance and support made it possible. I have tried to make this book one that they, too, can be proud of, and worthy of the hours of pleasure in their company which I have sacrificed.

1978 John B. Stenlake

Acknowledgements

Permission to reproduce the following Figures is gratefully acknowledged:

Fig. 10, D. H. Meadows and O. Jardetzky, and *Proceedings of the National Academy of Sciences*, USA; Figs. 32 and 43, B. B. Brodie and C. A. M. Hogben, and *Journal of Pharmacy and Pharmacology*; Fig. 33, T. R. D. Shaw, J. E. Carless, M. R. Howard and K. Raymond, and *The Lancet*; Fig. 34, A. T. Florence, E. G. Salole and J. B. Stenlake, and *Journal of Pharmacology*; Fig. 35, J. R. Marvel, D. A. Schlichting, C. Denton, E. J. Levy and M. M. Cahn, and *Journal of Investigative Dermatology*; Figs. 36 and 37, C. L. Gantt, N. Gochman and J. M. Dyniewicz, and *The Lancet*; Fig. 38, A. H. Beckett and A. C. Moffat, and *Journal of Pharmacy and Pharmacology*; Fig. 44, A. H. Beckett and M. Roland, and *Journal of Pharmacy and Pharmacology*.

Contents

Contents of Volume 1

1 The Characteristics of Drug–Receptor Interaction

SPECIFIC AND NON-SPECIFIC DRUG ACTION

Introduction

The action of drugs at the molecular level may be broadly classified in one of two ways depending upon whether it is determined essentially by the presence and arrangement of specific chemical groupings, or not. Drugs may, therefore, be classified according to their mode of action as either **structurally specific** or **structurally non-specific**.

Structurally specific action depends, as it implies, on the presence in the drug molecule of specific chemical groupings usually arranged in some specific spatial relationship with respect to each other. Structurally non-specific drug action, on the other hand, is a manifestation of a specific but purely physical property, which may arise in compounds of quite diverse chemical properties.

Structurally Non-specific Action

The majority of drugs have structurally specific actions, but a few groups of chemically-unrelated compounds are capable of inducing identical and sometimes intense pharmacological reactions, due to the possession of a common physical property unrelated to any one particular chemical structure. The action is therefore **structurally non-specific**. In contrast to the structurally specific compounds, small changes of chemical structure have little or no effect on the nature and intensity of the pharmacological response which they elicit. Also, their action is usually directly related to their **thermodynamic activity**, which is high, implying that their action is largely a function of concentration.

Ferguson's Principle and the Non-specific Action of General Anaesthetics

Ferguson (1939) reasoned that toxic concentrations of drugs are reached by a series of distributions between unspecified heterogeneous biophases, linking the external circumambient phase and the particular biophase which is the site of drug action. We thus have a series of compartments, unspecified in nature and number, interpolated between the external circumambient phase and the biophase, each one in thermodynamic equilibrium with the next. Since all the compartments are in equilibrium, the thermodynamic activity of the drug is the same in each compartment. Measurement of the thermodynamic activity of a drug in the external phase, therefore, provides a measure of its thermodynamic activity in the biophase. Thus, measurement of the concentration of a gaseous anaesthetic in alveolar air or of the solubility of a non-volatile hypnotic in blood plasma will provide a measure of its thermodynamic activity in the tissues of the central nervous system.

Ferguson (1939) postulated that the same degree of biological activity would be produced by the non-specific action of different compounds at the same level of thermodynamic activity, i.e. when the same relative saturation of the biophase is reached. He tested the toxicity of a variety of substances in wireworms and mice (**1**, 9), and showed that although the actual lethal concentrations varied by a factor of 10^4, the index of thermodynamic activity (P_t/P_s, where P_t is the partial vapour pressure at the toxic concentration, and P_s is the saturation vapour pressure) at lethal concentrations in the biophase approximated more closely to a constant figure. The same applies to the activity of general anaesthetics in man, taking the ratio of the anaesthetic vapour pressure in the inhaled gas mixture to the saturation vapour pressure as the measurement of thermodynamic activity. Thus, Eger and his collaborators (1965) have shown that measurement of equipotent alveolar concentrations of the anaesthetics, *Methoxyfluorane*, *Chloroform*, *Halothane*, *Ether*, *Cyclopropane*, xenon and *Nitrous Oxide*, correspond more closely to the oil/gas partition coefficient than any other physical constant. They have also shown (Eger, Saidman and Brandstater, 1965) that the minimum alveolar concentration of halothane and cyclopropane to prevent movement in dogs in response to painful stimulation varies linearly with temperature. Enthalpies of absorption calculated from the experimental results correlate well with enthalpies for the absorption of these anaesthetics by lipoprotein surface films. The results, therefore, are in agreement with Ferguson's principle.

The discovery that inert gases, such as nitrogen and helium, under pressure produced anaesthesia (Behnke and Yarbrough, 1938, 1939) led to the examination of the more highly fat-soluble inert gases, krypton and xenon, as potentially useful anaesthetics in mice (Lawrence, Loomis, Tobias and Turpin, 1946), and eventually to the demonstration of the anaesthetic activity of xenon in man by Cullen and Gross (1951). Administered with oxygen (20%), xenon gives rapid induction of anaesthesia, good muscle relaxation and rapid recovery, demonstrating as an inert gas the purely physical nature of the drug–biophase interaction.

A number of theories have been proposed concerning the mechanism of anaesthetic action. Wulf and Featherstone (1957) suggested that anaesthetic activity was due to displacement of the bimolecular phospholipid leaflets, which are thought to form the basic structure of cellular membranes (Danielli and Davson, 1935). Thus, it was suggested that any substance of molecular volume greater than that of oxygen or water, which normally separate the protein and phospholipid layers, is capable of physically disorientating the membrane structure, and in this way initiating anaesthetic action. The constant, *b*, which derives from the van der Waals equation

$$\left(P + \frac{a}{V^2}\right)(V - b) = RT$$

Table 1. Van der Waals Molecular Volume Constants (b) and Anaesthetic Activity

	Van der Waals Constant (b)[1] ($l\,mol^{-1} \times 10^2$)	Minimum Alveolar[2,3] Conc. (MAC) (Atmospheres)	Oil/gas partition[3] coefficient	MAC × Oil/gas[3]
Nitrous Oxide	4.4	1.88	1.4	2.63
Xenon	5.1	1.19	1.9	2.26
Cyclopropane	7.5	0.175	11.8	2.06
Chloroform	10.2	0.0077	265	2.08
Ether	13.2	0.030	65	1.95
Halothane		0.0087	224	1.95
Methoxyflurane		0.0023	970	2.23

[1] Wulf and Featherstone (1957)
[2] Minimum alveolar concentration (MAC) in atmospheres to prevent movement in response to painful stimulation (Eger, Brandstater, Saidman, Regan, Severinghaus and Munson, 1965)
[3] Eger, Lundgren, Miller and Stevens (1969)

gives a measure of molecular volume, and is greater in the case of most common anaesthetics (Table 1) than that of water ($3.047 \times 10^{-2}\,l\,mol^{-1}$) or oxygen ($3.183 \times 10^{-2}\,l\,mol^{-1}$), increasing approximately in order of the anaesthetic activity, as determined by Eger and his collaborators.

Pauling (1961) proposed an alternative theory in which it is suggested that the action of general anaesthetics is due to ordering of adjacent water molecules by the anaesthetic in the central nervous system. Anaesthetics are considered either to form hydrates or to promote the formation of clathrates in which the ions responsible for nerve conduction become trapped. Some doubt is cast on the theory by the failure of Eger and Shargel (1969) to demonstrate the formation of hydrates in aqueous anaesthetic mixtures containing *Methoxyfluorane*, *Halothane* or *Ether* at 0°, although cyclopropane forms two distinct hydrates under the same conditions. Furthermore, there is also a lack of correlation between hydrate dissociation pressures and the minimum alveolar concentration (MAC) of anaesthetics required to prevent movement in response to painful stimulation (Eger, Lundgren, Miller and Stevens, 1969). Similar reservations attend the iceberg theory (Miller, 1961), which proposes the formation of microcrystalline water around the anaesthetic molecules at the site of action. Eger, Lundgren, Miller and Stevens (1969), however, did find good agreement between MAC and the oil/gas partition coefficient of anaesthetics (Table 1) in agreement with the Ferguson principle.

Non-specific Bactericidal Activity of Long-chain Cationic Detergents

There is now substantial evidence to show that the bactericidal action of long-chain cationic detergents is structurally non-specific, and purely a function of their surfactant properties. Thus, the bactericidal activity increases with decreased critical micelle concentration (CMC) within a series of long-chain

Table 2. Bactericidal and Thermodynamic Activities of Long-chain Quaternary Ammonium Salts

Structure $(R-\overset{+}{N}R^1R^2R^3)$				CMC	Minimum Inhibitory Conc. ($\times 10^8$)		Thermodynamic Activity	
R	R^1	R^2	R^3	(N)	*Staph. aureus*	*E. coli*	*Staph. aureus*	*E. coli*
$C_{12}H_{25}$	Me	Me	Me	0.0228	7.50	7.50	0.033	0.033
$C_{12}H_{25}$	Me	Me	Et	0.0213	7.50	7.50	0.030	0.030
$C_{12}H_{25}$	Me	Me	Et	0.0199	7.50	7.50	0.038	0.038
$C_{12}H_{25}$	Et	Et	Et	0.0193	7.50	7.50	0.039	0.039
$C_{12}H_{27}$	Me	Me	Me	0.0112	2.50	2.50	0.022	0.022
$C_{14}H_{29}$	Me	Me	Me	0.0058	0.75	0.75	0.014	0.014
$C_{16}H_{33}$	Me	Me	Me	0.0015	0.75	0.75	0.050	0.050

quaternary ammonium salts (Cella, Eggenberger, Noel, Harriman and Harwood, 1952). In an extension of the concept that solutions in equilibrium with excess solid have the same thermodynamic activity, to the equilibrium between solutions with micelles, Ecanow and Siegel (1963) concluded that the thermodynamic activity should be equal to a constant fraction of the CMC. They have further shown that the ratio of minimum inhibitory concentrations against *Staphylococcus aureus* and *Escherichia coli* to the CMC is virtually constant (with one exception) for the group of long-chain quaternary compounds studied by Cella and his collaborators (1952), despite the fact that there is a fifteen-fold variation in their respective CMC's (Table 2).

The conclusion that the bactericidal action of long-chain quaternary ammonium salts is structurally non-specific has been confirmed by Weiner, Hart and Zografi (1965), and by Laycock and Mulley (1970).

Structurally Specific Action

Most drugs act by specific intermolecular interaction(s) with bio-receptor molecules. Specificity of action is determined by a precise combination and steric arrangement of chemical groups in the drug molecule which facilitate interaction by chemical bonding and physical interaction with appropriate bio-receptor molecules. The nature and relevance of such interactions is discussed in Chapter 2. Since the forces involved in chemical bonding and physical attraction between molecules are essentially short-range forces, the question of fit between the specific active groups of the drug and the complementary groups on the bio-receptor is highly relevant. Molecular size, overall molecular shape, spacing between essential structural features and the relative orientation of essential groups in a drug molecule, are just as important as criteria of potency as the availability or deficiency of electrons for reaction with the receptor molecule which they complement. The significance of such factors is considered in Chapter 3.

The Nature of Drug—Specific Receptor Interactions

The precise target molecule and the ultimate molecular interaction mechanism is known only in the case of a very few drugs. In some cases, it is possible to identify a triggering reaction, which is clearly linked to the observable pharmacological response produced by the drug, even when the subsequent sequence of reactions is not clear. In other cases, there is still only a confusing number of possible and plausible alternative sites and mechanisms of action. Ariens (1964), however, introduced the concepts of **affinity** and **intrinsic activity** to distinguish between molecular interactions which primarily result in binding of the drug to the receptor (affinity), and those which result in triggering a pharmacological response (intrinsic activity). For some drugs, the target molecule is an enzyme, but whether this is so or not, there is a close parallel between drug–receptor, enzyme–substrate, and enzyme–inhibitor reactions. For this reason, the nature and properties of enzymes will be considered (p. 12).

MATHEMATICAL ANALYSIS OF STRUCTURE–ACTION RELATIONSHIPS

Linear Physico-chemical—Activity Relationships

It is well-known that small changes in the chemical structure of structurally specific drugs lead to corresponding, and occasionally, marked changes in the pharmacological response which they elicit. Minor modifications of chemical structure have frequently been used to establish correlations between particular structural features of a drug and its pharmacological action. In this way, it is often possible to delineate the optimum molecular feature to produce a particular type of response. This approach, however, is empirical and fails to distinguish readily the relative influence of electronic, steric, and physical interactions of individual substituents. In order to overcome this difficulty, Hansch and Fujita (1964) adapted the Hammett equation (**1**, 11), which has been used extensively to establish correlations between structure and reactivity of aromatic compounds, to the analysis of structure–action relationships in molecular pharmacology.

In considering the action of a drug on a living system in which only two parameters can be measured, the dose administered and the response, Hansch and Fujita assume that only one reaction is rate-determining. This will be merely one of several processes which include absorption, protein binding, fat deposition and metabolism, in addition to that at the specific reaction site.

Thus, the rate at the critical reaction site can be expressed as:

$$\text{Rate of biological response} = \frac{\mathrm{d(response)}}{\mathrm{d}t} = ACK_{\mathrm{X}} \tag{1}$$

where A = the probability of a molecule reaching the specific reaction site in the time interval dt

C = the administered dose

K_{X} = the equilibrium constant for the rate-determining step.

The absorption and transport of drugs is controlled largely by lipid solubility, and transport rates in general correlate with the logarithm of their partition coefficient (log P) (Collander, 1954; Milborrow and Williams, 1968). Hansch and Fujita (1964) therefore expressed the probability term (A) as log P, the logarithm of the partition coefficient of the drug between an organic solvent (octanol) and water. Changes in the probability factor with substitution in a series of compounds can, therefore, be expressed as a **substituent constant**, π, defined by analogy with the Hammett substituent constant as

$$\pi_{\mathrm{X}} = \log\left[\frac{P_{\mathrm{X}}}{P_{\mathrm{H}}}\right] \tag{2}$$

where P_{X} = the partition coefficient of the substituted compound, and
P_{H} = the partition coefficient of the unsubstituted compound (i.e. X = H).

On the basis of evidence that the partition coefficients for a group of related compounds, such as those based on *Chloramphenicol* (Hansch, Muir, Fujita, Maloney, Geiger and Streich, 1963), normally exhibit an optimum partition coefficient for activity (P_0), and on the assumption that there is a normal Gaussian distribution of partition coefficients about the optimum, the probability factor A may be expressed as

$$A = f(\pi) = a\mathrm{e}^{-(\pi-\pi_0)^2/b} \tag{3}$$

where π_0 is the value of the substituent constant for optimum activity.

Hence, from eq. (1),

$$\frac{\mathrm{d(response)}}{\mathrm{d}t} = a\mathrm{e}^{-(\pi-\pi_0)^2/b}CK_{\mathrm{X}} \tag{4}$$

If the applied concentration of drug, C, be measured in terms of the concentration required to elicit a constant response in a fixed time interval, then from eq. (4)

$$\frac{\mathrm{d(response)}}{\mathrm{d}t} = 0$$

hence

$$\log\frac{1}{C} = k'\pi\pi_0 \quad k\pi^2 - k''\pi_0^2 + \log K_{\mathrm{X}} + k''' \tag{5}$$

π_0 at the optimum value (log P_0) of log P is a constant.

Assuming K_{X} depends on electron availability, substitution for K_{X} in the Hammett equation gives the general Hansch equation

$$\log\frac{1}{C} = k_1\pi - k_2\pi^2 + \rho\sigma + k_3 \tag{6}$$

where k_1, k_2 and k_3 are constants for the system determined by regression analysis, σ is the substituent constant and ρ the reaction constant as in the Hammett equation (**1**, 11).

Since log P and hence π (equation 2) are free energy terms, they are additive, and

$$\log P = \sum_{1}^{n} \pi$$

It is possible, therefore, to calculate log P and π values by addition of the corresponding values for their separate components (Hansch, Quinlan and Lawrence, 1968; Bird and Marshall, 1967). Thus, the π_X for the methylene unit can be calculated from the log P values of a pair of barbiturates, such as *Barbitone* (Barbital; log P 0.65) and *Butobarbitone* (Butethal; log P 1.65), which differ only in the constitution of one of the alkyl substituents.

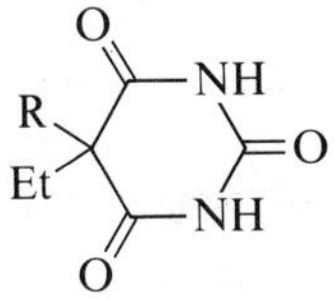

Barbitone R = $CH_3 \cdot CH_2$—
Butobarbitone R = $CH_3 \cdot CH_2 \cdot CH_2 \cdot CH_2$—

The difference in log P values (Δ log P) is 1.0 for two methylene units. The methylene (and methyl) group therefore has a π_X value of 0.5, and from this it is readily possible to calculate that the barbiturate ring has a π_X value of -1.35 [i.e. $0.65 - (4 \times 0.5)$].

Measurements of log P values have now been made on large numbers of different compounds from which tables of log P and π_X values have been compiled using methods similar to that illustrated above (Tute, 1971).

Applications of Hansch Analysis

The π^2 term in the general Hansch equation (6) is only essential in complex biological systems, where a large number of compartments (reaction steps) exist between the point of administration and the critical reaction site. In simpler systems, such as drug–protein binding, where transport from a remote site of administration is not involved, the best fit of data is obtained in equations which do not involve a π^2 term. Thus, in the binding of penicillins to human serum albumin (Bird and Marshall, 1967), an excellent fit of data for some 79 penicillins was obtained with equation (7), the correlation coefficient (r) being 0.924 and the variance about the mean value of log B/F, $S^2 = 0.66$

$$\log\left[\frac{\text{Bound penicillin}}{\text{Free penicillin}}\right] = 0.504\Sigma\pi - 0.665 \qquad (7)$$

The excellent fit with this equation, with but a single π term implies that whilst the extent of binding is dependent on the lipid–water partition coefficient of the penicillin, it relates to the hydrophobic character of the penicillin side-chain and its interaction with hydrophobic binding sites of the protein.

The significance of the π^2 term in determining the importance of transport processes for the concentration of a drug at its specific reaction site was demonstrated in an analysis of the distribution of a series of benzeneboronic acids in mice (Soloway, Whitman and Messer, 1960). The rates of accumulation of boron in brain tissue (C_b) and tumour tissue (C_t) within 15 min of injection were found to give the best fit in equations (8) and (9) respectively (Hansch, Steward and Iwasa, 1965a).

$$\log C_b = -0.540\pi^2 - 0.765\pi + 1.505 \tag{8}$$

$$\log C_t = -0.130\pi^2 - 0.029\pi - 0.405\sigma + 1.342 \tag{9}$$

The ideal partition coefficient ($\log P_0$) for penetration of brain tissue was found from equation (8) to be about 2.3. This compares closely with the value for $\log P_0$ found for optimum hypnotic activity (*ca* 2.0) in a large series of barbiturates and non-barbiturate hypnotics (Hansch, Steward, Anderson and Bently, 1967). In contrast, equation (9) which gives the best fit of data relating to tumour tissue, is significant in both π^2 and σ terms. From this, it was suggested that substituents with π values between -1.0 and -2.0 should favour concentration in tumour tissue. Additionally, dependence on the σ term in the equation indicates a degree of structural specificity which is favoured by the presence of electron-releasing substituents (negative σ values).

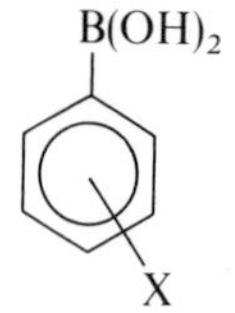

Benzeneboronic acids

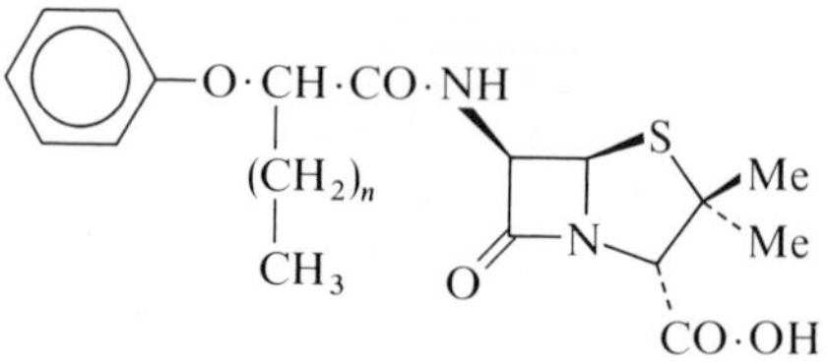

Phenoxymethylpenicillins

In contrast, an analysis of the minimum inhibitory concentrations of a series of phenoxymethylpenicillins against *Staphylococcus aureus* infections in mice (Gourevitch, Hunt and Lein, 1960), showed a more limited degree of dependence on the π^2 function, despite the complexity of the system. Three equations (10), (11) and (12) were generated by Hansch and Steward (1964) from computations of least squares fit of data.

	r(correlation coefficient)	s	
$\log \frac{1}{C} = 0.053\pi^2 - 0.610\pi + 0.019\sigma + 5.71$	0.918	0.192	(10)
$\log \frac{1}{C} = 0.055\pi^2 - 0.613\pi + 5.756$	0.918	0.187	(11)
$\log \frac{1}{C} = 0.445\pi + 5.673$	0.909	0.191	(12)

Comparison of equation (11) with equation (10) demonstrates that the σ term is unimportant, and hence that electronic effects of substituents in the phenoxy ring contribute little to the activity of the antibiotics except insofar as they influence the partition coefficient. The fact, however, that almost as good correlations are obtained with equation (12), which lacks the π^2 term and attributes an apparently linear relationship to the $\Sigma\pi$ values, and the biological response, must imply that the log P values for the series are relatively remote from that of the optimum (log P_0) and, hence, lie on that part of the distribution parabola which is virtually linear. The negative sign of the coefficient in π in equation (12) suggests that more active compounds would be obtained with substituents having negative values of π.

Other examples of Hansch analysis of linear relationships between lipophilic character and biological response have been summarised by Tute (1971) and Hansch and Dunn (1972).

Theoretical model-based equations have also been developed for the correlation of linear-free energy relationships with biological activity in ionisable substances (Martin and Hackbarth, 1976). The models, which comprise a series of aqueous and non-aqueous compartments, give rise to log $(1/C)$ *vs* log P curves which may be asymptotic, linear, or two-part consisting of two lines of unequal slope. Such equations show:

(a) whether the ion or neutral form of the drug is the active species,
(b) whether there is hydrophobic bonding to the receptor,
(c) the presence of inner compartments.

Parabolic Physico-chemical—Activity Relationships

A number of non-linear, parabolic relationships between lipophilic character and biological properties have also been analysed by Hansch and Clayton (1973). Non-linear effects arise for a variety of reasons, and may be due to kinetic (Penniston, Beckett, Bentley and Hansch, 1969) or thermodynamic factors (Higuchi and Davis, 1970). Other factors contributing to parabolic relationships include bulk tolerance when the active site is unable to accommodate the bulkier substituents of higher molecular weight members of a series; increasingly greater conformational distortions of the active site by successive members of a series due to physico-chemical interactions; micelle formation; limiting solubilities, and inconstant metabolism within a series.

Molecular Connectivity

More precise analysis of both linear and parabolic relationships between physical properties and biological activity has been made possible by invoking the concept of **molecular connectivity** Kier, Hall, Murray and Randic (1975). This employs the branching index devised by Randic (1975) to provide a means of analysing the relationships between the extent of molecular branching and properties which are critically dependent on molecular size and shape. Thus, this mathematically-derived branching index, renamed as the **molecular connectivity index**, χ, which is in reasonable agreement with the experimentally-derived Kováts index (Kováts, 1961), sums the additive and constitutive properties of any molecule to which it relates. Not only has the molecular connectivity index, χ, been shown to be correlated with a number of physical properties, including partition coefficients (Murray, Hall and Kier, 1975), water solubilities and boiling points (Hall, Kier and Murray, 1975), but it has also been shown to be correlated **linearly** with non-specific local anaesthetic activity (Kier, Hall, Murray and Randic, 1975), antifungal and butyrylcholinesterase activity (Kier, Murray and Hall, 1975). Murray, Kier and Hall (1976) have also examined a number of non-linear relationships using the molecular connectivity index, χ, in place of the physico-chemical parameter, log P, in which P is the octanol–water partition coefficient. Addition of a χ^2 term to the linear equation gave significant correlations at the 0.99 probability level between molecular connectivity and the antibacterial activity of long-chain quaternary ammonium salts, and also the hypnotic activity of barbiturates.

Free–Wilson Analysis

The approach to quantitative structure–action relationships (QSAR) proposed by Free and Wilson (1964) relates the occurrence of particular variable substituents directly to the level of a specified biological response shown by each member of the series. This method of analysis was illustrated by reference to the LD_{50} of the following group of compounds in mice.

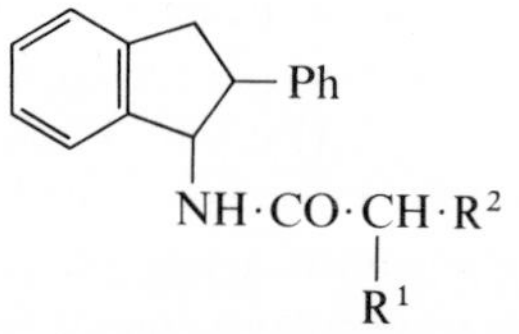

Ph

NH·CO·CH·R²

R¹

Substituent	LD_{50} values (mg/10g in mice)		
R^2	R^1 = H	R^2 = Me	Mean LD_{50}
NMe_2	2.13	1.64	1.885
NEt_2	1.28	0.85	1.065
Mean LD_{50}	1.705	1.245	1.475

From this, it is evident that the mean LD_{50} for the series is 1.475 and that the substituent group contributions at the R^1 position, $a[R^1]$, can be derived as follows.

$$\text{Substituent group contribution for H, } a[\text{H}] = 1.705 - 1.475 = +0.23$$

$$\text{Substituent group contribution for Me, } a[\text{Me}] = 1.245 - 1.475 = -0.23$$

Similarly, substituent group contributions at the R^2 position, $b\,[R^2]$, are given by

$$b[\text{NMe}_2] = 1.885 - 1.475 = +0.41$$

and

$$b[\text{NEt}_2] = 1.065 - 1.475 = -0.41$$

Likewise, the biological response for any member of the series can be derived from the expression:

Biological response = mean response (μ) + Σ substituent group contributions

Thus,

$$\text{LD}_{50} = \mu + a[R^1] + b[R^2]$$

Substituting, actual LD_{50} values, therefore produces four equations with **five** unknowns

$$\mu,\ a[\text{H}],\ a[\text{Me}],\ b[\text{NMe}_2],\quad \text{and}\quad b[\text{NEt}_2]$$

Since, however, the sum of all individual substituent group contributions at each position is zero,

$$a[\text{H}] = -a[\text{Me}]$$

$$b[\text{NMe}_2] = -b[\text{NEt}_2]$$

and the problem then reduces to the solution of **four** equations with **three** unknowns.

The value of this type of analysis lies in its ability to predict the minimum number of compounds of a given series which must be prepared and tested biologically to determine the mean response (μ) and each of the individual group substituent contributions. In a series with three variable substituents R^1, R^2 and R^3 with m variations in R^1, n variations in R^2 and p variations in R^3, there are $m \times n \times p$ possible compounds. Calculation of the mean response and individual group substituent contributions is possible with the synthesis and biological examination of $(m + n + p - 1)$ compounds. The number of unknowns is, however, reduced by the restriction that the sum of the substituent contributions at each position is zero. Thus, in an example where $m = 2, n = 3$ and $p = 3$, although there are eighteen possible compounds $(2 \times 3 \times 3)$, the nine unknowns $(\mu + 2R^1 + 3R^2 + 3R^3)$ are reduced by the restrictions to six $(\mu + R^1 + 2R^2 + 2R^3)$. Careful selection of six representative compounds

Matrix for a Series of Eighteen Compounds with Three Variable Substituents

	Substituents							
Compound	R^1		R^2			R^3		
	A	B	C	D	E	F	G	K
1	+		+			+		
2	+		+				+	
3	+		+					+
4	+			+		+		
5	+			+			+	
6	+			+				+
7	+				+	+		
8	+				+		+	
9	+				+			+
10		+	+			+		
11		+	+				+	
12		+	+					+
13		+		+		+		
14		+		+			+	
15		+		+				+
16		+			+	+		
17		+			+		+	
18		+			+			+

from the matrix of the eighteen possibles for synthesis and biological examination should, therefore, provide the required solution to the problem, provided always that each of the variables is additive and biological measurement is sufficiently precise. One representative selection of six compounds for synthesis might be compounds 1, 5, 9, 11, 15, 16.

ENZYMES

Introduction

Many proteins possess enzymic properties, either as such, or when in molecular association with specific coenzymes, such as nicotinamide adenine dinucleotide (NAD), thiamine pyrophosphate, pyridoxal phosphate or coenzyme A. They are characterised essentially by their ability to catalyse chemical reactions. This catalytic power of enzymes greatly exceeds that of comparable chemical reactions *in vitro*. Thus, chymotrypsin will catalyse the hydrolysis of certain peptides some 10^6 times more efficiently than either mineral acid or base. Many enzymic reactions frequently proceed at even greater rates, some as high as 10^{10}–10^{11} times the rate of the corresponding non-enzymic reactions. Hence, **turnover**

numbers (molecules of substrate reacting per molecule of enzyme per minute), which are usually at least 1000, can in some circumstances be as high as a million.

These exceptionally high rates of catalysis in enzymic reactions are achieved by the combination of a number of effects. Firstly, the versatility of proteins to combine with a large variety of compounds is associated with a conformational mobility, which enables the enzyme to hold and move reactants into close proximity with one another. Secondly, the amphoteric properties of proteins ensure the ready availability of acidic and basic groups for specific acid and base catalysis, whilst the widespread ability to co-ordinate metal ions, and a high degree of hydration, provide conditions which favour reactions subject to general acid-base catalysis. Furthermore, proximity of catalytic groups at active centres of enzymes favours reactions, which are rate-enhanced by concerted acid-base catalysis. Similarly, reactions subject to nucleophilic catalysis are catalysed by such powerful nucleophiles as the imidazole group of histidine and the sulphydryl group of cysteine, which are often found in enzymic proteins. Their acyl derivatives are also highly sensitive to nucleophilic attack, and hence are highly reactive intermediates for acyl transfer and its catalysis.

Specificity

The spectrum of enzymic reactions is wide, and includes hydrolysis, oxidation, reduction, addition, elimination, isomerisation and polymerisation. Within each class, however, enzymic reactions are highly substrate-specific. Thus, trypsin, chymotrypsin and pepsin are all proteases, but trypsin catalyses the hydrolysis of peptide links involving basic amino acids, such as lysine and arginine, whilst chymotrypsin and pepsin only hydrolyse peptides involving aromatic amino acids (phenylalanine and tyrosine).

Phe-Ala-Lys-Ala $\xrightarrow{\text{Trypsin}}$ Phe-Ala-Lys + Ala

Phe-Ala-Lys-Ala $\xrightarrow{\text{Chymotrypsin}}$ Phe + Ala-Lys-Ala

The Commission on Enzymes of the International Union of Biochemistry, 1961 has established a systematic nomenclature and classification for enzymes consisting of six groups, designated by class name and class number (1 to 6). Each class is divided into sub-classes and sub-sub-classes, designated by a second and a third number. Finally, each enzyme is designated by its own specific serial number. The classification, nomenclature and numbering system is illustrated in Table 3 (p. 36).

The majority of enzymes are not merely specific, but additionally **stereospecific**, catalysing reactions with only one enantiomer of a DL pair (Cushny, 1926; Beckett, 1959, 1962). Some, such as the L- and D-amino acid oxidases, are so completely stereospecific that they can be used to determine the level of optical purity (Karush, 1956). Others, however, are merely stereoselective. Thus, maltase (α-glucosidase) simply catalyses the hydrolysis of α-D-glucosides more rapidly than β-D-glucosides. This degree of stereospecificity demands a three-point attachment (**2**, 3) of the substrate to the enzyme surface (Ogston, 1948). A similar manner of substrate–enzyme attachment is also implied for dihydroxy-acetone phosphate in the specific exchange of one particular hydrogen atom for deuterium (in D_2O) which is catalysed by aldolase (Rose, 1958).

Substrate specificity is determined by the structure and orientation of the active site specific groups of the enzyme; that is, those groups which come into actual direct contact with the substrate. It is clear, however, from the identification of specific reactive centres in a number of enzymes, that the amino acid units which together constitute the active site are not necessarily adjacent, but merely in proximity to the substrate by virtue of the conformation of the protein. Thus, the glutamic acid residue, which constitutes the cationic-binding site on acetylcholinesterase, is remote from the nucleophilic serine group, which constitutes the esteratic site (**2**, 2). The existence of this type of structural and stereochemical organisation is in accord with the loss of enzymic activity, which so often accompanies denaturation of the enzyme, even when this is reversible.

The catalytic function of enzymes is often modified by interaction of the enzymic protein with small molecules or ions. Interactions of this sort may or may not occur directly at the active site. Many interactions with small molecules, however, also occur at secondary sites which are remote from the active centre; such interactons are described as **allosteric** effects (Monod, Changeux and Jacob, 1963; Monod, Wyman and Changeux, 1965). Allosteric effects either activate or inhibit enzymic activity. They do so by initiating essential changes, that alter either the conformation or the equilibrium between the quaternary protein polymer and its monomer units.

Two types of allosteric effect are recognised. Allosteric effects due to secondary binding of the enzyme's natural substrate are described as **homotropic** effects. These effects always activate the enzyme. Allosteric effects due to binding of ligands chemically unrelated to the substrate are described as **heterotropic** effects. These may either activate or inhibit the enzyme. Any one enzyme may, however, be under allosteric control by both homotropic and heterotropic mechanisms. Thus, many heterotropic effectors are the end product of a sequence of reactions, of which the enzyme allosterically-controlled forms one part. This type of allosteric reaction, therefore, constitutes a feedback control mechanism capable of switching the enzyme on or off in response to the supply and demand of the end product.

Identification of the amino acids which form covalent substrate–enzyme links at active sites is achieved either by the use of labelled substrates or specific binding reagents, followed by peptide hydrolysis and identification of the labelled fragments (Cohen, 1968; Vallee and Riordan, 1969). The procedures follow the methods described for the determination of the primary structure of peptides and proteins (**1**, 19).

The Distribution and Location of Enzymes

Enzymes are essential to virtually every aspect of intermediary metabolism (Fig. 1) in living tissue. They are, therefore, widely distributed throughout the blood, tissues and organs of the animal body. Some enzymes are soluble and occur in solution in blood plasma and in the cellular protoplasm of red and white blood cells and cellular tissue. Soluble enzymes appear in the supernatant fluid remaining after centrifugation of tissue homogenates. Other enzymes are present in cellular tissue within specialised bodies, such as mitochondria, lysosomes and microsomes, which are differentiated by size, density and function, and can be separated by tissue homogenisation and differential centrifugation. Much, but by no means all, the information available on sub-cellular distribution of enzymes is based on liver tissue. It is recognised, however, that this is by no means a guide to sub-cellular distribution in other tissues, which may be quite different.

Mitochondria are generally oval-shaped bodies usually ranging from 0.2 to 7μm in length, possessing a double lipoprotein membrane, the inner one of which is folded giving a characteristic striped appearance to the structure. Most mitochondrial enzymes are contained within this inner membrane, and consist principally, but not exclusively of those concerned with the processes of oxidative phosphorylation. These include the oxidative decarboxylation of pyruvate to acetate, and its entry as acetyl-S$\overline{\text{CoA}}$ into the tricarboxylic acid cycle (**1**, 13) which is responsible for the oxidation of sulphydryl-bound acetate to carbon dioxide and water with the release of energy for storage ultimately as adenosine triphosphate (ATP). Mitochondrial enzymes also catalyse the oxidation of fatty acids. This leads to the successive removal of a series of 2-carbon fragments as acetyl-$\overline{\text{CoA}}$, which feed the tricarboxylic acid cycle. Propionyl-$\overline{\text{CoA}}$, the residual 3-carbon fragment from enzymic oxidation of fatty acids with an odd number of carbon atoms, is homologated to succinyl-$\overline{\text{COA}}$, which also enters the tricarboxylic acid cycle. Proteolytic enzymes are also located in mitochondria in the liver. Characteristic enzymes of liver mitochondria, therefore, include cytochrome oxidase, glutamate, malate, isocitrate and succinate dehydrogenases, acetyl-$\overline{\text{CoA}}$ acetyltransferase and deoxyribonuclease.

The **lysosomes** are generally smaller (*ca* 0.2–0.8μm in length) and denser than the mitochondria, and separable from them by differential centrifugation. They contain mainly hydrolytic enzymes, such as lipases, phosphatases and peptidases, including, for example, carboxylesterase, cholinesterase, cholesterol esterase, alkaline phosphatase and glucose-6-phosphatase. Normally, hydrolysis occurs within the lysosomal membrane, with the hydrolytic products diffusing out through the membrane. When the cell dies, however, the lysosomal membrane breaks down, releasing the enzymes into the body of the cell, where they are then free to attack and degrade the normal cell constituents. This process is known as **autolysis**.

Still smaller bodies, known as **ribosomes** (*ca* 0.01–0.015μm) and **microsomes** (*ca* 0.01–0.25μm in diameter) are formed by fragmentation of the endoplasmic reticulum, a tubular lipoprotein network which extends from the cell wall

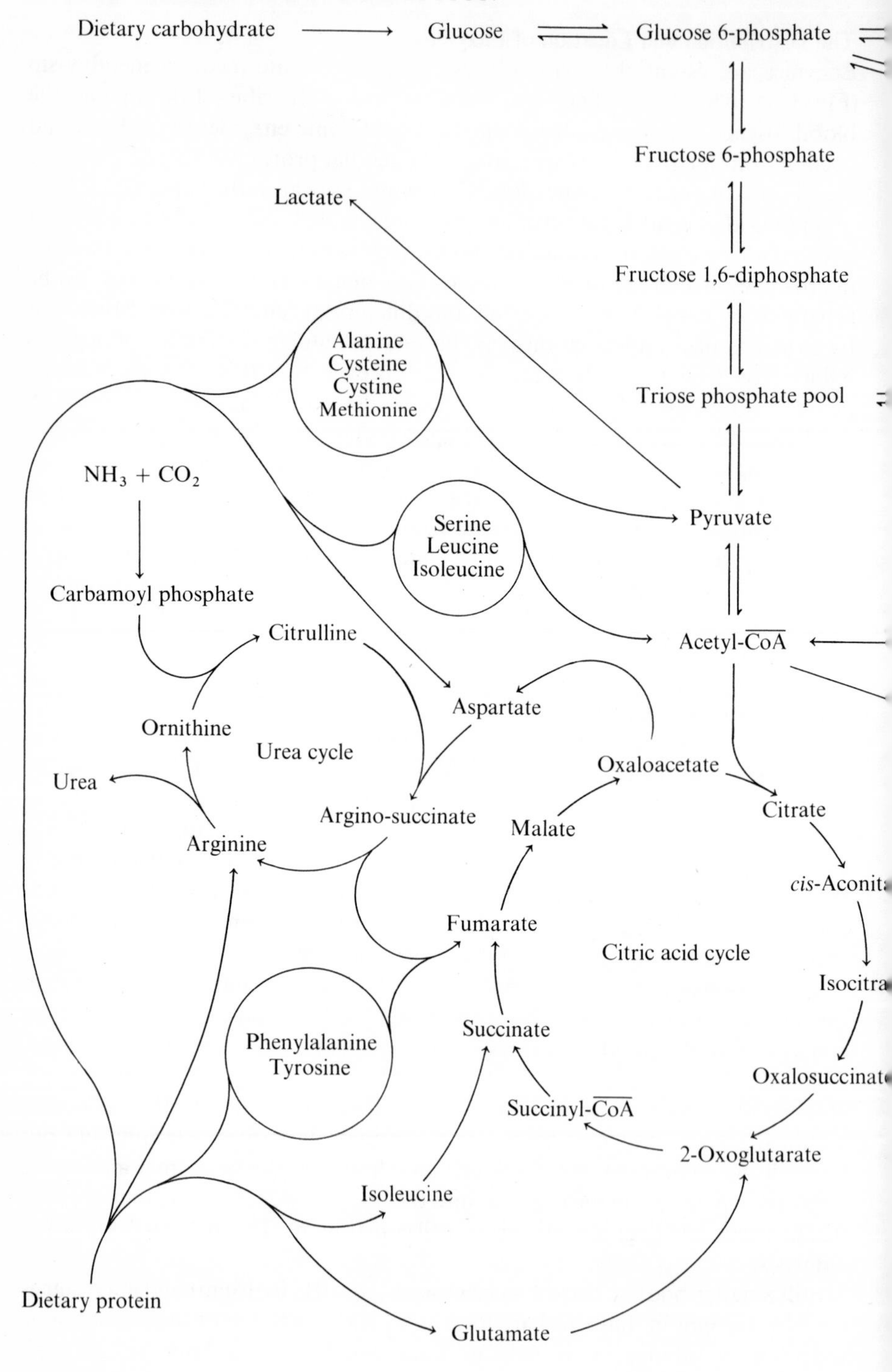
Dietary carbohydrate
Glucose
Glucose 6-phosphate
Fructose 6-phosphate
Lactate
Fructose 1,6-diphosphate
Alanine
Cysteine
Cystine
Methionine
Triose phosphate pool
$NH_3 + CO_2$
Serine
Leucine
Isoleucine
Pyruvate
Carbamoyl phosphate
Citrulline
Acetyl-$\overline{\text{CoA}}$
Aspartate
Ornithine
Urea cycle
Urea
Oxaloacetate
Argino-succinate
Citrate
Malate
Arginine
cis-Aconit
Fumarate
Citric acid cycle
Isocitra
Phenylalanine
Tyrosine
Succinate
Oxalosuccinat
Succinyl-$\overline{\text{CoA}}$
2-Oxoglutarate
Isoleucine
Dietary protein
Glutamate

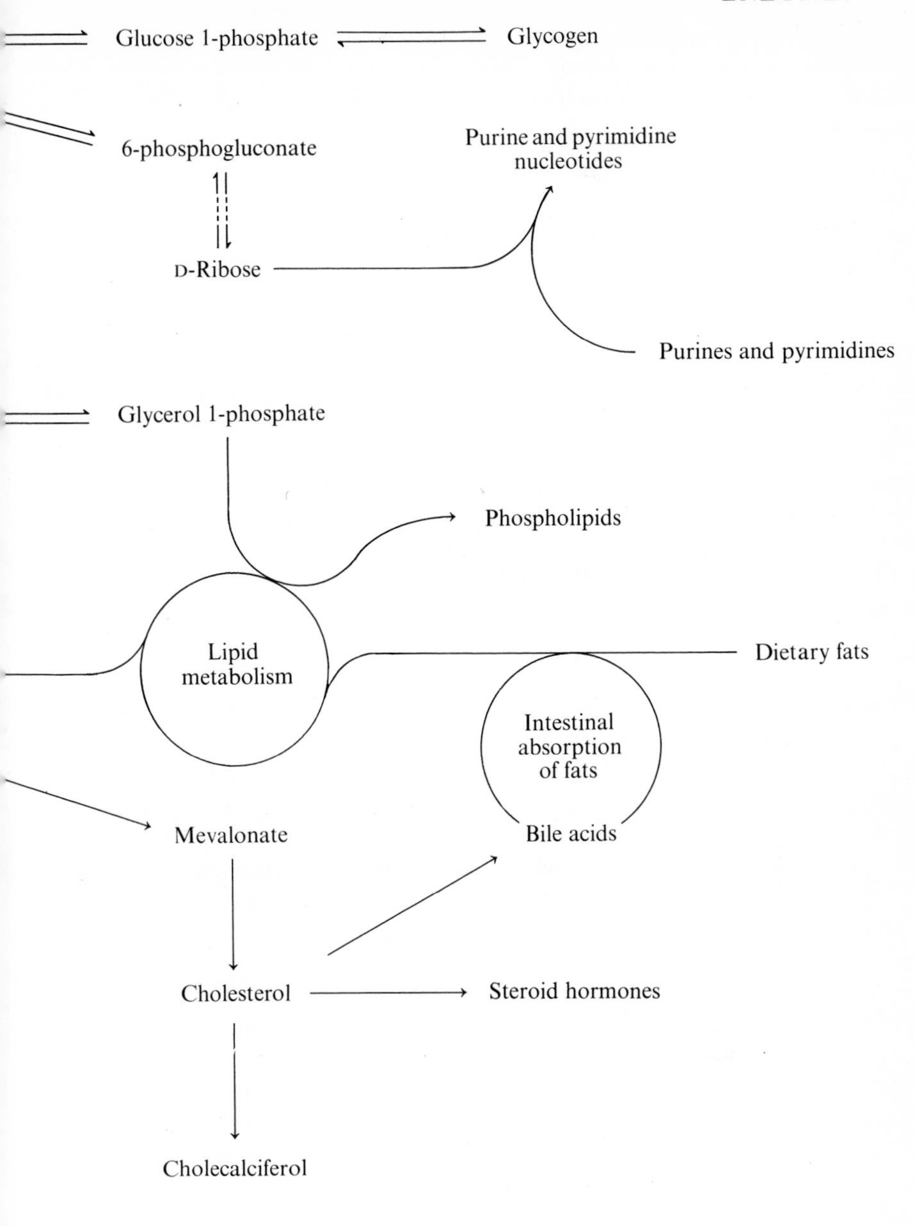

Fig. 1 Synopsis of mammalian intermediary metabolism

throughout the cytoplasm of liver cells. This is of two types, **rough** and **smooth** endoplasmic reticulum. Rough endoplasmic reticulum is characterised by the appearance of large numbers of small dense particles on its surface. These are the ribosomes which represent the seat of protein synthesis. Smooth endoplasmic reticulum has no such distinguishing feature, and breaks up when homogenised and forms microsomes, which contain the enzymes primarily responsible for the metabolism of foreign compounds. These enzymes, which consist of mixed-function oxidases and conjugating enzymes, are membrane-dependent, and lose activity if attempts are made to solubilise them (**2**, 5).

Small glycogen granules (*ca.* 0.1 μm in diameter) are also present in liver cells. These represent the main storage system for carbohydrate reserves, from which glucose 1-phosphate is released on demand by phosphorylase. Glucose-1-phosphate is isomerised in the body of the cell by phosphoglucoisomerase to glucose 6-phosphate. The latter is then either hydrolysed by hexokinase to yield glucose for transport across the cell wall and distribution in the blood to other tissues, or alternatively is oxidised (glycolysis) for energy release (**1**, 21).

The Role of Enzymes in Health and Disease

Because of their very high specificity, enzymes play a vital role in the metabolism of all life forms. In mammalian metabolism, proteases, lipases and glycosidases are essential for the digestion of the proteins, fats and carbohydrates of food. The primary products of digestion in turn become substrates for the enzymes of intermediary metabolism in reactions which are essential for creation of the energy-supply, and the synthesis of cell constituents. Other enzymes have specialised protective functions. These include the drug-metabolising enzymes essential for the detoxification of foreign compounds; also, lysozyme, which is present in lacchrymal secretion, blood, mucous membranes and spleen. This enzyme is capable of hydrolysing the glycosidic links of mucopolysaccharides, and on this account plays a part in the defence of mucous surfaces and other organs against invading micro-organisms.

Enzyme insufficiencies or enzymic defects may be either genetically-determined or pathological in origin. They can sometimes be overcome or compensated by the adoption of alternative metabolic pathways or even the development of new ones. Such adaptation is clearly an important determinant of drug resistance, and occurs in the development of resistance to antibiotics in pathogenic micro-organisms (**2**, 5), in the development of resistance to drug treatment in leukaemia and other forms of cancer (*Methotrexate*, **1**, 23), and may also account for the effects of tolerance which are shown to increasingly high doses of other drugs by some individuals.

A number of deficiency diseases, some fatal, are clearly attributable to genetically-determined enzyme deficiencies, usually described as inborn errors of metabolism (Garrod, 1909; Harris, 1971). For example, the genetic deficiency of phenylalanine hydroxylase, which blocks the normal conversion of phenylalanine to tyrosine in phenylketonurea, promotes the metabolism of phenylalanine by the alternative pathway leading to phenylpyruvic acid. Similarly, oxidation of tyrosine by tyrosinase to 3,4-dihydroxyphenylalanine, which is the biosynthetic precursor of the pigment, melanin, is deficient in albinism.

$C_6H_5-CH_2\cdot CH(NH_2)\cdot CO\cdot OH$ Phenylalanine → (Phenylalanine hydroxylase) → $HO-C_6H_4-CH_2\cdot CH(NH_2)\cdot CO\cdot OH$ Tyrosine

PHENYLKETONURIA

Phenylalanine ⇌ $C_6H_5-CH_2\cdot CO\cdot CO\cdot OH$ Phenylpyruvic acid

Tyrosine → (Tyrosinase) → $(HO)_2C_6H_3-CH_2\cdot CH(NH_2)\cdot CO\cdot OH$ 3,4-Dihydroxyphenylalanine

ALBINISM

Other genetic deficiencies, such as those of serum esterase which can give rise to *Suxamethonium* (Succinylcholine Chloride) apnoea, and liver acetyltransferase deficiency which influences the effectiveness of *Isoniazid*, can be important in the use of particular drugs (**2**, 5).

Monitoring of tissue and plasma enzyme levels is used routinely in diagnosis. For example, the detection of particular **isoenzymes**, such as the rise in the 'heart-type' lactic dehydrogenase (**1**, 19) in acute myocardial infarction (Wroblewski and Gregory, 1960), provides a useful means of confirming diagnosis (Wilkinson, 1970). Measurements of plasma enzyme levels are also of considerable value in monitoring the incidence of drug-induced effects.

Enzymes as Therapeutic Agents

There are relatively few well-documented examples of enzymes used in the treatment of disease, other than by replacement in cases of deficiency. Some proteolytic enzymes, such as trypsin and streptokinase (fibrinolysin), are used to disperse intravascular clots, and in the treatment of necrotic wounds and ulcers. *Colaspase* (L-asparagine amino hydrolase) is administered by injection in the treatment of lymphomas (Kidd and Todd, 1954; Broome, 1961) to limit the availability of asparagine. In this condition, the lymphoma cells are deficient in asparagine synthetase, and hence reliant on exogenous sources of the amino acid. Asparaginase from *E. Coli* or from *Erwinia chrysanthemi* (Crisantapase) has been used with some effect in the treatment of leukaemia (Whitecar, Bodey, Harris and Freireich, 1970) and corticosteroid-sensitive lymphosarcoma.

The so-called **spreading factor**, hyaluronidase, hydrolyses those glycosidic links of the mucopolysaccharide, hyaluronic acid, which involve the reducing group of its *N*-acetylglucosamine units. Hyaluronic acid is an important component of the tissue cement which binds tissues together, and its breakdown by hyaluronidase facilitates the speed of absorption and diffusion of drugs applied to the skin.

Enzymes in Analysis

The high specificity of carrier-bound enzymes provides the basis for their growing use in trace analysis (Orth and Brümmer, 1972). Thus, a chemically insolubilised apo-polyphenol oxidase from mushrooms has been shown to be a very selective and stable means of determining ng amounts of copper (Stone and Townshend, 1972).

Control of Enzymic Activity

The action of enzymes is subject to a number of indirect control mechanisms. Thus, the biosynthesis of enzymes, which is itself enzyme-catalysed, is subject to control by feed-back mechanisms involving both substrate and products. Such feed-back mechanisms often involve **allosteric** effects, that is effects brought about by the combination of small molecules at sites which are remote from the active centre of the enzyme. Such interactions often appear to effect alterations in the quaternary structure of the enzyme and the associated reversible association—dissociation phenomena between enzyme polymer, monomer and monomer sub-units. Thus, attachment of oxygen to one sub-unit of haemoglobin immediately increases the rate of attachement of oxygen to a second sub-unit, which in turn successively increases the rates of attachment to the third and fourth sub-units. This is an example of the natural substrate binding to the enzyme at an allosteric site; such effects are termed **homotropic** and always promote the enzymic reaction, i.e. they lead to more rapid metabolism of the substrate. Feedback control by factors other than the substrate, such as the products of enzymic reaction, is a **heterotropic** effect. Such effects may either assist or antagonise enzymic action, activating at low concentration and antagonising at high concentration (Monod, Wyman and Changeux, 1965).

Thus, cholic acid blocks its own biosynthesis by an allosteric inhibitory effect on cholesterol 7-hydroxylase (**1**, 22). Similarly, whilst the supply of the substrate, aspartate, co-operatively assists the activity of *E. coli* aspartate transcarbamylase, ATP, which functions as activator, and cytosine triphosphate which acts as inhibitor, exert allosteric control over the reaction.

Enzymic Reactions

As proteins, enzymes are readily denatured by heat, by extremes of pH and ionic strength, and by solvents other than water, with concomitant loss of activity. Within the limits of their stability, however, the rate of enzyme-catalysed reactions is dependent on both temperature and pH. Figure 2 shows a characteristic curve for the relationship between enzymic activity (expressed as reaction

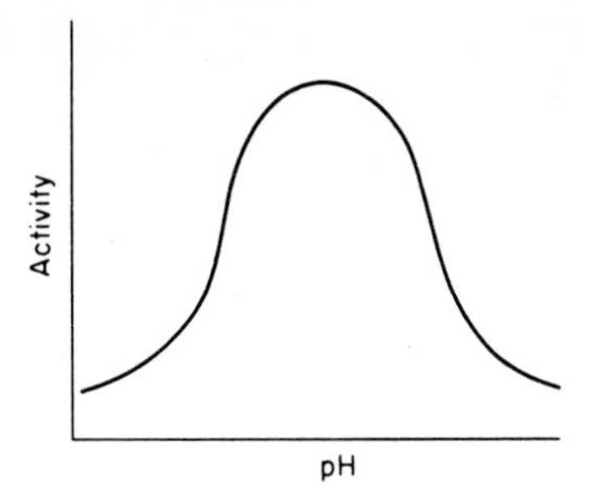

Fig. 2 Typical pH—activity profile for enzyme–substrate reactions

velocity) and pH. In general terms, such curves reflect the dissociation of ionisable groups of the enzyme or of the enzyme–substrate complex which directly influence enzymic activity. In some cases, as for example that of monoamine oxidase (Fig. 3), other factors intrude, such as the instability of the enzyme at high pH; in consequence, the curve is asymmetrical. Each enzyme, therefore, has an optimum pH, at which it reacts most rapidly with its substrate. However, even if temperature and pH are fixed at optimum values, the velocity is usually constant only in the initial stages of the reaction. The velocity of the reaction subsequently falls off exponentially with time as the substrate is consumed and the reaction products accumulate, due either to mass action or to feedback control.

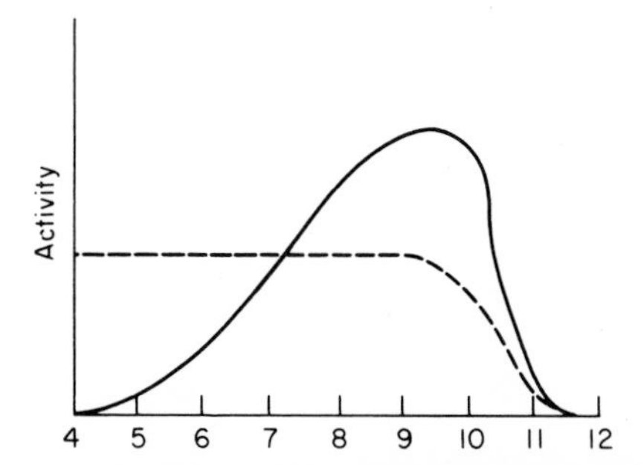

Fig. 3 pH—activity profile for monoamine oxidase
——— Activity at stated pH
------- Activity after 5 min at pH stated and subsequent adjustment to pH 7.3

Under optimum pH and temperature conditions, the velocity of an enzyme-catalysed reaction is initially proportional to both enzyme and substrate concentration. Provided low substrate concentrations are maintained, a straight-line relationship exists, but as substrate concentrations are increased, the reaction velocity tends to a maximum, and the velocity is then solely dependent on enzyme concentration (Fig. 4).

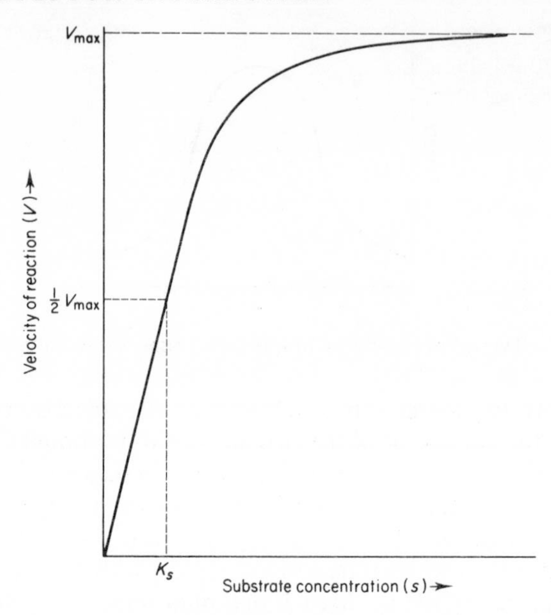

Fig. 4 Effect of substrate concentration on the velocity of an enzyme reaction

This departure from direct proportionality between substrate concentration and reaction velocity, which is theoretically demanded by the Law of Mass Action, is explained by application of the Michaelis and Menten equation (1913). This is based on the assumption that the reaction takes place in two stages involving firstly a reaction in which enzyme (E) and substrate (S) combine reversibly to form an enzyme–substrate complex (ES), and, secondly, irreversible decomposition in which ES breaks down to regenerate enzyme and release products.

$$E + S \underset{}{\overset{K_s}{\rightleftharpoons}} ES \tag{13}$$

$$ES \xrightarrow{k} E + \text{products} \tag{14}$$

Considering equation (13) when

e = total enzyme concentration
s = concentration of free substrate
p = concentration of enzyme–substrate complex

then

$$e - p = \text{concentration of free enzyme}$$

and

$$K_s = \frac{(e - p)s}{p} \tag{15}$$

where K_s, the dissociation constant of the enzyme–substrate complex, is known as the **Michaelis constant**.

Rearranging (15),

$$p = \frac{es}{K_s + s} \tag{16}$$

Considering now the reaction in equation (14), the velocity (v) is dependent on the concentration of the enzyme–substrate complex, p, and $v = kp$, where k is the velocity constant for the reaction. Substituting for p in equation (16),

$$v = \frac{kes}{K_s + s} \tag{17}$$

$$= \frac{ke}{\frac{K_s}{s} + 1} \tag{18}$$

As s increases in relation to the size of K_s, v will tend towards equality with ke, which is the maximum velocity (V_{max}) when the enzyme is saturated with substrate, and

$$v = \frac{V_{max}}{1 + \frac{K_s}{s}} \tag{19}$$

From this equation, it may be deduced that when s is equal to K_s, v is equal to $\frac{1}{2}V_{max}$.

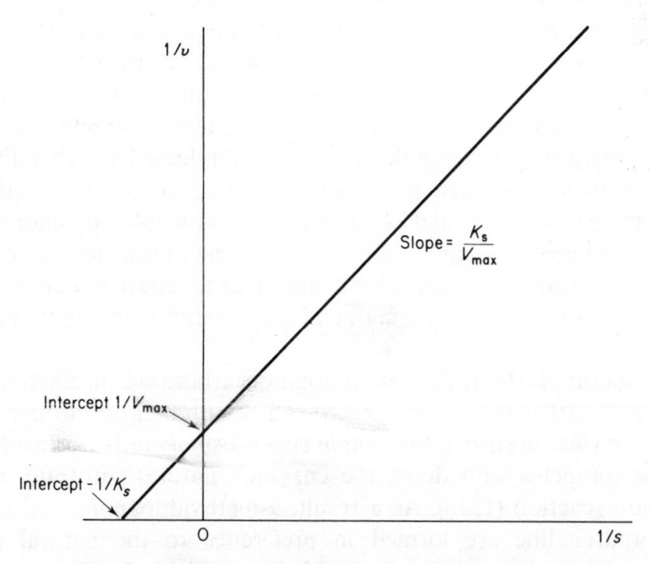

Fig. 5 Lineweaver and Burk plot of $1/v$ against $1/s$

The Michaelis constant (K_s) and maximum velocity (V_{max}) are important characteristics of an enzyme. They can be obtained from the plot of velocity against substrate concentration (Fig. 4). A more precise method is the Lineweaver and Burk (1934) plot of $1/v$ against $1/s$, derived by inverting equation (17), and writing V_{max} for ke as before:

$$\frac{1}{v} = \frac{K_s + s}{V_{max}\,s}$$

and

$$\frac{1}{v} = \frac{K_s}{V_{max}} \cdot \frac{1}{s} + \frac{1}{V_{max}} \qquad (20)$$

The plot of $1/v$ against $1/s$ gives a straight line, the slope of which is equal to K_s/V_{max} and the intercept on the ordinate equal to $1/V_{max}$ (Fig. 5)

ENZYME INHIBITION

Enzyme Inhibitors as Therapeutic Agents

The action of a number of important therapeutic agents is due to their ability to inhibit specific enzymes. For example, monoamine oxidase inhibitors (**1**, 16 and **2**, 5) depress the metabolism of natural adrenergic amines. This inhibition prolongs the life of released catecholamines, thereby prolonging their central nervous stimulant action. The production of stimulant effects in this way, however, may have certain dangers, in that the metabolism of other pressor amines, such as tyramine, which is normally present in certain foods, is also suppressed. Care should, therefore, be taken to avoid such foods as cheese, pickled herring, yeast products and Chianti wine, all of which contain tyramine, and have been known to cause hypotensive crises in patients receiving monoamine oxidase inhibitors. Similarly, anticholinesterases inhibit the action of acetylcholinesterase, depressing the hydrolysis of released acetylcholine (**1**, 11). As a result, anticholinesterases prolong the effect of acetylcholine and function as cholinomimetics (**1**, 21 and **2**, 2). Typical reversible anticholinesterases include *Physostigmine*, which is mainly used for its miotic action in the treatment of glaucoma, and *Neostigmine*, which is used to relieve muscular weakness in the treatment of *Myasthenia gravis*, and also to stimulate bowel and bladder movement.

The interaction of *Methyldopa* with dopa decarboxylase in the treatment of hypertension is rather more complicated. Thus, although *Methyldopa* inhibits dopa decarboxylase *in vitro*, inhibition *in vivo* is usually only transient. Instead, *Methyldopa* competes with dopa, the enzyme's natural substrate in the decarboxylation reaction (**1**, 18). As a result, α-methyldopamine and ultimately α-methylnoradrenaline are formed in preference to the natural products, dopamine and noradrenaline (norepinephrine), with consequent depletion of noradrenaline stores and replacement of the noradrenaline therein by α-methylnoradrenaline. Activation of sympathetic nerve impulses will then release α-methylnoradrenaline, which is similar in action to, but less potent than, the replaced noradrenaline, though, because of its resistance to destruction by monoamine oxidase, more prolonged in action. The overall effect is, therefore, a reduction in sympathetic transmitter activity.

Levodopa, the natural substrate of dopa decarboxylase, is used for the control of symptoms in Parkinson's disease. It achieves its effect by decarboxylation (**1**, 18) to dopamine within the brain. Much of the administered drug, however, is decarboxylated before it can enter the central nervous system, so that large doses are required to produce the required effect. This problem of excessive dosage is overcome by co-administration with a dopa decarboxylase inhibitor, *Carbidopa*, which is unable to cross the blood-brain barrier. This substantially increases the amount of *Levodopa* available for transport into the brain, and in favourable cases can lead to as much as a tenfold reduction in dosage.

Many enzymes are common to a wide range of living organisms. Nonetheless, treatment of some infectious diseases depends on drug-induced inhibition of a specific enzyme in the invading organism. The use of such agents, however, is only possible in cases where the enzyme under attack is absent from the host's metabolism, where it can be by-passed in the host by assimilation of the end-product from the diet, or where significant differences exist in the sensitivity of the key enzyme in host and invading parasite. For example, the effectiveness of sulphonamides in antibacterial chemotherapy depends on a major difference in the route of folic acid supply in bacterial and human metabolism. Thus, folic acid is not synthesised in mammals, but absorbed intact in the diet. In contrast, most bacteria and other invading parasites are unable to assimilate folic acid and need to synthesise it *de novo*. Sulphonamides act as competitive inhibitors of PABA, the natural substrate in the microbial biosynthesis of dihydropteroate, and hence dihydro- and tetrahydro-folic acids (**1**, 14). On the other hand, dihydrofolate reductase inhibitors, such as *Trimethoprim* (**1**, 14), which potentiate the action of sulphonamides when used in combination therapy in the form of *Co-trimoxazole*, act in both host and bacterial metabolism. Their action, however, is selective due to the fact that the bacterial enzymes are some 10^4 times more sensitive than their mammalian counterparts.

The selective action of *Sodium Stibogluconate* and *Stibophen* in the treatment of schistosomiasis is also due to substantial differences in host/parasite enzyme sensitivity to these drugs. They are potent inhibitors of schistosomal phosphofructokinase, but have only minimal effect on mammalian phosphofructokinase (**2**, 2). They act, therefore, by preventing conversion of fructose-6-phosphate to fructose 1,6-diphosphate in the parasite, thus blocking the anaerobic metabolism of glucose on which the survival of the organism largely depends.

Additional factors contribute to the selectivity of the antimalarial agent, *Pyrimethamine*, which is chemically-related to *Trimethoprim*, and derives its activity from its ability to block dihydrofolate reductase in the malarial parasite. Selectivity in this case is achieved, not only by significant differences in host/parasite enzymic sensitivity to the drug (Ferone, Burchall and Hitchings, 1969), but also by the high lipid solubility of *Pyrimethamine* arising from the presence of its phenyl and chlorophenyl substituents, which favour penetration of the lipid membrane of red cells infected with erythrocytic forms of the parasite. Such dihydrofolate reductase inhibitors, however, inevitably cause some folate depletion in the host, though this effect can be overcome by daily administration of folinic acid (N^5-formyltetrahydrofolic acid). This protects the patient's own metabolism by the supply of a pre-formed folic acid derivative, which because of its inability to penetrate the infective plasmodium (Ferone and Hitchings, 1966), does not modify the effect of the drug.

The use of certain sulphonamides, such as *Acetazolamide*, as diuretics is due to their action as carbonic anhydrase inhibitors in the renal tubular epithelium. Carbonic anhydrase is also present inside the red blood cells, where it controls the formation of carbonic acid, and is essential for the carriage of carbon dioxide from the tissues for discharge in the lungs. Its presence within the red cells, however, largely protects it from the action of the diuretic sulphonamides which do not penetrate the cells to any appreciable extent. In consequence, these inhibitors of carbonic anhydrase are effective *in vivo* only at the renal tubular epithelium. As a result, they function as diuretics causing the excretion of large volumes of alkaline urine containing Na^+, K^+ and HCO_3^- ions (**1**, 14).

Yet another form of selective drug action involving enzyme inhibition is seen in the action of a number of antibiotics which suppress bacterial cell-wall biosynthesis. Selectivity in this case depends essentially on the unique structure of bacterial cell walls, which consists of cross-linked peptidoglycan polymers. Their chemistry and biosynthesis is thus completely different from the phospholipid protein of mammalian cell membranes and, therefore, sensitive to attack by agents which are free from similar action to mammalian biosynthesis and metabolism. Bacterial cell-wall biosynthesis occurs essentially in three stages. These occur successively within the protoplasm of the cell, in attachment to the cell membrane, and in the existing cell wall exterior to the cell. The first stage, the synthesis of UDP-acetylmuramylpentapeptide within the cell, is suppressed by *Cycloserine* (**1**, 18 and 21). The antibiotic is a structural analogue of D-alanine, and hence a competitive inhibitor of L-alanine racemase and of D-alanine-D-alanine synthetase, both of which bind *Cycloserine* about 100 times more powerfully than their natural substrate (Roze and Strominger, 1966). The antibiotics, *Vancomycin*, *Ristocetin* and *Bacitracin*, affect various steps essential to the formation of a membrane-bound peptidoglycan in the second stage of cell-wall biosynthesis. Penicillins and cephalosporins act by competitive inhibition of a transpeptidase responsible for cross-linking of the cell-wall glycopeptide polymer units (Tipper and Strominger, 1968) in the third and final stage of cell-wall biosynthesis (**1**, 21, and **2**, 2).

Types of Enzymic Inhibition

Enzyme inhibitors may bring about either reversible or irreversible inhibition of the enzyme. The terms **reversible** and **irreversible** are purely descriptive of the process of interaction between enzyme and inhibitor, and irreversible inhibition does not imply destruction of the enzymic protein as may occur by the action of heat or simple protein denaturants (**1**, 19).

In general, the criterion for reversibility is the regeneration of full enzymic activity after dialysis of the inhibitor. Reversible inhibition implies a definite equilibrium between enzyme and inhibitor with a measurable inhibitor constant K_i, which is a quantitative expression of the affinity of the enzyme for the inhibitor. The degree of inhibition depends on inhibitor concentration and once equilibrium is reached is independent of time. Irreversible inhibition, on the other hand, increases with time, and the effectiveness of the inhibitor is determined by the rate constant for the reaction. The inhibition of acetylcholinesterase by *Dyflos* (Isoflurophate; Cohen, Oosterbaan, Jansz and Berends, 1959; Sanger, 1963) provides a good example of irreversible inhibition. The inhibitor forms a covalently-bonded phosphorus ester with the serine residue at the esteratic site of the enzyme in a reaction which is not reversible except by the intervention of a specific bond-breaking agent, such as *Pralidoxime* (**2**, 2).

Enzymic inhibition can be broadly classified as **competitive** or **non-competitive**. There are various refinements of these classes of inhibitor, and some fail to fall neatly into one group or the other and must, therefore, be classed as **mixed** (competitive–non-competitive) inhibitors. Essentially, the distinction between competitive and non-competitive inhibitors depends upon whether or not the inhibitor is bound at the usual substrate-binding site (competitive) or elsewhere on the enzyme (non-competitive) either at some part of the active site or alternatively at what is now described as the **allosteric site**.

Competitive Inhibition

Competitive inhibition occurs when the inhibitor resembles the natural substrate sufficiently strongly to ensure binding at the active site with formation of an enzyme–inhibitor complex (EI). This may either be totally incapable of breakdown to products (**dead-end** inhibitor) or alternatively may undergo conversion to an abnormal product. The inhibition of monoamine oxidase (Fig. 3) under acidic conditions provides an example of dead-end competitive inhibition by protons. The enzyme is stable to acid, and adjustment to neutrality after exposure to acid restores activity. Other examples of dead-end competitive inhibition of enzymes include the action of the monoamine oxidase inhibitor, *Tranylcypromine* (**1**, 16) and the competition between sulphonamides and *p*-aminobenzoic acid in bacterial folic acid biosynthesis (**1**, 14). The antimalarial, *Pyrimethamine*, also inhibits dihydrofolate reductase from *P. berghei* competitively (Rollo, 1955).

Competitive enzymic inhibition leading to abnormal products is exemplified by the inhibition of dopa decarboxylase by α-methyldopa (**1**, 18), which also leads to the slow formation of α-methyldopamine. The formation of thio-inosinate by inosinic phosphorylase from 6-mercaptopurine in place of hypoxanthine (Brockman, 1961) provides a further example of this type of inhibition.

In competitive inhibition there is usually a close chemical and structural resemblance between the normal substrate and the inhibitor, and the competition is represented by two competing reactions:

$$\mathrm{E} + \mathrm{S} \underset{}{\overset{K_s}{\rightleftharpoons}} \mathrm{ES} \tag{13}$$

and

$$\mathrm{E} + \mathrm{I} \underset{}{\overset{K_i}{\rightleftharpoons}} \mathrm{EI} \tag{21}$$

in which EI is the enzyme inhibitor complex.

Considering these two equilibria, in conjunction with the normal breakdown of ES to products (equation 14), and taking q as the concentration of EI, and i as the concentration of inhibitor, then

$$(e - p - q)s = K_s p \qquad \text{(a)}$$

$$(e - p - q)i = K_i q \qquad \text{(b)}$$

$$v = kp$$

from which, by substituting for q in eq. (a) and using the same procedure as for eqs (16)–(19),

$$v = \frac{V_{\max}}{1 + \dfrac{K_s}{s}\left(1 + \dfrac{i}{K_i}\right)} \tag{22}$$

From this, it will be apparent that the value of K_s, the Michaelis constant, is increased by a factor of $1 + i/K_i$, so that K_s is increased in direct proportion to the concentration of the inhibitor. The corresponding displacement of the Lineweaver–Burk plot is shown in Fig. 6A, in which the intercept on the baseline gives $1/kp$, where kp is $K_s(1 + i/K_i)$ the effective Michaelis constant in the enzyme–inhibitor interaction. The intercept on the ordinate $1/V_{\max}$ is unaffected by the inhibitor.

Uncompetitive Inhibition

This is a form of inhibition in which the inhibitor combines only with the already formed enzyme–substrate complex at a site remote from the active centre. This, therefore, is a form of **allosteric** inhibition, which occurs in accord with the equation:

$$\mathrm{E} + \mathrm{S} \rightleftharpoons \mathrm{ES} \xrightarrow{\mathrm{I}} \mathrm{ESI} \tag{23}$$

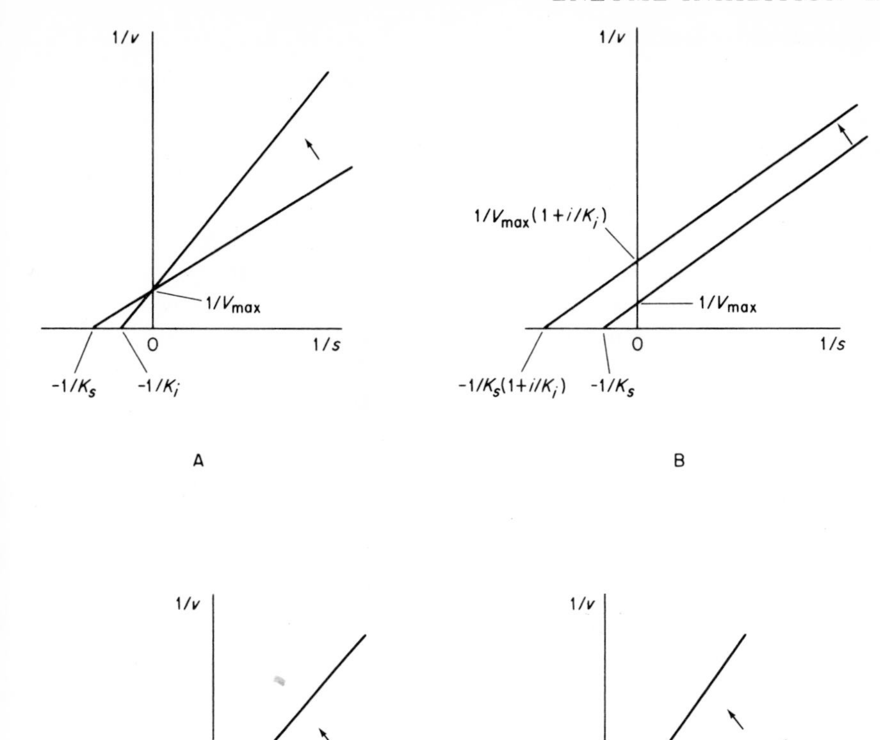

Fig. 6 Lineweaver and Burk plots of $1/v$ against $1/s$
A Competitive inhibition B Uncompetitive inhibition
C Non-competitive inhibition D Mixed inhibition

Inhibition of this sort is rare with single substrates, but is fairly common with multiple substrates. Formation of the ESI complex would appear to be facilitated by conformational changes in the enzyme which result from the formation of the enzyme–substrate complex (ES). The inhibition of the oxidised form of cytochrome oxidase by azide is uncompetitive (Nicholls, 1968), and is characterised by parallel displacement of the Lineweaver–Burk plot (Fig. 6B) in which both $1/V_{max}$ and $1/K_s$ are reduced by a constant factor $(1 + i/K_i)$.

Non-competitive Inhibition

In non-competitive inhibition, enzyme–substrate binding is unaffected, since the inhibitor is bound at an alternative site. There is, therefore, no effect on the Michaelis constant K_s. Competition arises through the complete failure of the enzyme–inhibitor–substrate complex (EIS) to break down. A reduction in V_{max} occurs, and any residual effect can be attributed to breakdown of the normal enzyme–substrate complex (ES). Alternatively, EIS may break down to give the usual products, but at a reduced rate compared with the rate for ES, and the overall V_{max} will be the sum of both. In the first case, which is the simpler, the reactions are represented by the following equations:

$$\mathrm{E + S} \xrightleftharpoons{K_s} \mathrm{ES} \tag{13}$$

$$\mathrm{ES} \xrightleftharpoons{k} \mathrm{E + products} \tag{14}$$

$$\mathrm{E + I} \xrightleftharpoons{K_i} \mathrm{EI} \tag{21}$$

$$\begin{matrix} \mathrm{ES + I} & \searrow & \\ & & \mathrm{EIS} \\ \mathrm{EI + S} & \nearrow & \end{matrix} \tag{24}$$

From this it may be deduced that

$$v = \frac{V_{max} \bigg/ \left(1 + \dfrac{i}{K_i}\right)}{1 + \dfrac{K_s}{s}} \tag{25}$$

The corresponding displacement of the Lineweaver–Burk plot is shown in Fig. 6C in which the intercept on the ordinate gives $1 + (i/K_i)/V_{max}$ corresponding to the reduction in V_{max}. The intercept on the baseline in this case is unchanged $(1/K_s = 1/K_i)$ in accord with the theory.

One of the most extensively studied examples of non-competitive inhibition is that of various esterases (e.g. acetylcholinesterase) and proteolytic enzymes (e.g. chymotrypsin) by *Dyflos* (DFP) and other phosphorus insecticides (**2**, 2). Their action is highly selective for a single serine residue present in chymotrypsin (Serine-195), which although one of many (Schaffer, May and Summerson, 1954), is the only one forming part of the active centre. Main (1964) has demonstrated that a reversible complex is formed initially between cholinesterase and malaoxon, and that acylation of the serine residue only occurs subsequently.

Other examples of non-competitive enzymic inhibition include the action of the hydrazide monoamine oxidase inhibitors of the *Iproniazid* type (**1**, 16), and the inhibition by thioinosinate of inosinate dehydrogenase and various other enzymes for which inosinate forms the substrate. Binding of *Methotrexate* to dihydrofolate reductase, which forms the basis of its action as an antitumour agent, is also non-competitive. The drug-bound enzyme is stabilised towards the action of proteolytic enzymes as a result of conformational changes brought about by complex formation (Hakala and Suolinna, 1966).

Mixed Competitive–Non-competitive Inhibition

The Lineweaver–Burk plot for mixed (competitive–non-competitive) inhibitors shown in Fig. 6D is intermediate between that of competitive (Fig. 6A) and non-competitive (Fig. 6C), with the intersection of plots for enzyme–substrate and enzyme–inhibitor interactions falling between the ordinate and the abscissa.

The inhibition of the amido-transferase responsible for the conversion of *N*-formylglycinamide ribonucleotide (FGAR) to *N*-formylglycinamidine ribonucleotide by azaserine and 6-diazo-5-oxo-L-norleucine (DON), which accounts for their effect on purine synthesis and hence for their antitumour and antibacterial activity (**1**, 18), is a mixed competitive–non-competitive action.

Levenberg, Melnick and Buchanan (1957) have shown that in the absence of glutamine, azaserine inactivates the enzyme irreversibly. Also that glutamine competes with azaserine for the active site, and delays, but does not prevent, irreversible inactivation. DON acts similarly, but is effective as an antibiotic at about 1/40th of the dose, and is equally more effective as an inhibitor of the enzyme. Buchanan and his collaborators conclude on the basis of studies with the purified enzyme and ^{14}C-labelled azaserine (French, Dawid, Day and Buchanan, 1963), that initially both the natural substrate (glutamine) and its inhibitor are bound competitively by their α-amino carboxyl groups, and that subsequently irreversible (non-competitive) inhibition of the enzyme occurs as a result of alkylation of the sulphydryl group which normally is responsible for the displacement of the amide group from glutamine. Azaserine does not react with sulphydryl groups under neutral conditions, but at the active site is first protonated and then attacked by the resulting cysteinyl sulphydryl anion. The amino acid sequence in the region of this particular cysteine residue has been partially identified, though the point of the initial competitive binding is not clear (French, Dawid and Buchanan, 1963). The structural specificity of the active centre, however, is high since neither the lower homologue of DON, 5-diazo-4-oxo-L-norvaline (French, Dawid, Day and Buchanan, 1963), nor the homologue of azaserine, *O*-α-diazopropionyl-L-serine, (Baker, 1967) are active. The following reaction sequence illustrates the proposed mechanism of action, which is based on the identification of a cysteinylacetylserine peptide formed by hydrolysis, and $O \rightarrow N$ acyl migration (Benoiton and Rydon, 1960).

$^{-}O{\cdot}CO{\cdot}CH(NH_3^{+}){\cdot}CH_2{\cdot}O{\cdot}\mathbf{CO{\cdot}CH{=}\overset{+}{N}{=}\overset{-}{N}}$ + H–S–(···-Ala-Leu-Gly-Val-Cy-···) $\longrightarrow$ $^{-}O{\cdot}CO{\cdot}CH(NH_3^{+}){\cdot}CH_2{\cdot}O{\cdot}\mathbf{CO{\cdot}CH_2{-}\overset{+}{N}{\equiv}N}$ + S^{-}–(···-Ala-Leu-Gly-Val-Cy-···)

$\longrightarrow$ N_2 + $^{-}O{\cdot}CO{\cdot}CH(NH_3^{+}){\cdot}CH_2{\cdot}O{\cdot}\mathbf{CO{\cdot}CH_2}$–**S**–(···-Ala-Leu-Gly-Val-Cy-···)

$\xrightarrow{H^{+}}$ ···-Ala-Leu-Gly-Val-NH·CH(CO·OH)·CH_2·**S·CH_2·CO**·NH·CH(CH_2OH)·CO·OH

The inhibition of acetylcholinesterase by carbamate esters, such as *Carbachol*, *Neostigmine* and *Pyridostigmine*, is also mixed competitive–non-competitive (**2**, 2).

DRUG–RECEPTOR INTERACTIONS

Affinity

There is a close parallel between the concept of enzyme–substrate and enzyme–inhibitor interactions on the one hand and that of drug (agonist)–bioreceptor and drug (antagonist)–receptor interactions on the other. This similarity is apparent in the high structural and stereochemical specificity of these reactions, and in the close analogy between agonist drug–receptor and enzyme–substrate reactions, which may be represented as:

$$\text{Drug} + \text{Receptor} \underset{k_2}{\overset{k_1}{\rightleftharpoons}} \text{Drug–Receptor Complex} \qquad (26)$$

$$\text{Drug–Receptor Complex} \longrightarrow \text{Pharmacological Response} \qquad (27)$$

Apart from the experimental difficulty of accurately measuring the concentration of drug in the biophase, and of assessing the 'concentration' of receptors, it is not possible to equate the pharmacological response even in an isolated tissue with the concentration of a product, as in enzyme–substrate reactions. Nevertheless, the close similarity of dose–response curves (Fig. 7) to those for enzyme–substrate interactions (Fig. 4) permits accurate measurement of the affinity constant (K_{aff}) of the tissue for any particular drug (Ariens *et al.*, 1964). This corresponds to the association–dissociation constant, postulated by Paton (1961) as the rate-determining factor of drug–receptor interactions.

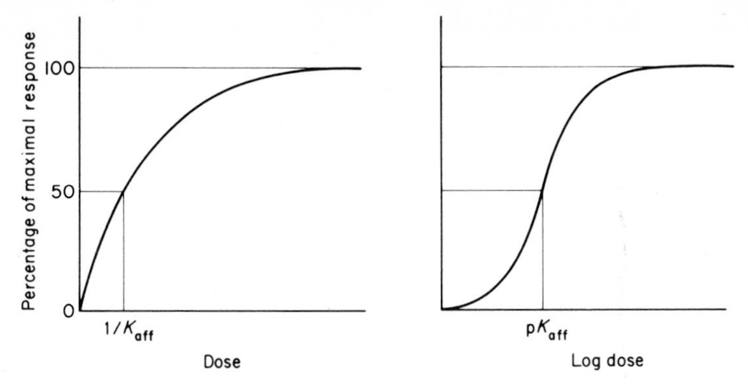

Fig. 7 Dose–response and log dose–response curves for drug–receptor interaction

The affinity constant is an association constant which merely relates to the equilibrium attained in the formation of the drug–receptor complex. An expression for it can be derived in the same way as for the Michaelis constant.

If, v_1 = velocity of forward reaction
v_2 = velocity of reverse reaction
r = total concentration of all receptors
D = concentration of drug in the biophase
C_D = concentration of drug–receptor complex

then

$$r - C_D = \text{concentration of free receptors}$$

at equilibrium

$$v_1 = k_1 D(r - C_D)$$

and

$$v_2 = k_2 - C_D$$

since $v_1 = v_2$

$$\frac{k_1}{k_2} = K_{aff} = \frac{C_D}{(r - C_D)D} \tag{28}$$

If now it is assumed that the pharmacological response is proportional to the extent to which receptors are occupied by the drug, then a 50% response will be obtained when half the receptors form the drug–receptor complex. This will correspond to a situation in which

$$\frac{C_D}{r - C_D} = 1$$

and equation (28) becomes

$$K_{aff} = \frac{1}{D_{50}}$$

in which

K_{aff} is the affinity constant

and

D_{50} is the dose which will produce a 50% response.

Intrinsic Activity

In practice, the observed effect on an isolated tissue is not always linearly proportional to the stimulus, since some drugs, called **partial agonists**, produce less than the maximum response. In such cases, the response does not depend solely on the proportion of receptors occupied, but on the **efficacy** (Stephenson, 1956) or **intrinsic activity** (Ariens, 1954), which is a measure of the effectiveness with which the drug–receptor complex is able to bring about its pharmacological response. The **intrinsic activity** (α) is equal to 1 when all the receptors are occupied by an active agonist ($D_r = 1$). For drugs which exhibit a lower maximal response, α is less than unity and this is determined experimentally by comparing the maximal response with that of a highly active agonist.

Drug Antagonism

The parallel with enzymic reactions extends to antagonist–receptor interactions. Similarly, drug antagonism may also be classified either as **competitive**, **uncompetitive**, **non-competitive**, or **mixed competitive–non-competitive** exactly in the same way as for enzymic inhibition. The equilibria in competitive drug antagonism may be represented as:

$$\text{Drug} + \text{Antagonist} + \text{Receptor} \rightleftharpoons \text{Drug–Receptor Complex} + \text{Antagonist–Receptor Complex} \quad (29)$$

$$\text{Drug–Receptor Complex} \rightleftharpoons \text{Pharmacological Response} \quad (30)$$

If, as before, r = total concentration of all receptors
D = concentration of drug in the biophase
C_D = concentration of drug–receptor complex

and

A = concentration of antagonist in the biophase
C_A = concentration of drug–antagonist complex
K_D = affinity constant of drug–receptor complex
K_A = affinity constant of drug–antagonist complex

then

$$K_D = \frac{C_D}{(r - C_D - C_A)D}$$

$$K_A = \frac{C_A}{(r - C_D - C_A)A}$$

Rearranging and equating these expressions, with the elimination of C_A, gives equation (31) in which the affinity constant is effectively increased by a factor of $(1 + K_A \cdot A)$ compared with that of the drug–receptor complex in the absence of antagonist (equation 28).

$$K_D = \frac{C_D}{(r - C_D)} \cdot \frac{(1 + K_A \cdot A)}{D} \tag{31}$$

A typical series of log-dose–response curves (Fig. 8A) for competitive antagonism shows a series of parallel displacements to the right with increasing antagonist concentration. Examples of competitive drug antagonism include the antagonism of acetylcholine by *Tubocurarine* and tubocurarine-like neuromuscular blocking agents at the motor-end-plate (Ariens, Simonis and van Rossum, 1964), and the antagonism between morphine and antagonists, such as *Nalorphine* (Cox and Weinstock, 1964; Martin, 1967).

In contrast, the typical log-dose–response curve pattern for non-competitive antagonism shows a series of diverging displacements to the right with increasing antagonist concentration (Fig. 8B). Antagonism of this type is seen in the action of the depolarising neuromuscular blocking agents, such as *Decamethonium* and *Suxamethonium*, which in contrast to the tubocurarine-like blockers, are not displaced from the receptor by high concentrations of acetylcholine.

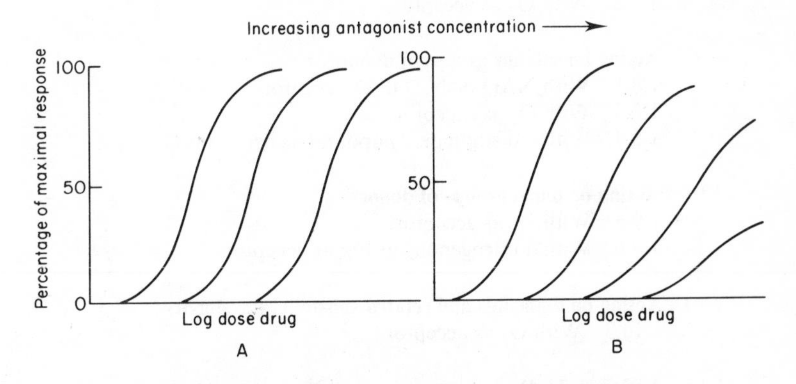

Fig. 8 Log dose–response curves for drug antagonism

A Competitive antagonism B Non-competitive antagonism

Table 3. Classification and Nomenclature of Enzymes

1. OXIDOREDUCTASES

1.1 **Acting on the CHOH group of donors**
- 1.1.1 With NAD or NADP as acceptor
- 1.1.2 With a cytochrome as acceptor
- 1.1.3 With O_2 as acceptor

1.2 **Acting on the aldehyde or keto group of donors**
- 1.2.1 With NAD or NADP as acceptor
- 1.2.2 With a cytochrome as acceptor
- 1.2.3 With O_2 as acceptor
- 1.2.4 With lipoate as acceptor

1.3 **Acting on the CH—CH group of donors**
- 1.3.1 With NAD or NADP as acceptor
- 1.3.2 With a cytochrome as acceptor
- 1.3.3 With O_2 as acceptor

1.4 **Acting on the CH—NH_2 group of donors**
- 1.4.1 With NAD or NADP as acceptor
- 1.4.3 With O_2 as acceptor

1.5 **Acting on the C—NH_2 group of donors**
- 1.5.1 With NAD or NADP as acceptor
- 1.5.3 With O_2 as acceptor

1.6 **Acting on NADH or NADPH as donor**
- 1.6.1 With NAD or NADP as acceptor
- 1.6.2 With a cytochrome as acceptor
- 1.6.4 With a disulphide compound as acceptor
- 1.6.6 With a nitrogenous group as acceptor

1.7 **Acting on other nitrogenous compounds as donors**
- 1.7.3 With O_2 as acceptor

1.8 **Acting on sulphur groups of donors**
- 1.8.1 With NAD or NADP as acceptor
- 1.8.3 With O_2 as acceptor
- 1.8.4 With a disulphide compound as acceptor

1.9 **Acting on haem groups of donors**
- 1.9.3 With O_2 as acceptor
- 1.9.6 With a nitrogenous group as acceptor

1.10 **Acting on diphenols and related substances as donors**
- 1.10.3 With O_2 as acceptor

1.11 **Acting on H_2O_2 as acceptor**

1.98 **Enzymes using H_2 as reductant**

1.99 **Other enzymes using O_2 as oxidant**
- 1.99.1 Hydroxylases
- 1.99.2 Oxygenases

2 TRANSFERASES

2.1 **Transferring one-carbon groups**
- 2.1.1 Methyltransferases
- 2.1.2 Hydroxymethyl-, formyl-, and related transferases
- 2.1.3 Carboxyl- and carbamoyl-transferases

2.2 **Transferring aldehydic or ketonic residues**

2.3 **Acyltransferases**
- 2.3.1 Acyltransferases
- 2.3.2 Aminoacyltransferases

2.4 **Glycosyltransferases**
- 2.4.1 Hexosyltransferases
- 2.4.2 Pentosyltransferases

2.5 **Transferring alkyl or related groups**

2.6 **Transferring nitrogenous groups**
- 2.6.1 Aminotransferases
- 2.6.2 Amidinotransferases
- 2.6.3 Oximinotransferases

2.7 **Transferring phosphorus-containing groups**
- 2.7.1 Phosphotransferases with an alcohol group as acceptor
- 2.7.2 Phosphotransferases with a carboxyl group as acceptor
- 2.7.3 Phosphotransferases with a nitrogenous group as acceptor
- 2.7.4 Phosphotransferases with a phospho-group as acceptor
- 2.7.5 Phosphotransferases apparently intramolecular
- 2.7.6 Pyrophosphotransferases
- 2.7.7 Nucleotidyltransferases

2.8 **Transferring sulphur-containing groups**
- 2.8.1 Sulphurtransferases
- 2.8.2 Sulphotransferases
- 2.8.3 CoA-transferases

3. HYDROLASES

3.1 **Acting on ester bonds**
- 3.1.1 Carboxylic ester hydrolases
- 3.1.2 Thiolester hydrolases
- 3.1.3 Phosphoric monoester hydrolases
- 3.1.4 Phosphoric diester hydrolases
- 3.1.5 Triphosphoric monoester hydrolases
- 3.1.6 Sulphuric ester hydrolases

Table continued overleaf

3.2 **Acting on glycosyl compounds**
- 3.2.1 Glycoside hydrolases
- 3.2.2 Hydrolysing *N*-glycosyl compounds
- 3.2.3 Hydrolysing *S*-glycosyl compounds

3.3 **Acting on ether bonds**
- 3.3.1 Thioether hydrolases

3.4 **Acting on peptide bonds (peptide hydrolases)**
- 3.4.1 α-Aminopeptide aminoacidohydrolases
- 3.4.2 α-Carboxypeptide aminoacidohydrolases
- 3.4.3 Dipeptide hydrolases
- 3.4.4 Peptide peptidohydrolases

3.5 **Acting on C—N bonds other than peptide bonds**
- 3.5.1 In linear amides
- 3.5.2 In cyclic amides
- 3.5.3 In linear amidines
- 3.5.4 In cyclic amidines

3.6 **Acting on acid anhydride bonds**
- 3.6.1 In phosphoryl-containing anhydrides

3.7 **Acting on C—C bonds**
- 3.7.1 In ketonic substances

3.8 **Acting on halide bonds**
- 3.8.1 In C-halide compounds
- 3.8.2 In P-halide compounds

3.9 **Acting on P—N bonds**

4. LYASES

4.1 **Carbon–carbon lyases**
- 4.1.1 Carboxy-lyases
- 4.1.2 Aldehyde-lyases
- 4.1.3 Ketoacid-lyases

4.2 **Carbon–oxygen lyases**
- 4.2.1 Hydro-lyases

4.3 **Carbon–nitrogen lyases**
- 4.3.1 Ammonia-lyases
- 4.3.2 Amidine-lyases

4.4 **Carbon–sulphur lyases**

4.5 **Carbon–halide lyases**

5. ISOMERASES

5.1 **Racemases and epimerases**
- 5.1.1 Acting on aminoacids and derivatives
- 5.1.2 Acting on hydroxyacids and derivatives
- 5.1.3 Acting on carbohydrates and derivatives

5.2 ***cis–trans*-Isomerases**

5.3 **Intramolecular oxidoreductases**
- 5.3.1 Interconverting aldoses and ketoses
- 5.3.2 Interconverting keto- and enol-groups
- 5.3.3 Transposing C=C bonds

5.4 **Intramolecular transferases**
- 5.4.1 Transferring acyl groups
- 5.4.2 Transferring phosphoryl groups

5.5 **Intramolecular lyases**

6. LIGASES

6.1 **Forming C—O bonds**
- 6.1.1 Aminoacid–RNA ligases

6.2 **Forming C—S bonds**
- 6.2.1 Acid–thiol ligases

6.3 **Forming C—N bonds**
- 6.3.1 Acid–ammonia ligases (amide synthetases)
- 6.3.2 Acid–aminoacid ligases (peptide synthetases)
- 6.3.3 Cyclo–ligases
- 6.3.4 Other C—N ligases
- 6.3.5 C—N ligases with glutamine as *N*-donor

6.4 **Forming C—C bonds**

2 Bonding and Biological Activity

Introduction

In ultimate molecular terms, the action of a drug on living tissue results from either a physical interaction, a chemical reaction, or more often than not a combination of physical and chemical interactions between the drug and some natural receptor molecule. In this, the drug may compete with the natural substrate for the active or allosteric site of an enzyme, thereby inhibiting the enzyme (**2**, 1). Alternatively, it may react with an essential metabolic intermediate or with some natural macromolecule such as a structural, cytoplasmic or enzymic protein, a nucleic acid, or lipopolysaccharide. However, irrespective of the nature of receptor molecule, the first step is one of association with the drug. This is promoted by physical forces including van der Waals forces, hydrogen bonding, charge-transfer complexing, electrostatic and hydrophobic interactions, and not infrequently, is stabilised by the formation of one or more chemical bonds which may be ionic, covalent or co-ordinate-covalent. A multiplicity of physical and chemical interactions is almost invariably involved in any one drug–receptor interaction, but in many instances one particular type of interaction or bond formation has a decisive influence.

VAN DER WAALS FORCES

The fundamental attractive forces between one molecule and another are known as **van der Waals forces** or London (1930) dispersion forces. They arise from zero point energy, i.e. the energy which all molecules possess even in their lowest energy state. This energy manifests itself as a result of the internal vibration of the constituent atoms of the molecule with respect to each other. The result is the continuous creation of a large number of transient dipoles within the molecule. The summation of all these local temporary dipoles results in averaging, so that the net molecular dipole is zero. Nonetheless, individual localised temporary dipoles are able to induce similar dipoles in phase with themselves in adjacent molecules, irrespective of whether the latter are chemically of the same species or different. This creates a series of small net attractive forces between molecules, wherever they approach sufficiently closely to each other. Since these forces have their origins in atomic vibrations, each individual dipole has only transient existence with a life of about 10^{-6} s in duration. They also occur only at very short range, their intensity being inversely proportional to (approximately) the sixth power of the distance separating the two molecules. There is, however, a minimum intermolecular distance beyond which one molecule is unable to

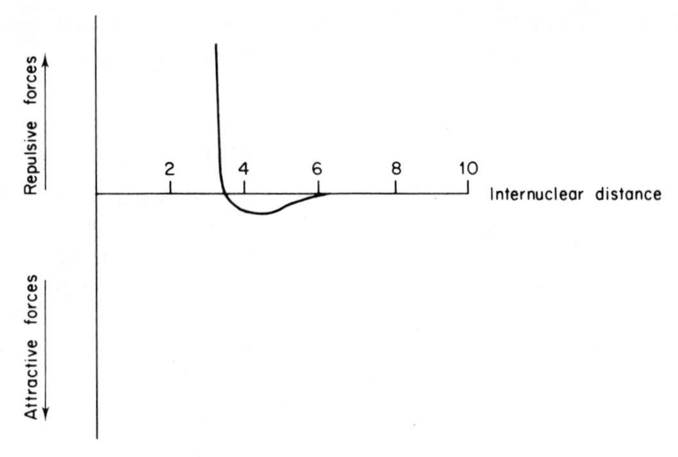

Fig. 9 Variation of intermolecular attraction and repulsion forces with internuclear separation

approach more closely to another that sets an upper limit to the size of van der Waals forces. This is due to both steric and electrostatic repulsion forces, which increase rapidly and exceed the attraction force once the critical intermolecular separation distance is reached (Fig. 9).

This critical intermolecular separation distance is given by the sum of the van der Waals radii of the two atoms concerned. The van der Waals radius of an atom is always greater than its covalent radius (Table 6, **1**, 2), and provides a measure of the minimum effective radius of an atom. Thus, two unbonded atoms in a single molecule or in two different molecules will not normally approach more closely to each other than the distance represented by the sum of their van der Waals radii. Any closer approach would require an expenditure of energy, giving rise to steric compression, and may result in bond shortening, or more likely, bond bending, since the energy requirement for the latter is less than for the former. These factors combine, therefore, to limit the effective distance within which van der Waals forces operate to within about 4–6Å. Within these limits, typical van der Waals forces range from about 1.9 down to 0.3 kJ mol^{-1}.

The influence of van der Waals attraction forces increases with molecular complexity, as the points of contact between molecules increase in number. Such forces, therefore, clearly contribute to the stabilisation of the charge-transfer complexes formed, for example, by polycyclic aromatic carcinogens with their bioreceptors (**1**, 22); by antimalarial acridines and various carcinolytic agents with DNA (p. 57); and by *Procaine* with thiamine phosphate (p. 59). The enhancement of the antibacterial activity of acridines and related bases with increasing surface area is similarly a reflection of increased van der Waals attraction.

Salem (1962) has shown that for association of long-chain hydrocarbons, where the distance between the hydrocarbon chains is small compared to the total chain length, van der Waals attraction forces are increased, and are inversely proportional approximately to [intermolecular distance]^5. He calculates that in a monomolecular film of stearic acid, the attraction force between the hydrocarbon tails of the acid, which are 4.8Å apart, is 33.5 kJ mol^{-1}. In isostearic acid films, however, where the intermolecular chain separation is increased to 6.0Å due to chain-branching, the intermolecular attraction force is reduced to 12 kJ mol^{-1}.

Such forces also play a significant rôle in the binding of steroids to plasma proteins, and are seen in the enhancement of steroid solubilities by such simple amino acids as valine and leucine (Miller, 1972). They also account for the enhancement of fat solubility, protein binding and hypnotic activity of hydrocarbon substituents in barbiturates (**1**, 17; **2**, 1 and 4).

Van der Waals interaction forces are also directly related to the polarisability of the atoms and groups concerned. Polar aromatic substituents markedly enhance van der Waals interactions between haptens and antibodies (Pauling and Pressman, 1945), in accordance with the following expression in which molar refractivities are a direct measure of polarisation:

$$\Delta W = \frac{15.9 \times 10^5}{r^6}(R_A - R_H)\text{kJ mol}^{-1}$$

where ΔW = change of van der Waals energy produced by substituting a group A for hydrogen in the hapten molecule

r = intermolecular (hapten–antibody) distance

R_A = molar refractivity of substituent A

R_H = molar refractivity of hydrogen

Table 4 illustrates the change (ΔW) in van der Waals energy which can occur as a result of aromatic substitution by particular substituents at various typical

Table 4. Effect of Polar Substituents on van der Waals Interaction Energies of Aromatic Compounds

		ΔW in kJ mol^{-1} at internuclear separations of		
Substituent A	$R_A - R_H$	4Å	5Å	6Å
CH_3CONH—	13.0	5044	1320	444
O_2N—	6.58	2550	668	225
I—	12.98	5035	1318	443
Br—	7.79	3020	790	265
Cl—	5.0	1940	506	170
F—	−0.14	−42	−14	−4.8
CH_3O—	6.86	2660	698	235
CH_3—	4.92	1910	500	168
HO—	1.81	710	184	62
H_2N—	4.44	1720	450	151

intermolecular separation distances. Even at 6Å, the energy change is seen in most cases to be the equivalent of a covalent bond (150–600 kJ mol^{-1}), whilst at 4Å, the interactions are very substantial indeed. It follows by analogy, therefore, that the enhanced affinity of halogenated anaesthetic hydrocarbons, such as *Chloroform* ($CHCl_3$), for lipid tissue is, therefore, partly due to the polarity of the halogen. The negative value of ΔW for fluorine in the replacement of hydrogen is of interest in considering its effect in fluorinated steroids, and in trifluoromethyl-substituted compounds, such as *Trifluoperazine*, particularly in view of the fact that the van der Waals radius of fluorine (1.33Å) is little more than that of hydrogen (1.2Å). These factors are also clearly of considerable significance as determinants of polarisation and polarisability in such interhalogen compounds as the anaesthetics, *Halothane* ($CHBrCl \cdot CF_3$) and *Methoxyfluorane* ($CHCl_2 \cdot CF_2 \cdot O \cdot CH_3$), and in the aerosol inhalation propellants, *Dichlorodifluoromethane* ($CCl_2 \cdot F_2$) and *Dichlorotetrafluoroethane* ($CClF_2 \cdot CClF_2$).

HYDROGEN BONDING

Water and Aqueous Solubility

Hydrogen bonding, the fundamentals of which have been briefly described (**1**, 1), plays an essential role in the stabilisation of many important physiological systems, as well as in their interactions with drugs. Water, the fundamental solvent of all biological systems, is itself extensively hydrogen-bonded, and hence shows evidence of a partially ordered structure. Neutron diffraction studies and the Raman spectrum of water in D_2O show evidence of both hydrogen-bonded and non-hydrogen-bonded water molecules, which accounts for the fact that water exhibits both elastic (liquid-like) and non-elastic (solid-like) properties. The rôle of hydrogen bonding in the structuring of water is apparent from crystallographic studies, which show that in the solid state of ice-I, each oxygen atom

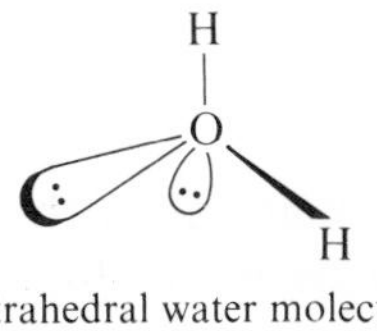

Tetrahedral water molecule

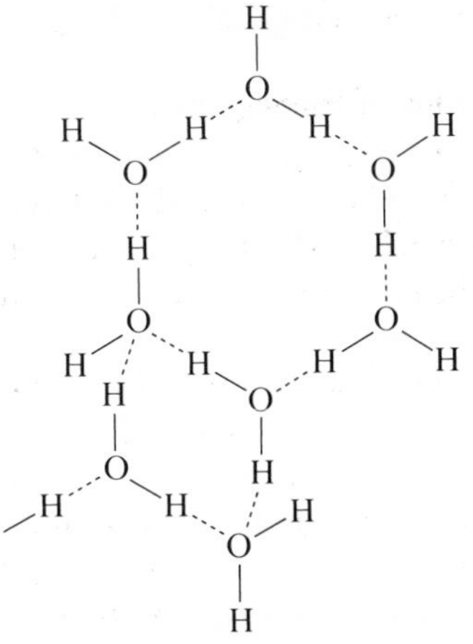

Hydrogen bonding of water molecules in ice-I

is bonded tetrahedrally through hydrogen to four other oxygen atoms. This type of open lattice has a series of hexagonal holes which are capable of accommodating other molecules, either with or without bond distortion.

The structuring of liquid water is an important factor in hydrophobic interactions (p. 51), and probably accounts for the occasional, quite marked difference in properties of otherwise closely-related substances with which water is capable of hydrogen bonding (Franks, 1969). For example, the significantly lower partial molal volume of β-methyl glucoside compared to its α-anomer is attributed to the much better fit of the former with the characteristic oxygen atom separation (4.75Å) in ice-I.

Hydrogen bonding leading to hydration attributes water solubility to π-deficient N-heterocyclics, such as pyridine (miscible) and its derivatives, *Nicotinamide* (Niacinamide; 1 in 1) and *Nikethamide* (miscible), which are markedly more soluble than the analogous benzenoid aromatics, benzene (1 in 1200) and salicylamide (1 in 500). Water solubility may be increased as the number of N-hetero-atoms is increased, but other contrary factors frequently intervene as, for example, intramolecular hydrogen bonding in *Pyrazinamide*, which is significantly less soluble (1 in 60) than *Nicotinamide*. In contrast, π-excessive N-hetero-aromatics in which the nitrogen-lone pair is less readily available for hydrogen bonding have quite low water solubilities (**1**, 23).

The solvating effect of hydrogen bonding to water, however, is often heavily outweighed by the adverse influence of large hydrophobic groups as, for example, the hydrocarbon skeleton in hydroxy- and keto-steroids. Even so, the extent of hydroxylation of the cardiotonic glycosides, *Digitoxin*, *Digoxin* and *Ouabain*, influences both water and lipid solubility, and hence contributes significantly to their absorption and excretion characteristics (Caldwell, Martin, Dutta and Greenberger, 1969). Thus, *Digitoxin*, the least hydroxylated, is the most readily absorbed and gives the highest blood levels, whilst *Ouabain*, the most heavily hydroxylated, is only slowly absorbed, but rapidly excreted giving a very short duration of action (Table 73, **1**, 22).

Stabilisation of Biopolymers

Hydrogen bonds stabilise the tertiary structure of proteins as in the α-helix (**1**, 19; Fig. 27D) proposed by Pauling and Corey. This is stabilised by the formation of N—H---O hydrogen bonds between certain of the amide hydrogens and carboxyl groups situated in adjacent turns of the helix. Not all amide hydrogens, however, are involved in inter-chain hydrogen bonding; others are available for association with solvent water or for the formation of intermolecular hydrogen bonds with drug molecules. The latter exchange readily with deuterium oxide, whereas inter-helix hydrogen-bonded hydrogens exchange much less readily (Blout, de Losé and Asadourian, 1961). Measurement of HEAH (hard to exchange amide hydrogen) gives a measure of intramolecular hydrogen bonding. Denaturation of proteins with urea or other denaturants breaks the intramolecular bond and permits unravelling of the ordered helical or sheet structure. Stabilisation of the double helix of RNA and DNA is also largely the result of multiple hydrogen bonding between the purine and pyrimidine base pairs (**1**, 23).

Enzyme Inhibition

Inhibition of ribonucleases by cytidine 3′-phosphate has been shown to be accompanied by increases of pK_a and downfield shifts of the C_2H PMR resonance of two of the four histidine residues present in the enzyme. Thus, histidine 119 shows an increase of pK_a from 5.8 to 7.4 and a marked shift of 60Hz due to hydrogen bonding with the nucleotide phosphate group, whilst histidine 12 undergoes a similar increase of pK_a from 6.2 to 8.0, but a somewhat smaller downfield shift of its C_2H due to interaction with ribose (Fig. 10, Meadows and Jardetzky, 1968). At the same time, the doublet due to the proton at C_6 of the nucleotide is broadened and shifted downfield due to the ring current from phenylalanine 120, which is adjacent to it. The enzyme–inhibitor complex may be represented as shown in Fig. 11 (Meadows, Roberts and Jardetzky, 1969).

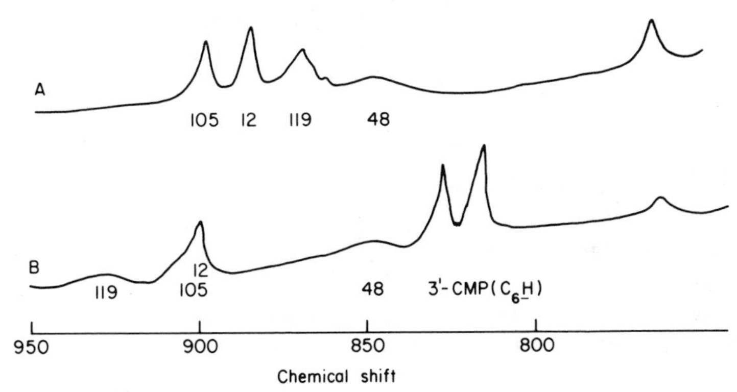

Fig. 10 PMR spectra of ribonuclease (A) and ribonuclease—cytidine complex (B)

Hydrogen Bonding in Drug Action

Steroid–Protein Binding

Many reactions of hormones and drugs with biopolymers and metabolites at molecular level are dependent on hydrogen bond formation to secure the correct juxtaposition of drug and receptor. Hydrogen bonds are weak (Table 5) in contrast to covalent bonds, and irrespective of the particular donor or acceptor, lie within fairly narrow limits (7-40 kJ mol^{-1}). The extent, therefore, to which any one drug–receptor hydrogen bond enhances the stability of the complex is questionable, since most drugs capable of forming hydrogen bonds may well be hydrated in aqueous media. The transfer of a hydrogen bond from water to bioreceptor is unlikely, therefore, to be accompanied by any appreciable energy change. This is clearly illustrated by the very weak protein binding characteristics

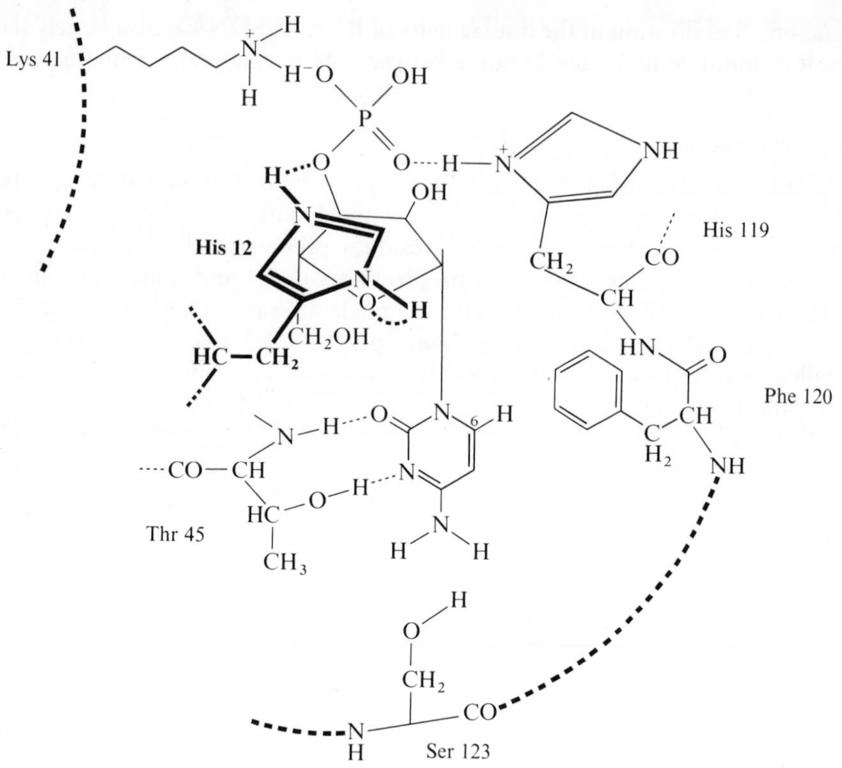

Fig. 11 Ribonuclease—cytidine 3′-phosphate complex

of polyhydroxy compounds, such as sugars, which are heavily hydrated in aqueous solution (Bruce, Giles and Jain, 1958; Giles and McKay, 1962).

In contrast, keto-steroids with but one or two hydrogen bond acceptor groups are more strongly bound to plasma proteins, due no doubt to the hydrophobic interactions of their not unsubstantial hydrocarbon skeleton. Nevertheless, whilst androstane-17-one and androstane-3,17-dione, which differ only in the 3-oxo group of the latter, are both markedly more soluble in bovine serum albumin (5%) than in a standard buffer solution at pH 7.3, the solubility of the dione is 3250 μmol l^{-1} of albumin solution compared with only 140 μmol l^{-1}

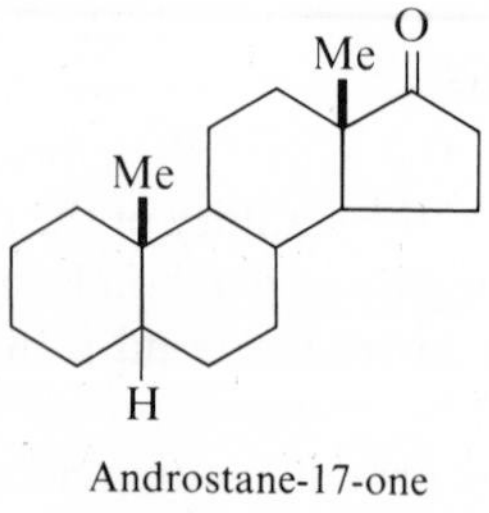

Androstane-17-one

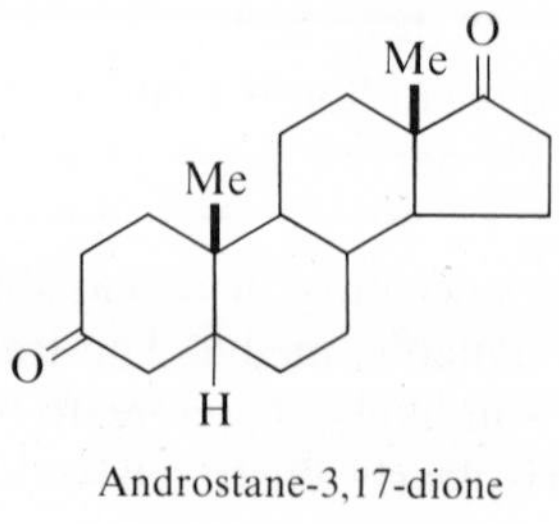

Androstane-3,17-dione

for androstane-17-one (Eik-nes, Schellman, Lumry and Samuels, 1954). This substantial difference is no doubt due to hydrogen bonding.

Similarly, the solubility of 17-methyltestosterone in aqueous solutions of synthetic peptides (angiotensin amide) is significantly greater than that of testosterone (Miller, 1972). This accords with the enhanced hydrogen bond acceptor properties of the former as a result of electron donation from the methyl group, though enhanced van der Waals attraction may also make a significant contribution to binding. A similar enhancement of steroid binding to human serum albumin by a suitably placed methyl substituent has also been demonstrated by Westphal and Ashley (1958) for both 2α- and 4-methyl substituted Δ^4-3-ketosteroids (Table 6), which seems more likely, therefore, to implicate enhancement of hydrogen bonding acceptor properties than van der Waals attraction.

Actinomycin D–DNA Complex

The *Actinomycin D*–DNA complex, which acts as an inhibitor of DNA-dependent RNA polymerase (Kersten, Kersten and Rauen, 1960; Harbers and Müller, 1962), has been shown to be formed by intercalation of the amino-phenoxazine nucleus between the base-pairs of the double helix, immediately adjacent to a guanine–cytosine base-pair (Müller and Crothers, 1968). Helix distortion occurs due to hydrogen bonding of the D-valyl-NH groups of the actinomycin cyclic peptide units with DNA. PMR studies of *Actinomycin D* (Lackner, 1970) show that in the stable conformation of the peptide, these are internally hydrogen bonded to the sarcosine carbonyl groups (Fig. 28, **1**, 19). In the DNA complex, however, the cyclic peptides undergo conformational changes which permit interaction by hydrogen bonding with the DNA backbone.

Since a guanine-free double helical nucleic acid containing a 2,6-diaminopurine instead of adenine also binds actinomycin (Cerami, Reich, Ward and Goldberg, 1967), it is suggested that there is hydrogen bonding of the actinomycin quinone oxygen to an amino group in the 2-position of the purine. Accordingly, Müller and Crothers (1968) have postulated that the complex is further stabilised by hydrogen bonding between the D-valyl-NH group (acting as donor) on the benzenoid side of the actinomycin nucleus, and the phosphate oxygen between guanine and the base on the other side of the intercalated actinomycin. The acceptor for the D-valyl-NH hydrogen bond on the quinonoid side of the actinomycin nucleus is the keto oxygen of cytosine in the guanine–cytosine base-pair serving as the binding site (Fig. 12). The complex is stereospecific, since the optical antipode of *Actinomycin D*, in which the configuration of all the asymmetric centres has been reversed, neither forms a complex nor shows biological activity.

Dimethicone–Silica

The intrinsic antifoaming properties of polydimethylsiloxanes, such as *Dimethicone* 1000, which find application in their use as antiflatulents, are limited by their

Table 5. Intermolecular Hydrogen Bonds

Hydrogen bond	Donor	Acceptor	Solvent	$-\Delta H^\circ$ (kJ mol^{-1})
—O—H···O=C	Acetic acid	Acetic acid	Vapour phase	30.5
	Phenol	Acetone	CCl_4	19.5
	Cholesterol	Glyceryl triacetate	CCl_4	18
—O—H···O<	Methanol	Methanol	Vapour phase	32
	Methanol	Dibutyl ether	Vapour phase	19.5
	Water	Dioxan	CCl_4	14.5
	Phenol	n-Dibutyl ether	CCl_4	24
	Phenol	Dioxan	CCl_4	21
—O—H···S<	Phenol	n-Dibutyl sulphide	CCl_4	17.5
—O—H···N≤	Methanol	Triethylamine	Vapour phase	32
	Phenol	Triethylamine	CCl_4	35
—O—H···N(=)—	Phenol	Pyridine	CCl_4	27
—O—H···X (halogen)	Phenol	Cyclohexyl fluoride	CCl_4	15.5
	Phenol	Cyclohexyl chloride	CCl_4	9.2
	Phenol	Cyclohexyl bromide	CCl_4	8.8
	Phenol	Cyclohexyl iodide	CCl_4	7
—O—H···π	Phenol	Acetonitrile	CCl_4	18
	Phenol	Hexamethylbenzene	CCl_4	7

$-S-H\cdots N\lessgtr$	Thiophenol	Pyridine	CCl_4	10
$-N-H\cdots O=C\lessgtr$	γ-Butyrolactam	γ-Butyrolactam	CCl_4	14.5
	α-Pyridone	α-Pyridone	CCl_4	18.5
$-N-H\cdots S=C\lessgtr$	γ-Thiobutyrolactam	γ-Thiobutyrolactam	CCl_4	12
	α-Thiopyridone	α-Thiopyridone	CCl_4	14.5
$-N-H\cdots O\lessgtr$	Aniline	Tetrahydrofuran	C_6H_{12}	12.5
	Thiocyanic acid	Tetrahydrofuran	CCl_4	26.5
$-N-H\cdots S\lessgtr$	Thiocyanic acid	n-Butyl sulphide	CCl_4	15
$-N-H\cdots N\lessgtr$	Aniline	Aniline	C_6H_{12}	7
$-N-H\cdots N\lessgtr$	Aniline	Pyridine	C_6H_{12}	14.5
	Indole	Pyridine	CCl_4	15
	Pyrrole	Pyridine	CCl_4	13.5
$-N-H\cdots\pi$	Aniline	Benzene	C_6H_{12}	7

Table 6. Influence of 2α-Methyl and 4-Methyl Substituents on the Binding of Δ^4-3-ketosteroids to Human Serum Albumin (Westphal and Ashley, 1958)

Steroid	Δε (Human serum albumin)[1]
Testosterone	13.2
2α-Methyltestosterone	16.5
4-Methyltestosterone	16.0
11-Deoxycorticosterone	12.6
2α-Methyl-11-deoxycorticosterone	17.6
Cortisone	7.3
2α-Methylcortisone	13.5
Cortisol	5.7
2α-Methylcortisol	9.2
19-Nortestosterone	13.6
4-Methyl-19-nortestosterone	18.2

[1] $\Delta\varepsilon = \dfrac{100(\varepsilon_w - \varepsilon_p)}{\varepsilon_w}$

where ε_w = molecular extinction coefficient of steroid in buffer
ε_p = molecular extinction coefficient of steroid in presence of protein at $\lambda_{max}^{H_2O}$ of the protein-free solution

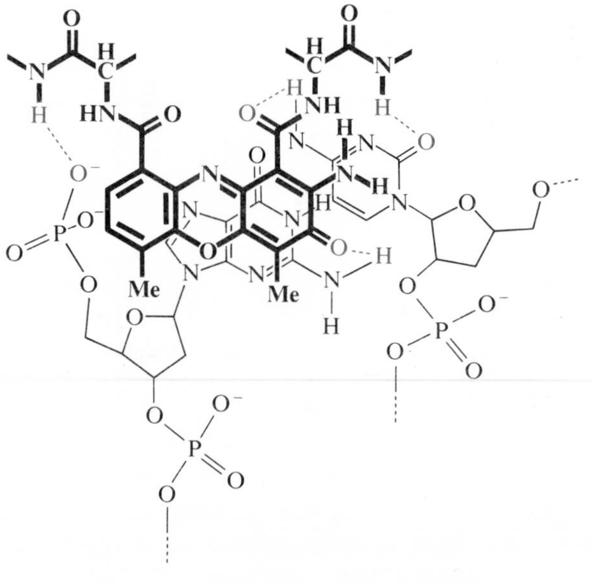

Fig. 12 *Actinomycin D*—DNA complex

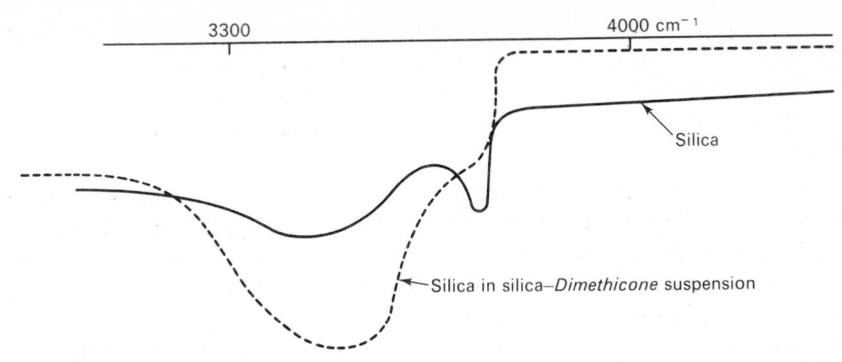

Fig. 13 Infrared absorption spectra (hydroxyl region) of silica, and silica in silica–*Dimethicone* suspension.

high viscosity and hydrophobic properties. These antifoaming properties are enhanced by mixing with finely divided silica or aluminium hydroxide when they become spread out and strongly bound as a hydrophobic film on the solid support particle surface. Despite various suggestions to the contrary, there is no chemical reaction in the accepted sense of the term between the two components, but merely strong intermolecular hydrogen bonding of the *Dimethicone* and the silica surface (Buist, Burton and Elvidge, 1973). This is demonstrated by the enhanced hydrogen bonded OH absorption at 3450 cm^{-1} and a complementary fall in the free OH absorption at 3700 cm^{-1} of silica in silica–*Dimethicone* mixtures (Fig. 13). Presentation of the liquid *Dimethicone* as a film, coating the solid surface of the finely divided silica particles, provides an easy means of exposing a large surface of the silicone polymer to the foam. This enhances dispersability of the *Dimethicone* throughout the foams, and thereby considerably increases its antifoaming efficiency (Birtley, Burton, Kellett, Oswald and Pennington, 1973).

HYDROPHOBIC INTERACTIONS

Principles

Hydrophobic interactions arise in hydrocarbon structures essentially from the inability of the carbon–hydrogen bond to enter into hydrogen bonding. This means that hydrocarbons do not become solvated in water and other hydroxylic solvents. As a result, adjacent solvent molecules in hydrocarbon–water mixtures become more ordered in an attempt to assist accommodation of the hydrocarbon molecules within the body of the solvent. This localised increase in the ordered structure of the 'solvent' results in a loss of entropy, and hence in an increase in the free energy of the system. At the same time, formation of an interface between water-insoluble non-polar groups and the aqueous medium also requires the input of energy to oppose the interfacial tension. In consequence, hydrocarbon structures undergo conformational changes. These result in chain contraction which, together with the mutual association of like molecules, leads to a reduction of the overall hydrophobic surface area exposed to water. This reduction in interfacial area is automatically accompanied by a decrease in the structured component of the solvent, and consequently results in a regain of entropy.

The overall effect is readily apparent when a liquid hydrocarbon such as cyclohexane is shaken with water. The oil, which is dispersed into droplets on shaking, experiences a large increase in the oil–water interfacial area, but once shaking is stopped, it immediately coalesces in pools and eventually completely separates into two phases. More refined experiments with model systems show that hydrophobic interactions are directly related to the number of interacting hydrophobic groups, and furthermore, being typical mutual solubility effects, they increase with temperature. Thus, binding constants of carboxylate ions with vinylpyrrolidone–alkylpyridinium co-polymers increase with temperature

Table 7. Binding Constants of Carboxylate Ions with Polyvinylpyrrolidone–hexylpyridinium 1:23—Co-polymer

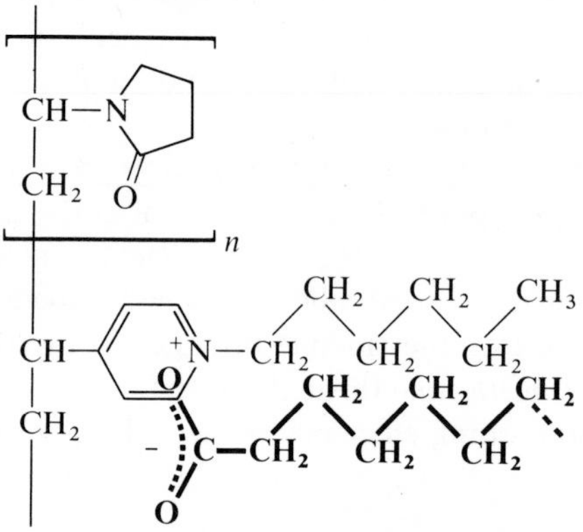

Carboxylate Ion	Temperature (°C)	Binding Constant (K in $l\,mol^{-1}$)
$CH_3{\cdot}CH_2{\cdot}CO{\cdot}O^-$	28	15.92
	37	23.57
$CH_3{\cdot}CH_2{\cdot}CH_2{\cdot}CO{\cdot}O^-$	28	22.35
	37	37.03
	45	57.48
$CH_3{\cdot}CH_2{\cdot}CH_2{\cdot}CH_2{\cdot}CO{\cdot}O^-$	28	30.82
	37	46.20
	45	98.26
$CH_3{\cdot}CH_2{\cdot}CH_2{\cdot}CH_2{\cdot}CH_2{\cdot}CO{\cdot}O^-$	28	49.64
	37	75.61
	45	170.73

and with anionic chain length (Nagwekar and Muangnoicharoen, 1973; Table 7).

Tanford (1962) has calculated the magnitude of typical hydrophobic interactions from the relative solubilities of amino acids in water and other solvents. The contribution per methylene group is quite small, being about 3.4 kJ mol^{-1}. Even so, the summation of a large number of such interactions can clearly amount to a considerable binding force. Conformational changes, which shorten the hydrocarbon chain when exposed to hydrophilic conditions, have been demonstrated in conductivity experiments with polymethylene-bis-quaternary ammonium salts (Elworthy, 1963b; **1**, 3).

Drug–Protein Binding

The presence in protein molecules of amino acid units, such as valine, leucine and isoleucine, which incorporate significant hydrocarbon fragments, provides an inbuilt capability to enter into hydrophobic bonding. This manifests itself in two ways. Firstly, it permits intramolecular hydrophobic bonding in the large flexible molecules of high molecular weight proteins, which promotes and stabilises chain-folding (**1**, 19) and creates hydrophobic pockets or clefts. Similar, but intermolecular, interactions also play a part in the association of peptides and proteins, which gives rise to quaternary structure. Secondly, intermolecular interactions between the hydrophobic areas of drug molecules and those of natural macromolecules also play a significant part in drug–macromolecular complex formation and stabilisation. Such interactions are a prominent feature of drug–protein binding.

Hydrophobic interactions account for the carrier capacity of proteins for lipid-soluble drugs. For example steroid–protein binding is, at least in part, assisted by hydrophobic effects, though the results of some experiments are

Table 8. Association Free Energy ($\Delta G°$) and Apparent Association Constants (K_a) for Interaction of Methyltesterone with Angiotensin II Amide and its Constituent Amino Acids (Stenlake, 1974)

	K_a (M^{-1})		$\Delta G°$ (kJ mol^{-1})	
Solute	4°	37°	4°	37°
Angiotensin II amide	200	1600	−2.92	−4.55
Arginine	7	4.7	−1.08	−0.95
Proline	8.4	3.5	−1.17	−0.77
Aspartic acid	14	12	−1.45	−1.53
Histidine	6.8	8.8	−1.06	−1.34
Isoleucine	4.2	17	−0.79	−1.75
Phenylalanine	14	8.8	−1.45	−1.34
Valine	5.6	14	−0.95	−1.63
Tyrosine	214	55	−2.95	−2.47

Table 9. Binding of Barbiturates to Plasma Proteins (Goldbaum and Smith, 1954)

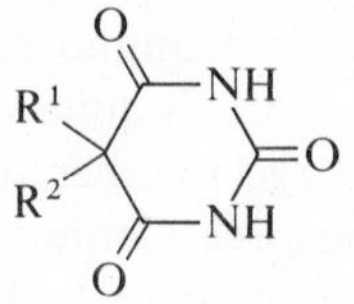

R^1	R^2	Fraction bound at pH 7.4
Ethyl	Ethyl	0.05
Ethyl	iso-Propyl	0.13
Ethyl	n-Butyl	0.28
Ethyl	iso-Amyl	0.35
Ethyl	1-Methylbutyl	0.37
Ethyl	n-Hexyl	0.65

capable of alternative interpretations. Thus, the large positive entropy of testosterone–serum albumin binding is considered to be due to changes in protein conformation rather than in the structure of associated water (Schellman, Lumry and Samuels, 1954). Nevertheless, it is evident from experiments with model systems, such as the octapeptide, angiotensin II amide and its constituent amino acids, that hydrophobic bonding to amino acid chains contributes significantly to the solubility of methyltestosterone (Table 8; Miller, 1972; Stenlake, 1974). Thus, the significant increase in ΔG° values and association constants with temperature between 4 and 37° for the interactions of methyltestosterone with isoleucine and valine indicates involvement of these amino acid residues of the peptide in hydrophobic bonding. It is also of note that the strong association of methyltestosterone with tyrosine, and the fall in both ΔG° and the association constant with rise in temperature between 4 and 37°, is indicative of hydrogen bonding to the tyrosyl hydroxyl group.

Hydrophobic interactions also play an important part as secondary factors in the protein binding of some polar molecules, such as barbiturates and fatty acids. Barbiturates, all with pK_a's about 7.6, show increased binding as the substituent chain length increases (Goldbaum and Smith, 1954), reaching 65% with hexobarbitone (Table 9).

Homologous series of fatty acids also show an increase in the binding to bovine serum albumin as chain length increases (Teresi and Luck, 1952). Similar results have been obtained for natural long-chain fatty acids (Goodman, 1958), from which it is apparent that at normal blood levels (0.5 mEq l^{-1}), less than 0.01% is unbound. Albumin, therefore, functions as an efficient medium for the transport of unesterified fatty acids; some fats and fatty acids are also carried by the α- and β-globulins.

The binding of digitalis glycosides to plasma proteins is also influenced by hydrophobic interactions (Lukus and DeMartino, 1969). Thus, *Digitoxin*, which is 97% bound in human plasma, is bound almost exclusively to albumin at a single binding site. The reaction is endothermic ($\Delta G = -33.5$ kJ mol^{-1}; $\Delta S = 142$ J mol^{-1} °K^{-1}) and with an entropy change together indicative of hydrophobic bonding accompanied by a conformational change in the protein structure (Kauzmann, 1959). In contrast, *Digoxin*, as a consequence of its more highly hydroxylated structure, is only about 23% bound to albumin in human plasma. This accounts for high plasma concentrations and slower excretion usually observed with *Digitoxin* as compared to *Digoxin* in man. It also accounts for the observation (Lukus and DeMartino, 1969) that *Digitoxin* migrates with albumin on continuous paper electrophoresis in plasma, whereas *Digoxin* does not accompany albumin. Moreover, although irrelevant to consideration of hydrophobic bonding, it is worthy of note that *Digoxin* actually fails to move from the origin in electrophoresis on starch gels, indicating a very strong affinity between the tri-digitose unit of the glycoside and the polysaccharide gel. This property may, therefore, be a factor of some importance in promoting the ready concentration of glycosides in skeletal and cardiac muscle, and in other tissues highly dependent on carbohydrate energy sources.

CHARGE-TRANSFER

Principles

Charge-transfer complexes (Mulliken, 1952) are formed by the transfer of charge from the molecules of one compound (the donor) to those of another (the acceptor). The forces involved in typical charge-transfer interactions are usually no more than about 30 kJ mol^{-1}, which is not surprising in view of their 'long range' character (*ca* 3.0–3.4Å) compared with covalent bond lengths. Charge-transfer occurs from the highest occupied molecular orbital (HOMO) of the donor to the lowest empty molecular orbital (LEMO) of the acceptor molecule, the extent of transfer being determined by whether the complex is in the ground or excited state. In the ground state, the usual intermolecular physical forces apply and the extent of charge-transfer is comparatively small. The complex, however, may be raised to the excited state by the absorption of light energy of appropriate wavelength when an electron is wholly transferred from donor to acceptor. Complex formation is then accompanied by spectroscopic changes with the appearance of new absorption bands, characteristic of the charge-transfer transitions.

The usual donor sources are either π-electrons or non-bonded electrons, so that typical donor molecules are either unsaturated compounds (ethylenic or acetylenic), aromatics with electron-donating substituents, π-excessive hetero-aromatics, or groups with unshared electron pairs, such as alcohols, ethers, thiols, sulphides, halides and amines. π-Electron donors usually give rise to weaker complexes (association constant 0.2–20 l mol^{-1}) than those derived from non-bonded electron donors (association constant up to $10^4 \, \mathrm{l\,mol^{-1}}$). Typical acceptors are π-deficient systems, particularly benzenoid aromatics with electron-withdrawing substituents, or hetero-aromatics.

The interaction between cyclohexene and iodine is well-established as a simple example of charge-transfer interaction. Thus, whilst iodine merely dissolves in cyclohexane to give a violet solution, in cyclohexene it forms a brown complex which slowly fades as addition of iodine to the double bond occurs to give *trans*-1,2-di-iodocyclohexane.

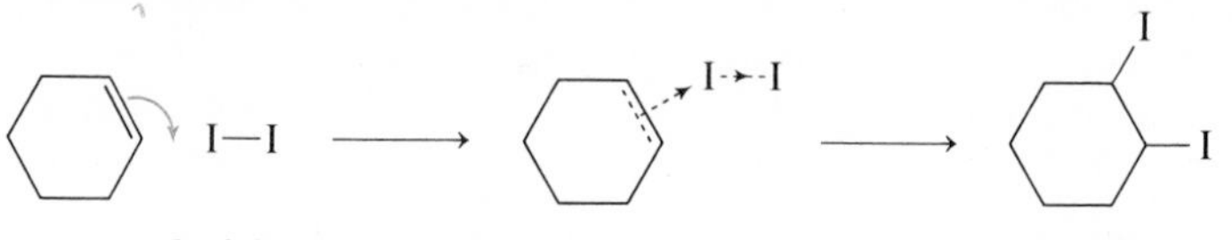

I—I ⟶ I···I ⟶ I, I

Cyclohexene–iodine complex (brown)

1,2-Di-iodocyclohexane (colourless)

Cortisone and other unsaturated keto-steroids have similarly been shown to form charge-transfer complexes with iodine (Sgent-Györgyi, 1960). The formation of yellow complexes between ethylenic compounds and tetranitromethane is also due to charge-transfer (Iles and Ledwith, 1969).

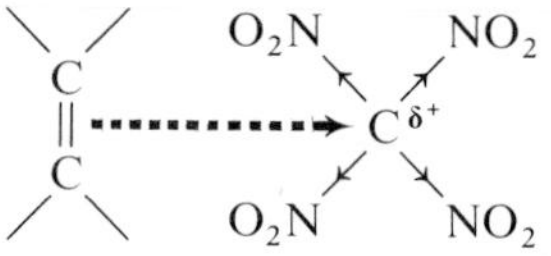

C=C ········→ $C^{\delta+}(NO_2)_4$

Picric acid and trinitrobenzene are also able to form relatively stable charge-transfer complexes with other aromatic hydrocarbons, such as mesitylene (1,3,5-trimethylbenzene) and polycyclic aromatics, such as anthracene. Complex formation in such systems involves the transfer of an electron from an electron donor (π-excessive) molecule (mesitylene) to an electron-deficient molecule (picric acid) as a result of orbital overlap. X-ray studies of charge-transfer bonding between aromatic molecules have shown that this results in a plane-to-plane orientation in the crystal lattice.

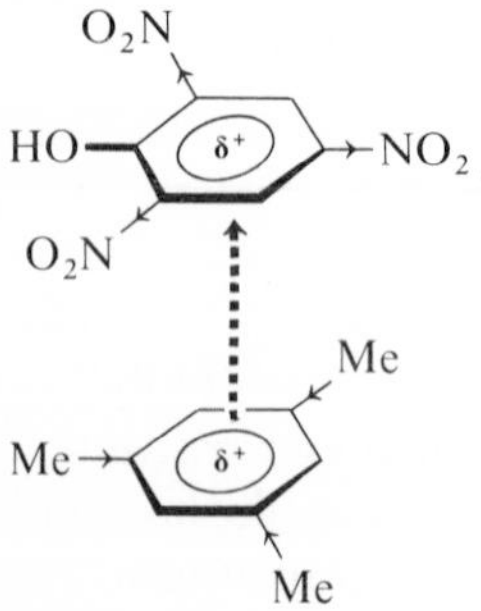

O_2N, HO, NO_2, O_2N, δ^+, Me, Me, δ^+, Me

Mesitylene–picric acid complex

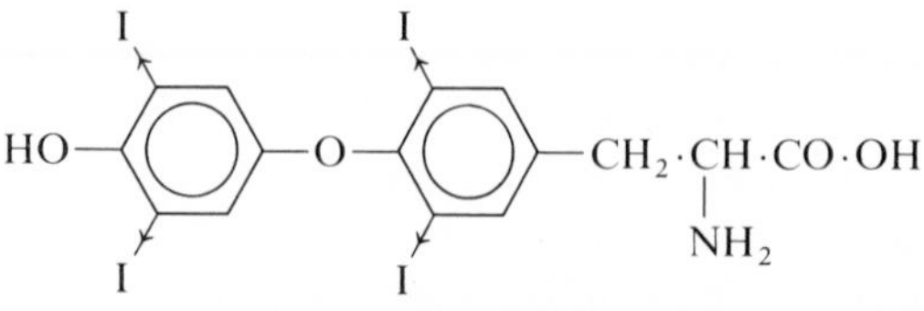

I, I, HO, O, I, I, $CH_2 \cdot CH \cdot CO \cdot OH$, NH_2

Thyroxine

Sgent-Györgyi (1960) has postulated the significance of this type of bonding in biological systems, and suggests that the powerful electron-attracting properties of 2,4-dinitrophenol and *Thyroxine* account for their ability to uncouple oxidative phosphorylation processes. Although there is strong evidence that charge-transfer complexing of polycyclic aromatic hydrocarbons is the primary event in their carcinogenic action (**1**, 22), more recent studies show that certain polyhydroxy metabolites are actually covalently bound to RNA (Nakanishi, Kasai, Harvey, Jeffrey, Jennette and Weinstein, 1977).

Proflavine–DNA

There is reasonable evidence to indicate that the antibacterial action of aminoacridines is linked to their ability to intercalate in the DNA polymer chain and form charge-transfer complexes with the purine bases (Peacocke and Skerrett, 1956). This correlates with the ability of aminoacridines to stain nucleoproteins *in vivo*, and to inhibit bacteriophage nucleoprotein synthesis in *E. coli*. (Foster, 1948).

Complexing of *Proflavine* (3,6-diaminoacridine) with DNA has been demonstrated by viscosity and sedimentation measurements of the complex in solution, and by X-ray diffraction (Lerman, 1961, 1963), to cause some distortion and unwinding of the helix. Hydrogen bonding between the base pairs is apparently unaffected, and consequently the lateral spacing between the helix strands is unaffected. There is evidence, however, for some unwinding of the backbone polymer chain with extension of the helix pitch as the mass per unit length is lowered. As a result of intercalation, chemical reactivity of the functional groups of the aminoacridine is decreased, as shown by the decrease in its diazotisation rate. The importance of the flat aromatic ring system for intercalation and complexing was demonstrated by Peacocke and Skerrett (1956), who showed that hydrogenation of one of the aromatic rings of 9-aminoacridine, which destroys the flatness of the ring system, also abolishes strong interaction with DNA, and reduces antibacterial activity to negligible proportions. A possible structure for the complex is shown in Fig. 14.

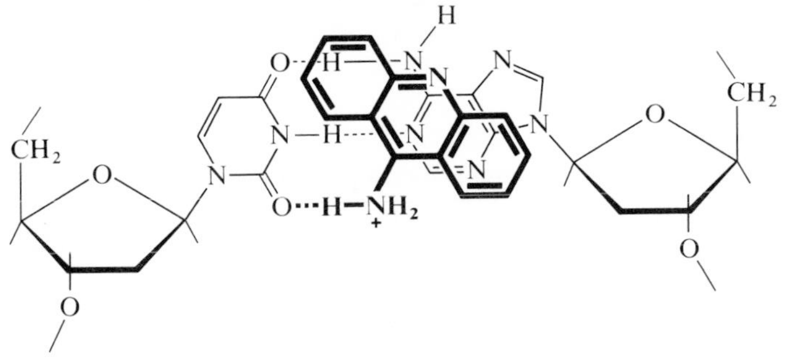

Fig. 14 9-Aminoacridine—DNA complex

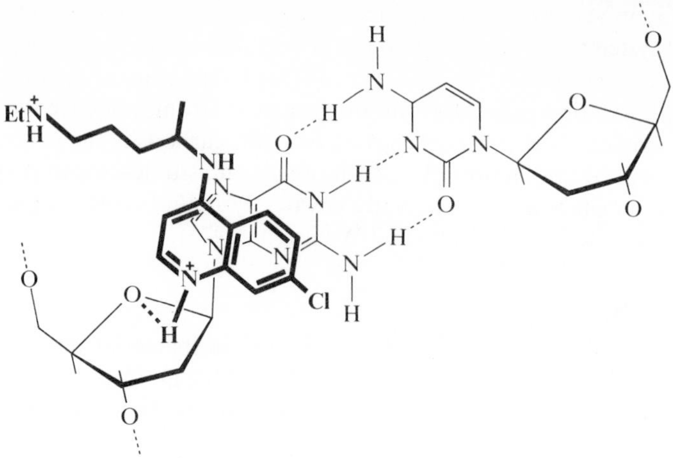

Fig. 15 *Chloroquine*—DNA complex

Chloroquine–DNA

The quinoline and acridine antimalarials, *Chloroquine* and *Quinacrine*, are also considered to act as a result of intercalation between a proportion of the base-pairs of the DNA double helix, stabilised by charge-transfer interactions. As discussed elsewhere (p. 65), *Chloroquine* is linked primarily by ionic bonding of its diamino side-chain across the minor groove of DNA. This, however, permits intercalation of the heteroaromatic nucleus which occurs preferentially adjacent to guanine or adenine (Allison, O'Brien and Hahn, 1965; Cohen and Yielding, 1965). PMR studies (Sternglanz, Yielding and Pruitt, 1969) have established charge-transfer interactions between *Chloroquine Phosphate* and both AMP and GMP, which support the view that DNA complexes may be similarly stabilised. Thus, both AMP and GMP show upfield shifts in the H_8, H_2 and H'_1 (ribose) proton signals in the presence of *Chloroquine*, accompanied by similar upfield shifts due to shielding of the *Chloroquine* ring proton signals. No shifts in the *Chloroquine* side-chain proton signals were seen with either AMP or GMP, but addition of ATP caused the side-chain methyl protons to coalesce from a sharp quartet to a broad triplet, indicative of electrostatic interaction with the additional charged phosphate residues. A possible structure for the *Chloroquine*–DNA complex is shown in Fig. 15.

Actinomycin D–DNA

Charge-transfer interaction, which accompanies intercalation of *Actinomycin D* (= Actinomycin C_1) (Dactinomycin) in DNA (p. 47), has been demonstrated spectroscopically. Complex formation is accompanied by a shift in the ultra-violet absorption maximum of the antibiotic at 425 nm to longer wavelengths. The binding of *Actinomycin D* to DNA is, therefore, measured by the decrease in optical density at 425 nm on the addition of DNA (Reich, Goldberg and Rabinowitz, 1962; Fig. 16).

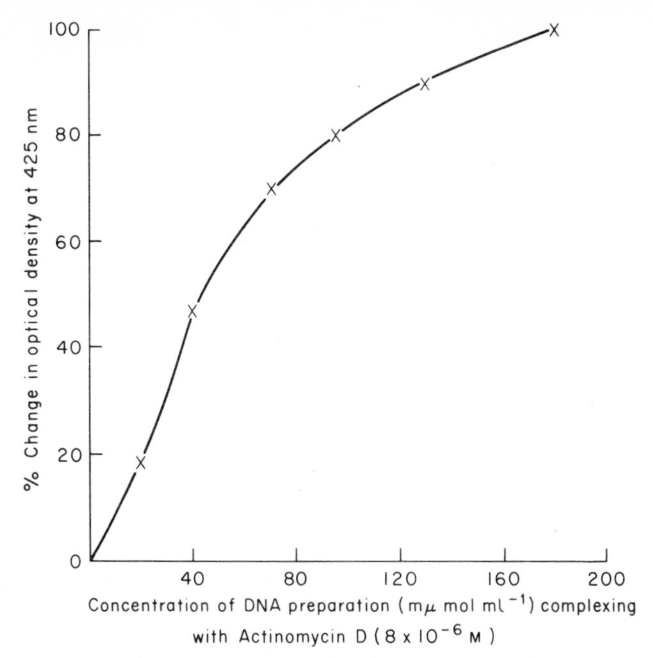

Fig. 16 Spectroscopic changes accompanying the formation of *Actinomycin D*—DNA complex

Local Anaesthetic–Thiamine

Eckert (1962) has shown that *Procaine* and other local anaesthetics form π-electron charge-transfer complexes with thiamine, which appears on the basis of a considerable body of evidence to have a specific rôle in nerve conduction processes. Thus, electrical stimulation of nervous tissue causes the release of thiamine (Cooper, Roth and Kini, 1963), and *Pyrithiamine* irreversibly blocks the electrical activity of nerve due to displacement of thiamine (Armett and Cooper, 1965). It has also been shown by fluorescence microscopy that thiamine is localised in the nerve membrane rather than in the axon (Tanaka and Cooper, 1968), and is present mainly (*ca* 80%) as thiamine pyrophosphate (Itokawa and Cooper, 1969). Furthermore, thiamine pyrophosphatase (TPPase) is also similarly localised specifically in nerve membrane structures. Studies with ^{35}S-labelled thiamine have shown that thiamine is released from frog and rat spinal cord by drugs which are known to affect ion movement in nervious tissues, including acetylcholine (at 10^{-6}M), *Carbachol* (at 10^{-5}M), *Ouabain* (at 10^{-6}M) and the local anaesthetic, *Lignocaine* (Lidocaine; at 10^{-2}M). It is particularly significant, too, that whereas thiamine is normally membrane-bound as thiamine pyrophosphate, it is released largely as thiamine monophosphate.

The formation of a 1 : 1 charge-transfer complex between *Procaine* (donor) and thiamine (acceptor) has been postulated on the basis of a shift in the ultraviolet absorption maximum for *Procaine* from *ca* 280 nm to 340 nm (Eckert,

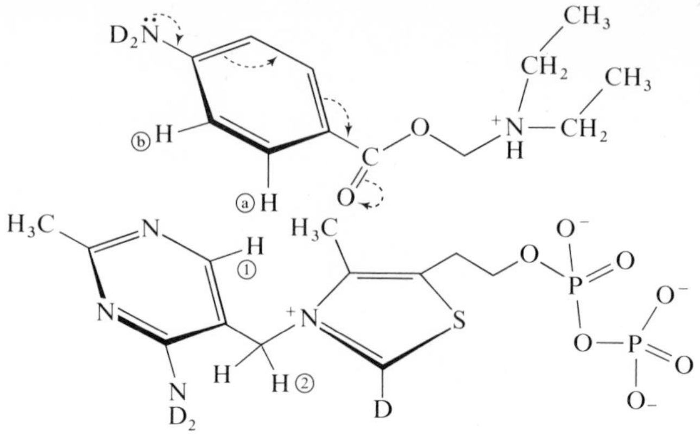

Fig. 17 *Procaine—Thiamine Pyrophosphate* complex

1962) has been confirmed by PMR studies (Thyrum, Luchi and Thyrum, 1969). Examination of mixtures of thiamine pyrophosphate and *Procaine* at various mole ratios in deuterium oxide and pD 7.00 showed that the chemical shifts of protons H^a and H^b of *Procaine* and H^1 and H^2 of thiamine are ratio-dependent showing upfield shifts due to shielding, which are consistent with off-set stacking of the pyrimidine and benzene rings. Examination of molecular models, based on the X-ray crystallographic analysis of thiamine (Kraut and Read, 1962), which show it to be folded with a dihedral angle of 76°, demonstrates that the proposed off-set stacking is consistent not only with the proposed charge-transfer complexing, but also with ion–dipole interaction between the procaine ester-carbonyl and thiazolinium ion, and ionic bonding of the procaine diethylamino-cation and thiamine phosphate anion (Fig. 17).

Histamine Methyl Transferase Inhibition

Histamine methyl transferase, the enzyme responsible for transfer of a methyl group from *S*-adenosylmethionine to histamine, is extremely widely distributed in animal tissue. The enzyme possesses a high degree of substrate specificity and is primarily responsible for inactivation of histamine in most species. The majority of antihistamines are potent inhibitors of histamine methyl transferase at histamine concentrations less than 10 μM. At higher histamine concentrations, however, they **enhance** enzyme activity to an extent which correlates significantly with their antihistamine activity (Taylor and Snyder, 1972).

Calculations show that enzyme inhibition by tryptamine and promazine derivatives is related to their electron donor activity (Merril, Snyder and Bradley, 1966; Table 10), and that the highest fractional electron density in the promazines is centred on sulphur.

Table 10. Electron Donor Activity of Phenothiazines and Inhibition of Histamine Methyl Transferase (Merril, Snyder and Bradley, 1966)

Phenothiazine	HOMO[1]	Highest fractional electron density[3]	LEMO[2]	$K_i \times 10^6$
Aminopromazine	0.3154	0.4463	−0.6114	13
Chlorpromazine	0.3330	0.4382	−0.5912	16
Promazine	0.3365	0.4352	−0.5856	20
Acepromazine	0.3514	0.4061	−0.3819	27

[1,2] HOMO (highest occupied molecular orbital) and LEMO (lowest empty molecular orbital) are expressed in β-units, which are approximately equal to −80 kJ. Hence, the smaller the value for the HOMO, the greater is the donor activity, and the smaller the value of the LEMO, the greater is its electron-acceptor properties.

[3] Highest fractional distribution of electron density in the HOMO, in all cases on sulphur.

ELECTROSTATIC INTERACTION

Electrostatic interaction may occur between drug and receptor as a result of one of the following types of interaction between

(a) two ions of opposite charge (ionic bonding)

$$[Y]^+[X]^-$$

(b) an ion and a molecular dipole

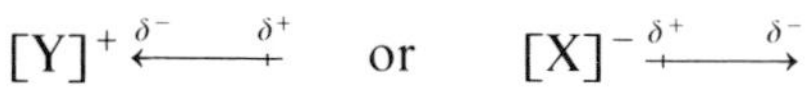

$$[Y]^+ \overset{\delta^- \qquad \delta^+}{\longleftarrow\!\!\!+} \quad \text{or} \quad [X]^- \overset{\delta^+ \qquad \delta^-}{+\!\!\!\longrightarrow}$$

(c) two molecular dipoles

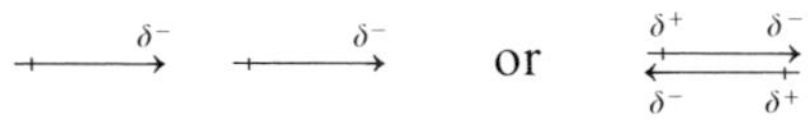

$$\overset{\delta^-}{+\!\!\!\longrightarrow} \quad \overset{\delta^-}{+\!\!\!\longrightarrow} \quad \text{or} \quad \underset{\delta^- \qquad \delta^+}{\overset{\delta^+ \qquad \delta^-}{\substack{+\!\!\!\longrightarrow \\ \longleftarrow\!\!\!+}}}$$

(d) an ion and an induced dipole

These various types of electrostatic interaction occur with decreasing intensity in the order given, i.e. from (a) to (d).

Ionic Bonding

The characteristic ionic bond is one formed between two ions of opposite charge, as, for example, between the ion of a Group I or Group II metal and a halogen or complex anion, such as nitrate (NO_3^-), sulphate (SO_4^{2-}) or phosphate (PO_4^{3-}). The electrostatic interaction energy (E) is given by the expression:

$$E = \frac{q_1 q_2}{Dr} \tag{1}$$

where q_1 and q_2 = the charges on the ions
r = the distance separating the two ions
D = the dielectric constant of the medium separating the ions

Ion-pairing only occurs when E is sufficiently great to overcome the energy due to the random thermal motion of the ions. Since most inorganic ions are normally hydrated in aqueous solution, and since also the dielectric constant of water is relatively high (*ca* 80), salts formed from inorganic ions are generally completely ionised in dilute aqueous solution, and hence also in physiological environments.

Many inorganic ions, including Na^+, K^+, Mg^{2+}, Ca^{2+}, Cl^-, HCO_3^-, PO_4^{3-}, and $H_2P_2O_7^{2-}$, have important physiological rôles which are influenced by their ability to combine to form stable (ionic) salts. Thus, idiopathic hypercaluria, in which the daily urinary excretion of calcium exceeds 300 mg, can be controlled by the administration of sodium cellulose phosphate, which binds Ca^{2+} ions preferentially as calcium cellulose phosphate. Since cellulose and its derivatives are not degraded in the human gastro-intestinal tract, the cellulose phosphates are not absorbed from the human intestine, and their excretion reduces the amount of dietary calcium available (Connolly, 1970).

Magnesium ions (Mg^{2+}) are believed to be involved in the bonding of noradrenaline to ATP. It has been suggested that the use of Li^+ ions in the treatment of manic depression may involve complexation with noradrenaline in place of Mg^{2+}. This could be a reflection of the diagonal relationship of Li to Mg in the Periodic Table of elements, since Li^+ and Mg^{2+} have almost identical ionic radii. Lithium, however, has a co-ordinate number of only four, whereas that of magnesium is six. Lithium may, therefore, decrease free noradrenaline in the brain by interfering with its normal co-ordination with Mg^{2+}, thus affecting the nature and degree of binding of noradrenaline (**2**, 3, Fig. 66).

In drug action, we are more often concerned with interactions between organic ions and ionised acceptor molecules in the host. Natural macromolecules, which more often than not possess ionic centres at physiological pH, are frequently involved. Thus, the terminal basic groups of lysine and arginine and the free carboxyl groups of aspartic and glutamic acid units of peptides and proteins are all completely ionised giving rise to cationic and anionic centres capable of ionic bonding with ionised drug molecules. Other amino acid units, such as histidine and tyrosine, which possess more weakly basic and acidic centres are also involved, though to a lesser degree. Similarly, the ionised phosphate residues of the polyribose phosphate backbone in nucleic acids, and sulphate groups of mucopolysaccharides, give rise to polyanionic structures which provide bonding sites for organic cations.

Strong Acids and Bases

Compounds which are fully ionised in solution irrespective of pH include the quaternary ammonium compounds, acetylcholine chloride, *Tubocurarine Chloride* and *Suxamethonium Bromide* (Succinylcholine Bromide), and related synthetic curariform neuromuscular blocking agents. Other examples of completely ionised compounds are the cationic and anionic detergents, *Cetylpyridinium Bromide* and *Sodium Lauryl Sulphate*. All such compounds readily associate with natural macromolecules via ionic bonds. Thus, the anionic site of acetylcholinesterase, which is responsible for binding the cationic head of acetylcholine prior to initiating its hydrolysis, appears to be the carboxylate ion of a glutamate or aspartate residue in the enzyme protein, so that enzyme and substrate are linked by this electrovalent bond.

$CH_2{\cdot}CH_2{\cdot}O{\cdot}CO{\cdot}CH_3$
N^+
Me Me Me

Acetylcholine

O
O

$CH_2{\cdot}CH_2{\cdot}CH{\cdot}CO{\cdot}NH$---
NH

Glutamate residue of acetylcholinesterase

The permanence of the bond depends on the extent to which stabilisation is achieved in the resulting 'salt' (i.e. the affinity constant), and the ease with which displacement can be effected by other ions. Most ionic bonds formed between oppositely-charged ions have bond strengths between 4 and 8 kJ mol^{-1}. For example, the strength of the ionic bond formed in the inhibition of antigen–antibody reactions by the $p(p'$-hydroxyphenylazo)-phenyltrimethyl ammonium ion has been found experimentally to be about 6.3 kJ mol^{-1} (Pressman, Grossberg, Pence and Pauling, 1946). The strength of the bond (equation 1) is determined both by the closeness of approach and by the effective dielectric constant of the medium separating the two ions. The two factors are interrelated according to the relationship (2) devised by Schwarzenbach (1936).

$$D = 6r - 11 \qquad (2)$$

where D = effective dielectric constant
r = the distance between the two charges in Å

Measurement of the variation of dissociation constant between pairs of dibasic acids (Schwarzenbach, 1936) gave an effective dielectric constant (D) of 31 which has been widely used in calculations of inter-ionic distances. Thus, the ionic bond in the inhibition of the antigen–antibody reaction by quaternary ammonium compounds was found to be about 7Å in length (Pressman, Grossberg, Pence and Pauling, 1946).

Weak Acids and Bases

The majority of acidic and basic drugs are either weak acids or weak bases. They are, therefore, only partially dissociated into ions in solution to an extent determined by the dissociation constant (pK_a) of the acid (or base-conjugate acid) and the pH of the solution (**2**, 4). Thus, at physiological pH (7.4), acids of pK_a 6.9 are 76% ionised, whilst acids of pK_a 6.5 or less are 89% or more in the anionic form; similarly, bases of pK_a 8.0 and above are 80% or more in the cationic form. For example, *Physostigmine* (eserine; pK_a 6.12) strongly inhibits acetylcholinesterase at pH 5.9 when the drug is substantially present as a cation (61%). Inhibition, however, decreases with increasing pH to about 20% at pH 10.1, when the percentage of cation is negligible, the residual effect being attributed to the unionised base (Wilson and Bergmann, 1950).

Ionic bonding forms the essential mechanism for the association of a number of weakly acidic or weakly basic drug substances with receptor ions, as in the action of sulphonamides (**1**, 14), sympathomimetics and adrenergic blocking agents (**1**, 16), and the protein binding of organic acids and bases (**2**, 4). Salt formation also accounts for the reaction between strongly basic antibiotics, such as *Neomycin*, *Streptomycin* and *Kanamycin* with mucopolysaccharides, which limits their absorption from the gastro-intestinal tract (**1**, 21). It also plays an important rôle in the action of antimalarials, local anaesthetics and acridine antiseptics.

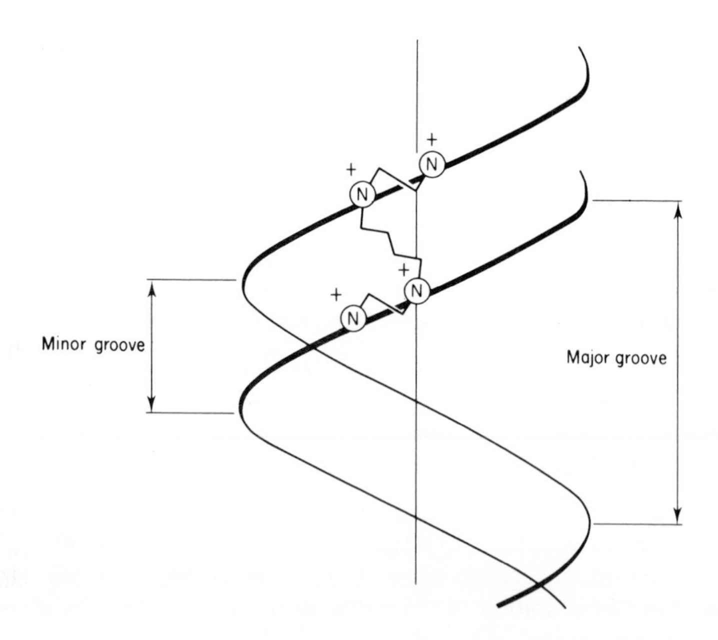

Fig. 18 Spermine—DNA complex

Antimalarials

Chloroquine and *Mepacrine* (Quinacrine) inhibit plasmodial DNA synthesis (Schellenberg and Coatney, 1961) by stabilising the DNA double helix against thermal 'melting' (separation of the double helix strands). Spermine, which is equally effective (Tabor, 1962), has been shown by X-ray diffraction (Suwalsky, Traub and Shmueli, 1969) to complex with DNA across the minor groove by electrostatic attraction between the protonated amino groups and the DNA-phosphate ions (Fig. 18).

According to O'Brien and Hahn (1965), the *Chloroquine*–DNA complex is similarly stabilised by ionic bonding of the diamino side-chain across the minor groove to complementary strands of the DNA helix. The degree of stabilisation of similar complexes with aliphatic diamines depends on aliphatic chain length. The most effective diamines are 1,4-diaminobutane and 1,5-diaminopentane, which are structurally analogous to the chloroquine side-chain (Mahler and Mehrotra, 1963). Introduction of double and treble bonds into the chloroquine side-chain, which shorten the inter-nitrogen distance and place constraint on the flexibility of the chain, reduce antimalarial activity (Singh, Stein, and Biel, 1969).

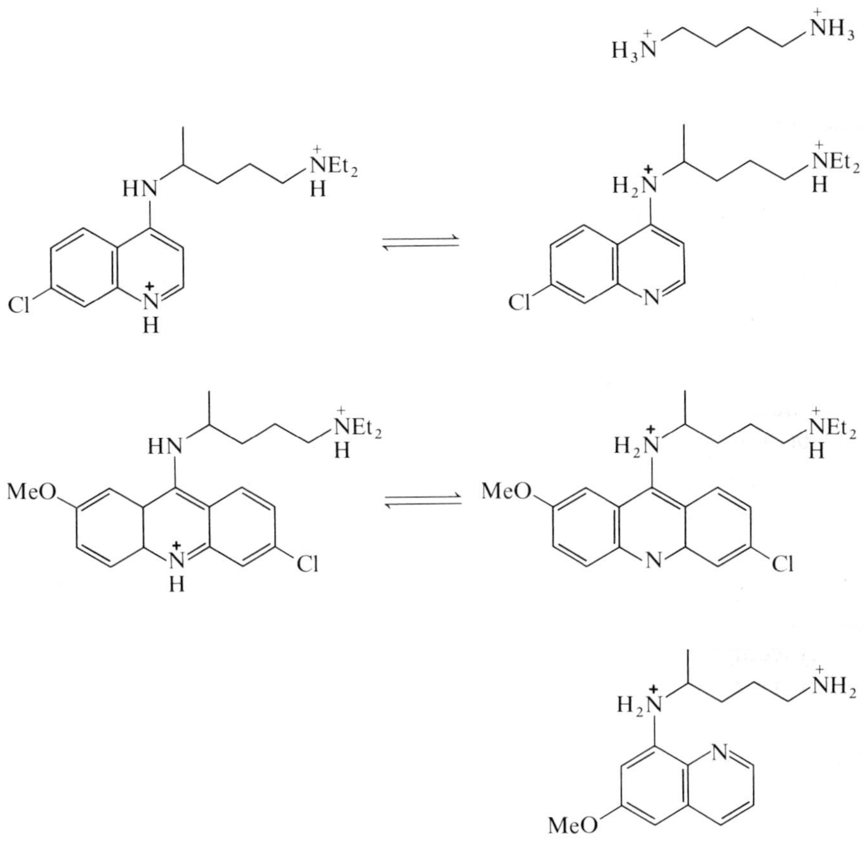

$H_3\overset{+}{N}$ $\overset{+}{N}H_3$

HN $\overset{+}{N}Et_2$ H Cl $\overset{+}{N}$ H ⇌ $H_2\overset{+}{N}$ $\overset{+}{N}Et_2$ H Cl N

HN $\overset{+}{N}Et_2$ H MeO $\overset{+}{N}$ H Cl ⇌ $H_2\overset{+}{N}$ $\overset{+}{N}Et_2$ H MeO N Cl

$H_2\overset{+}{N}$ $\overset{+}{N}H_2$ N MeO

It is of interest that the 8-aminoquinoline, *Primaquine*, which causes destruction of the **exo**-erythrocytic phase in *P. vivax* malaria, is unable to compete with *Chloroquine* for DNA binding sites (Morris, Andrew, Whichard and Holbrook, 1970) in equilibrium dialysis experiments at pH 6.0. At this pH, 8-aminoquinolines, which have pK_a *ca* 3.2 (8-amino) and 10.2 (terminal NH_2), occur as the singly protonated cation. 4-Aminoquinolines, such as *Chloroquine*, which has pK_a 8.1 (4-amino) and 10.2 (terminal NEt_2), exist almost exclusively as the diprotonated cation. Contrary to expectation, however, even when the pH of the dialysis medium is varied up to pH 8.0, the results in these binding experiments are unaltered, since no change occurs in the protonation of the 8-aminoquinolines, and although the proportion of diprotonated *Chloroquine* is reduced from 98 to 40%, the duration of the experiment (20 hr) gives ample time for equilibrium to be re-established as the diprotonated drug is bound.

Chloroquine–DNA binding is also significantly reduced by Mn^{2+} and Mg^{2+} ions (Cohen and Yielding, 1965), which are bound preferentially to the negatively-charged phosphate groups of DNA, and known to prevent the interaction of dyestuffs such as rosaniline (Cavalieri, Angelos and Balis, 1951).

Local Anaesthetics

It is generally held that local anaesthetics penetrate nervous tissue as the free base, and that the cationic base-conjugate acid blocks action potentials from inside the membrane. This has been demonstrated in experiments on the giant axons of the squid, *Loligo pealei*, perfused externally and internally with solutions at pH 8.0 and 7.0 respectively (Narahashi, Yamada and Frazier, 1969). The local anaesthetic activity of two lignocaine derivatives, A(pK_a 6.3) and B(pK_a 9.8) were measured when applied externally and internally (Table 11).

The results show that when the proportion of free base applied externally is substantially enhanced (compound B), a significant increase in local anaesthetic activity results. Similarly, internal pH changes which substantially reduce the proportion of cation present inside the axon significantly reduce local anaesthetic activity (compound A). In support of these conclusions, it was also shown that quaternary ammonium compounds with local anaesthetic activity block excitability of the nerve more strongly when applied inside rather than outside the nerve membrane. Furthermore, internally-applied quaternary ammonium compounds have local anaesthetic potencies which are independent of pH.

Acridine Antibacterials

Albert and his collaborators (Albert, Rubbo and Goldacre, 1941; Albert, Rubbo, Goldacre, Davey and Stone, 1945; Albert and Goldacre, 1948) have demon-

N NH_2 + H^+ $\rightleftharpoons$ N $\overset{+}{N}H_3$

Table 11. Effect of External and Internal pH on the Potency of Local Anaesthetics applied to Giant Squid Axon

2,6-Me₂-C₆H₃-NH·CO·CH₂·N(Me)·CH₂·CH₂·OMe

pK_a 6.3

	% base	% cation	Maximum rate of rise of action potential per cent[1] local anaesthetic applied externally	internally
pH 7.0	83	17	56	28
pH 8.0	98	2	—	56
pH 9.0	99.8	0.2	54	—

2,6-Me₂-C₆H₃-NH·CO(CH₂)₃·NEt₂

pK_a 9.8

	% base	% cation	externally	internally
pH 7.0	0.2	99.8	88	38[2]
pH 8.0	1.6	98.4	—	48[2]
pH 9.0	13.7	86.3	27	—

[1] High values indicate low local anaesthetic potency
[2] Difference not significant

strated that the antibacterial action of acridine antiseptics is both pH- and pK-dependent, such that this activity is due to the ionised form of the drug (Table 12).

The importance of the ionised form of the drug has been confirmed using *E. coli* as the test organism, which is unaffected by pH in the range 5.5–8.5 (Albert, 1968). A plot on a logarithmic scale of the minimum bacteriostatic concentration of various acridines against pH (which is also a logarithmic expression) gives a straight-line relationship (Fig. 19) only when the concentration is expressed in terms of acridine cation. The negative slope of the graph is a clear demonstration of direct competition between acridinium and hydrogen ions. For this reason, it has been recommended that treatment of wounds with *Aminacrine* (9-aminoacridine) to prevent sepsis should be preceded by washing with sodium bicarbonate (Poate, 1944).

Table 12. Ionisation and Bacteriostatic Action of Acridine and Related Bases (Albert, 1968)

Substituent	Minimum Bacteriostatic Conc.[1]	pK_a (37°)	% Ionised
1-Amino	1 in 10 000	5.7	2
2-Amino	10 000	5.6	2
3-Amino	80 000	7.7	73
4-Amino	5000	4.2	<1
9-Amino (*Aminacrine*)	160 000	9.6	100
2,7 Diamino	20 000	5.8	3
3,6 Diamino (*Proflavine*)	160 000	9.3	99
3,7 Diamino	160 000	7.8	76
3,9-Diamino	160 000	11.1	100
4,5-Diamino	<5 000	3.8	<1
4,9-Diamino	80 000	9.0	98

[1] *Strept.pyog.* after incubation for 48 hr at 37° in 10% serum broth, pH 7.3

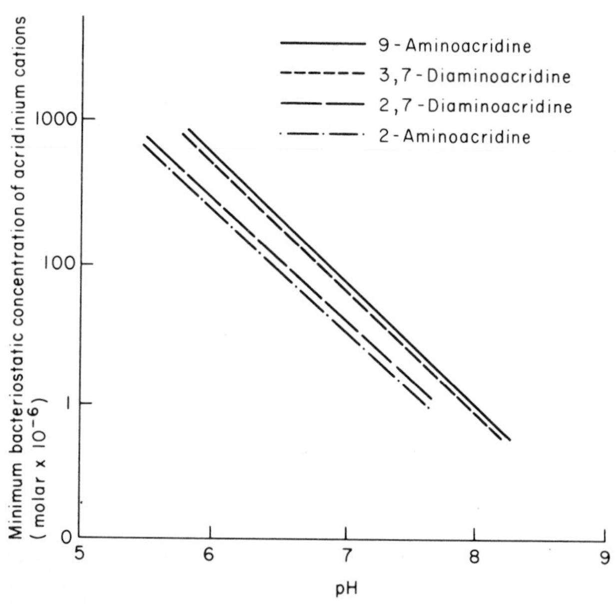

Fig. 19 Plot of minimum bacteriostatic concentration of acridinium cations for *E. coli* against pH

Ion–Dipole and Dipole–Dipole Interactions

Principles

Electrostatic interactions other than the formation of discrete ionic bonds also contribute to drug–receptor interactions. These are all much weaker than ionic bonds, and result from ion–dipole (*ca* 2 kJ mol^{-1}), dipole–dipole (*ca* 0.5 kJ mol^{-1}) and ion–induced dipole (*ca* 0.5 kJ mol^{-1}) interactions. Dipoles exist in a wide range of organic compounds as a result of electronegativity differences between adjacent atoms.

Molecular dipoles are expressed as **dipole moments**. When a molecule with an unsymmetrical charge distribution in placed in an electric field, it will tend to align itself so that its positive pole is oriented towards the negative electrode and its negative pole is towards the positive electrode. The turning moment to which the molecule is subjected is the **dipole moment**. Alignment of the molecules is, however, temperature-dependent, since it is opposed by thermal agitation which increases with temperature.

Dipole moments are in effect the product of the overall molecular charge and its distance from the effective centre of the molecule. They are expressed in **Debye** units (D) which are of the order of 10^{-20} esu. m, since the unit electric charge is 4.8×10^{-10} esu and the usual order of molecular dimensions is in Å (10^{-10} m). The dipole moment is actually the vector summation of all the individual **bond moments** in the molecule. In consideration of organic molecules, it is, however, more convenient to use functional group moments wherever these are appropriate (Table 13).

Symmetrical groups, such as —H, —X (halogen), —CN and $—NO_2$ have **linear** group moments, i.e. acting along the axis of the central atom and the adjoining carbon atom. The sum of two equal and opposite linear group

Table 13. Bond and Group Dipole Moments

Bond Dipole Moments

Bond	Bond Moment (+⟶)	Bond	Bond Moment (+⟶)	Bond	Bond Moment (+⟶)
H—C	0.40	C—N	0.22	C=O	2.3
H—S	0.68	C—O	0.74	C=S	2.6
H—Br	0.78	C—S	0.90	C≡N	3.5
H—Cl	1.08	C—I	1.19	N=O	2.0
H—N	1.31	C—Br	1.38		
H—O	1.51	C—F	1.41		
H—F	1.94	C—Cl	1.46		

Group Dipole Moments

Group	Group Moment (+⟶)	Group	Group Moment (+⟶)
—C(H)=O	2.7	$—\overset{+}{N}O_2^{-}$	3.9
$—\overset{+}{S}—\overset{-}{O}$	3.9		

(From Smyth, *Dielectric Behaviour and Structure*, McGraw Hill, 1963)

moments is zero provided the vectors are either collinear or parallel, as for example in *p*-dichlorobenzene and *trans*-1,2-dichloroethylene.

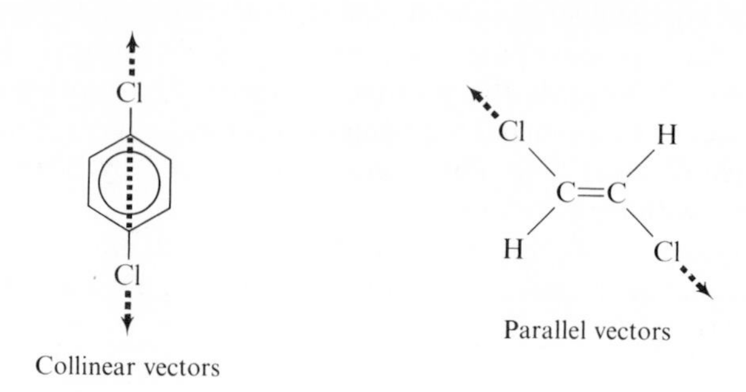

Collinear vectors

Parallel vectors

Molecules with unequal, but opposite linear vectors will have net dipole moments resulting primarily from simple arithmetic addition of the competing group dipole moments. Discrepancies between observed and calculated dipole moments are probably due to the formation of induced dipoles.

p-Chloronitrobenzene
$\mu_{calc} = 2.4D$
$\mu_{obs} = 2.6D$

p-Dimethoxybenzene
transoid ($\mu_{calc} = 0$) cisoid ($\mu_{calc} = 2.25D$)
$\mu_{obs} = 1.7D$

Unsymmetrical groups, such as OH, OMe, NH_2 and CO · OH have **non-linear** moments, and give a resultant dipole when two such identical groups are opposed to each other. Thus, the observed dipole moment of *p*-dimethoxybenzene is the sum of the moments of the two extreme planar (for maximum resonance) conformations.

Vector addition of non-linear arrangements of both linear and non-linear groups can be accomplished either graphically or mathematically using the expression:

$$\mu^2 = \mu_1^2 + \mu_2^2 + 2\mu_1\mu_2 \cos\theta$$

Typical non-linear arrangements of opposed dipoles are seen in *m*-dichlorobenzene and *cis*-1,2-dichloroethylene.

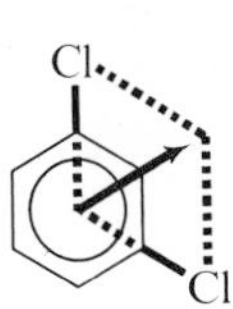

m-Dichlorobenzene (μ = 1.48D)

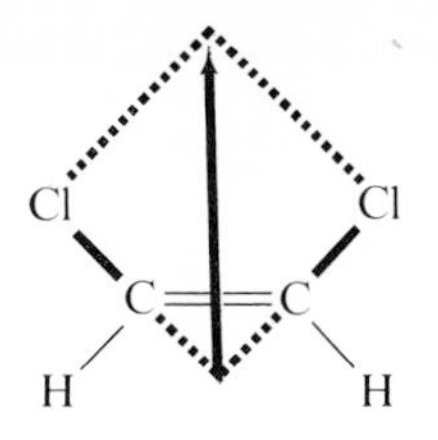

cis-1,2-Dichloroethylene

The electrostatic energy in an ion–dipole interaction is given by the expression (Gill, 1965):

$$E = \frac{Ne\mu \cos \phi}{D(r^2 - d^2)} \tag{3}$$

where N = Avogadro's number
e = electronic charge
μ = dipole moment
ϕ = the angle between the line joining the ion to the centre of the dipole and the direction of the dipole
D = the effective dielectric constant of the medium
r = the distance between the ion and the centre of the dipole
d = the length of the dipolar bond

Local Anaesthetics

Ion–dipole interactions are generally much smaller than those involved in the formation of ionic bonding. There are few well-documented examples of ion–dipole interactions, but there is good evidence that the local anaesthetic potency of *Procaine* and its derivatives is directly related to the dipolar character of the ester carbonyl group (Buechi and Perlia, 1971). Thus, as early as 1928 (Kindler, 1928, 1929) had demonstrated a direct relationship between the local anaesthetic activity of a series of *Procaine* analogues and electron availability at the ester carbonyl, as shown by the pK_a of the corresponding carboxylic acid (Table 14).

Amino- and methoxyl-substituents capable of donating electrons from the para position of the aromatic ring by the mesomeric effect promote polarisation of the carbonyl group, and this is associated with a high level of local anaesthetic activity.

R—C_6H_4—$CO{\cdot}O{\cdot}CH_2{\cdot}CH_2NEt_2$

Table 14. Potency of Local Anaesthetics and Bond Order of the Ester Carbonyl Group

$O \cdot CH_2 \cdot CH_2 \cdot NEt_2$ $\longleftrightarrow$ $O \cdot CH_2 \cdot CH_2 \cdot NEt_2$

$H_2\ddot{N}$—C—O $\quad$ $H_2\overset{+}{N}$=C—O^-

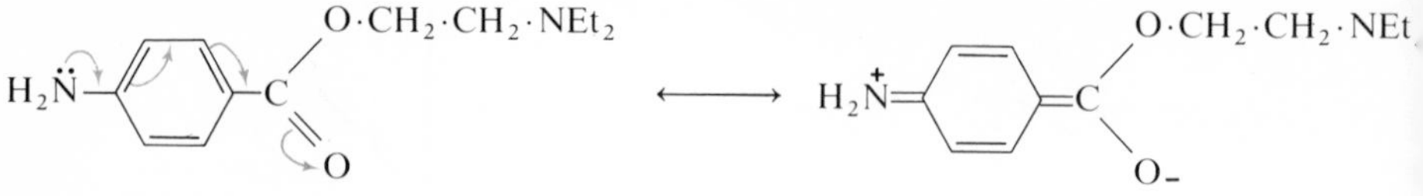

Substituent (R)*	pK_a (corresponding acid)	Infrared frequency of $>C{=}O$ (cm^{-1})	Infiltration Anaesthetic ED_{50} (mmol/100 ml)
CH_3O—	4.96	1708	0.060
H_2N—	4.49	1711	0.075
H—	4.22	1727	0.630
Cl—	4.03	1729	0.090
O_2N—	3.40	1731	0.740

* Structure, page 71. (from Galinsky, Gearien, Perkins and Susina, 1963)

The actual structure will be a resonance hybrid of the extreme forms (Table 14), the extent of polarity being capable of measurement in terms of the infrared absorption frequency of the ester carbonyl group (Galinsky, Gearien, Perkins and Susina, 1963) and the dissociation constant of the corresponding carboxylic acid (Kindler, 1928, 1929).

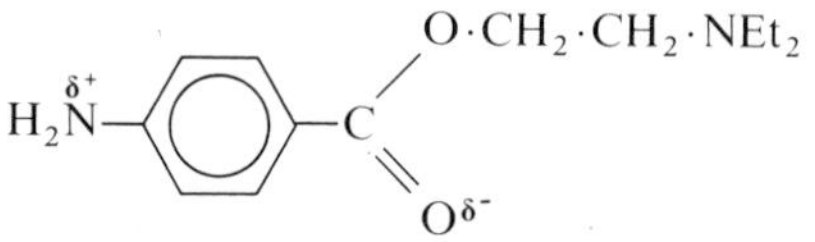

$O \cdot CH_2 \cdot CH_2 \cdot NEt_2$

$H_2\overset{\delta+}{N}$—C—$O^{\delta-}$

In contrast, *p*-substitution with electron-withdrawing substituents increases the double bond character of the ester carbonyl, leading to reduction of local anaesthetic activity.

It cannot, of course, be assumed that the sole function of the aromatic ring in *Procaine* is the transmission of electron density from the substituent to create a pharmacologically active $\overset{\delta+}{C}$—$\overset{\delta-}{O}$ dipole, since it also contributes to the overall lipophilicity of the base, a property which is important for its transmission across the nerve membrane (**2**, 4). There is, for example, evidence that the aromatic ring is not essential, since a number of simple aliphatic esters of diethylaminoethanol have significant local anaesthetic activity (Cano and Ranédo, 1920; McElvain and Carney, 1946). It is essential, however, that such compounds should combine lipid solubility with the requisite degree of dipolar character and aromatic compounds, such as *Procainamide*, which lack this combination, have little anaesthetic effect. In this compound, amide resonance competes with the amino group resonance, so that C—O dipole formation is inhibited. Also, its water solubility (597 mmol/l) is about 100 times that of

Procaine (5.74 mmol/l) and in consequence its oil/water partition coefficient (0.03) is low compared with that of *Procaine* (0.98). Its local anaesthetic activity is, therefore, negligible (Kindler, 1928, 1929). By contrast, the methyl substituents in *Lignocaine* enhance solubility, and assist the amide resonance and formation of a C—O dipole, though this effect is minimised by steric hindrance and the consequent reduction of molecular planarity.

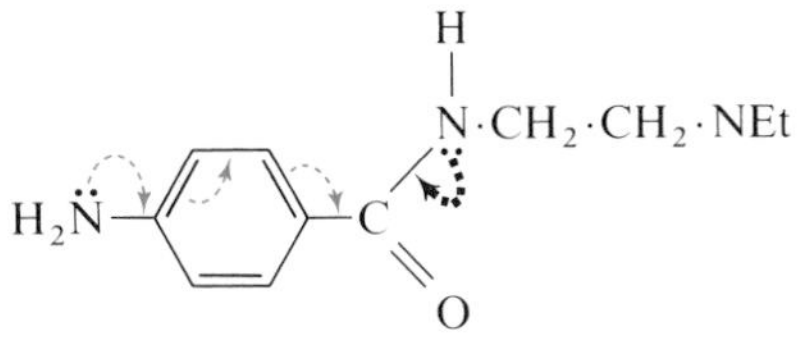

Procainamide 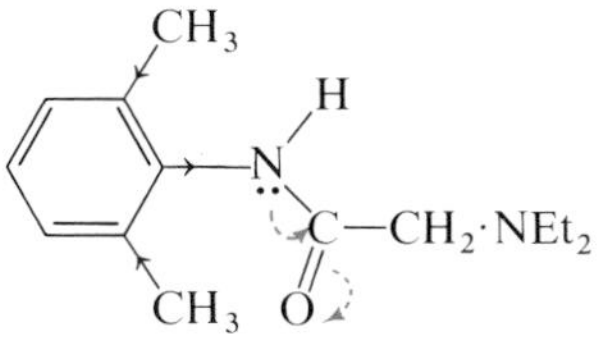 *Lignocaine*

Ion–induced-dipole Interactions

Ions are capable of inducing polarisation in adjacent neutral molecules, creating an attraction force, which is dependent on the polarisability of the molecule in which the dipole is induced. In general, ion–induced dipole interactions are considered to be quite small and of the order of about 0.5 kJ mol^{-1} (Adams and Whittaker, 1950).

Dipole–Dipole Interactions

Individual dipole–dipole interactions also undoubtedly contribute to drug–macromolecular interaction. The interaction energy, which like ionic interactions may assist or oppose drug–receptor attachment, depending upon dipole polarity, is determined by the strength, separation and orientation of the paired dipoles. The hydrogen bond can perhaps be considered to be a special case of a dipole–dipole interaction.

COVALENT BONDING

The great majority of drugs appear to combine with their receptors by weak molecular interactions which individually are readily reversible, but collectively create a strong link. A few, however, are known to enter into combination by the formation of covalent bonds which individually are very much stronger and are much less readily broken. Most covalent bonds have bond energies of the order of 140–800 kJ mol^{-1}. Thus, they are some 10–40 times stronger than an ionic bond. A single covalent bond, therefore, provides the same degree of stability in the drug–receptor complex as several ionic bonds.

Non-specific Covalent Bonding

The gas-sterilising agents, formaldehyde and ethylene oxide, exert a non-specific killing effect on bacteria by covalent bond formation. Formaldehyde is believed to act by combination with the free amino group of proteins to form methylene–imines, causing protein denaturation.

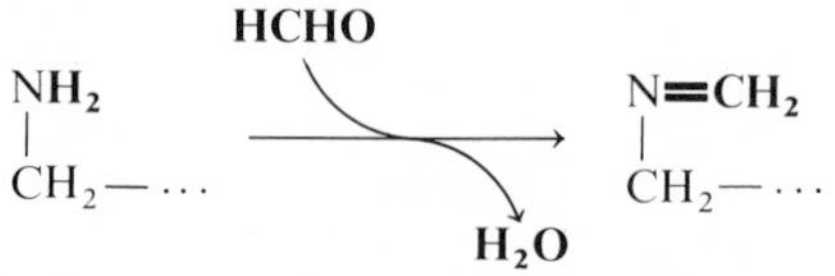

$$\begin{array}{c} NH_2 \\ | \\ CH_2-\cdots \end{array} \xrightarrow[\;-H_2O\;]{HCHO} \begin{array}{c} N{=}CH_2 \\ | \\ CH_2-\cdots \end{array}$$

The reactivity of ethylene oxide, because of ring strain in the 3-membered ring, is such that it is capable of attack at a much wider range of reactive groups, including terminal free primary and secondary amino groups, serine hydroxyls and the phenolic hydroxyl groups of tyrosine residues (**1**, 7).

Organic Arsenicals

The selective trypanocidal and amoebacidal action of organic arsenicals (Ehrlich, 1909) is due to the reaction of trivalent aresenoxides with sulphydryl groups (SH), which are believed to be associated with a respiratory enzyme. It must be emphasized that organic arsenicals are only effective as trypanocides in the trivalent arsenoxide form as present in *Oxophenarsine*. Such compounds, however, have relatively high host toxicity, and the pentavalent compounds, *Acetarsol* (Acetarsone) and *Tryparsamide*, which have much lower host toxicity, are used much more widely, The pentavalent compounds are slowly reduced *in vivo* to trypanocidal and amoebacidal effective arsenoxides.

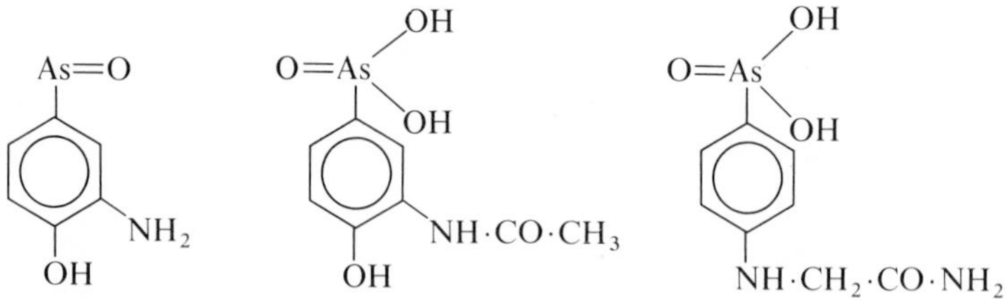

Oxophenarsine: benzene ring with As=O; NH_2; OH

Acetarsol: benzene ring with O=As(OH)(OH); $NH\cdot CO\cdot CH_3$; OH

Tryparsamide: benzene ring with O=As(OH)(OH); $NH\cdot CH_2\cdot CO\cdot NH_2$

Oxophenarsine *Acetarsol* *Tryparsamide*

The primary reaction of the arsenoxides is considered to be the formation of two As—S covalent bonds to the receptor protein:

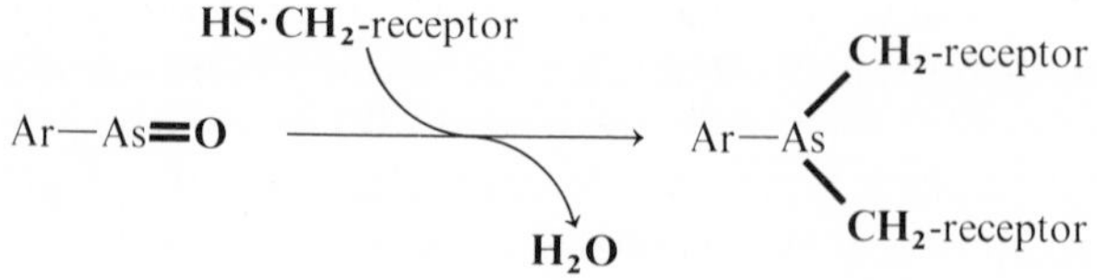

$$Ar-As{=}O \xrightarrow[\;-H_2O\;]{HS\cdot CH_2\text{-receptor}} Ar-As\begin{array}{l} \diagup CH_2\text{-receptor} \\ \diagdown CH_2\text{-receptor} \end{array}$$

The electron structure of trivalent arsenic is:

K	L	M	N (shells)		
			s	*p*	*d*
2	2,6	2,6,10	↑↓	↑ ↑ ↑	vacant

The *p*-orbitals in the N shell are filled in covalent bond formation in trivalent compounds by pairing with an electron (↓) from each of the three carbon atoms forming the three bonds:

K	L	M	N		
			s	*p*	*d*
2	2,6	2,6,10	↑↓	↑↓ ↑↓ ↑↓	vacant

This arrangement is to be compared with that in pentavalent arsenic, in which one of the *s* electrons has been promoted to a vacant *d*-orbital:

			s	*p*	*d*
2	2,6	2,6,10	↑	↑ ↑ ↑	↑

and leads to the following pairing of electrons in pentavalent compounds:

2	2,6	2,6,10	↑↓	↑↓ ↑↓ ↑↓	↑↓

A similar electron arrangement is considered to be formed in the reversal of drug–receptor formation which occurs on the addition of the sulphydryl compounds, cysteine or *Dimercaprol* (BAL).

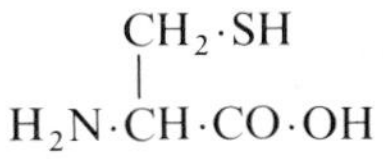

$$\begin{array}{c} CH_2\cdot SH \\ | \\ H_2N\cdot CH\cdot CO\cdot OH \end{array}$$

Cysteine

$$\begin{array}{l} CH_2\cdot SH \\ | \\ CH\cdot SH \\ | \\ CH_2\cdot OH \end{array}$$

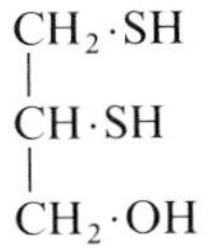

Dimercaprol

According to Albert (1968), this is possible because the vacant *d*-orbitals of the trivalent arsenic receptor complex are capable of accepting a lone pair of electrons from sulphur to give a new resonance-stabilised transition complex, with an electron structure similar to that in pentavalent arsenic compounds, i.e.

			s	*p*	*d*
2	2,6	2,6,10	↑↓	↑↓ ↑↓ ↑↓	**↑↓**

(electron pair from antidote)

The new receptor–drug–antidote complex is of low stability, and because of resonance there is an equal chance of any one of the three As—S bonds being broken. Hence, in the presence of excess of the free sulphydryl antidote, the drug–receptor covalent bond is readily broken by a simple mass action effect.

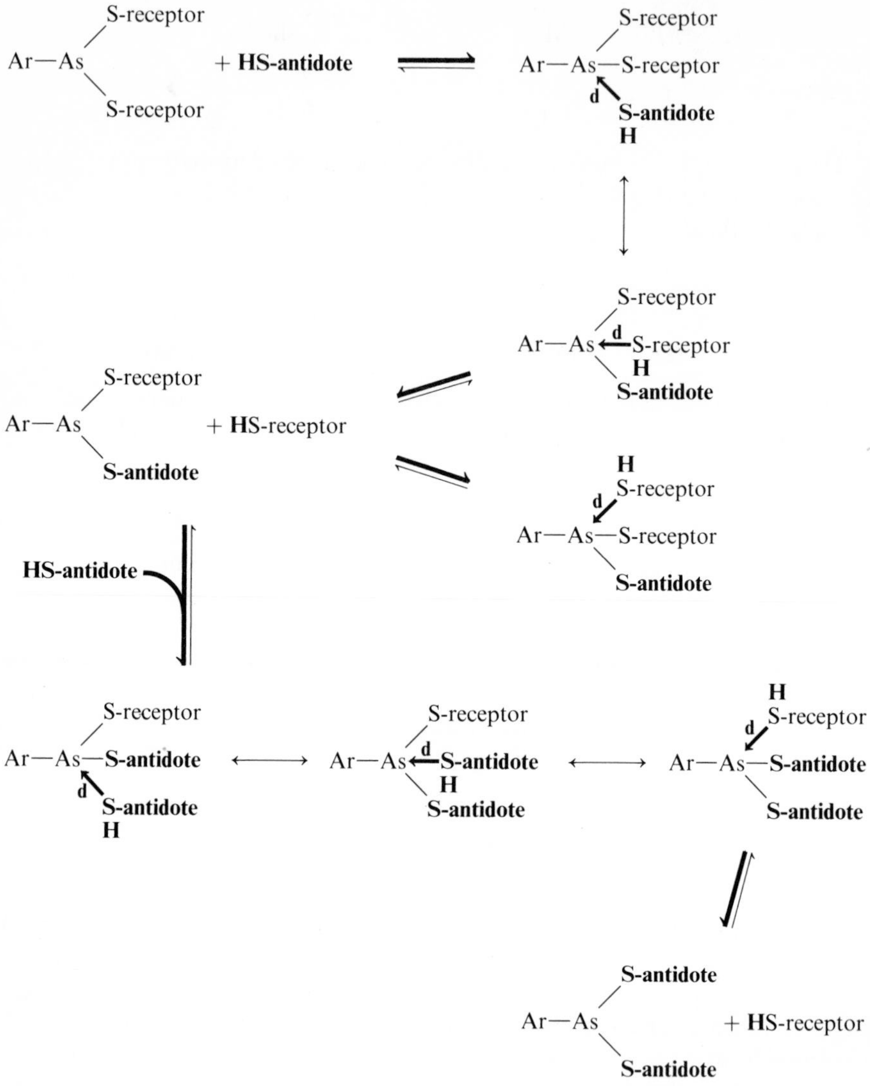

Ar—As(S-receptor)(S-receptor) + **HS-antidote** ⇌ Ar—As(S-receptor)—S-receptor ← d **S-antidote H**

↕

Ar—As(S-receptor) ←d S-receptor H, **S-antidote**

Ar—As(S-receptor)(**S-antidote**) + **H**S-receptor ⇌ (the two structures above and below)

H S-receptor →d Ar—As—S-receptor, **S-antidote**

HS-antidote ⇌ (downward)

Ar—As(S-receptor)—**S-antidote** ← d **S-antidote H** ⟷ Ar—As(S-receptor) ←d **S-antidote H**, **S-antidote** ⟷ **H** S-receptor →d Ar—As—**S-antidote**, **S-antidote**

⇌

Ar—As(**S-antidote**)(**S-antidote**) + **H**S-receptor

Substances, such as cysteine and *Dimercaprol*, therefore, act as efficient antidotes for arsenic poisoning. The action of the war-gas, Lewisite ($ClCH{=}CH \cdot AsCl_2$) is analogous to that of the arsenical trypanocides, its action being due primarily to covalent combination with the thiol groups of pyruvate oxidase (Peters, 1936, 1948). The reaction is completely reversed by the action of *Dimercaprol*.

Organic Antimonials

Trivalent organic antimonials used in the treatment of schistosomiasis, or bilharzia, a parasitic worm infection transmitted by water snails, similarly act by combination with the sulphydryl groups of phosphofructokinase. This enzyme, which effects the conversion of fructose-6-phosphate to fructose-1,6-diphosphate, controls a key step in the glycolysis of carbohydrate in schistosomes. The antimonial drugs, *Stibophen* and *Sodium Stibogluconate*, inhibit the enzyme selectively without affecting phosphofructokinase in the host, and hence cause the parasite to accumulate the substrate fructose-6-phosphate (Bueding and Mansour, 1957).

SO_3Na SO_3Na

Sb

NaO_3S NaO SO_3Na

Stibophen

S—H

Phosphofructokinase

$+ Mg^{2+}$

Fructose 6-phosphate + ATP ⇌ Fructose 1,6-diphosphate + ADP + H^+

SO_3Na

Sb—S-phosphofructokinase

NaO_3S

Organic Mercurials

Organo-mercury diuretics such as *Mersalyl Acid* are believed to be metabolised to mercuric ions (Hg^{2+}) in the kidney. According to Weiner, Levy and Mudge (1962), the mercuric ions are the effective diuretic agent, and function by inhibition of a sulphydryl enzyme. The antibacterial agents, *Phenylmercuric Nitrate* and *Phenylmercuric Acetate*, similarly act by combination with essential sulphydryl groups (Fildes, 1940).

Biological Acylating Agents

Phosphorus Insecticides

Phosphorus insecticides, such as *Dyflos* (Isoflurophate; di-isopropylphosphofluoridate), inactivate esterases, the enzymes which normally accomplish ester hydrolysis. They inhibit acetylcholinesterase, non-specific plasma esterases and lipase; they also block the peptidases, trypsin and chymotrypsin. The enzymes are inactivated by phosphorylation of the serine residue, which is considered to be the primary nucleophile in the normal hydrolytic reaction (Schaffer, May and Summerson, 1954; Cohen, Oosterbaan, Jansz and Berends, 1959; Sanger, 1963).

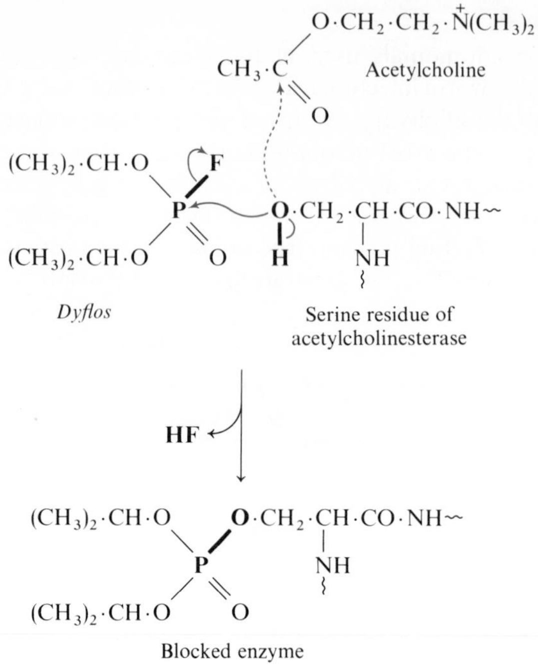

$CH_3 \cdot C(=O) \cdot O \cdot CH_2 \cdot CH_2 \cdot \overset{+}{N}(CH_3)_2$ Acetylcholine

$((CH_3)_2 \cdot CH \cdot O)_2P(=O)F$ *Dyflos*

$HO \cdot CH_2 \cdot CH(NH\sim) \cdot CO \cdot NH\sim$ Serine residue of acetylcholinesterase

$\xrightarrow{-HF}$

$((CH_3)_2 \cdot CH \cdot O)_2P(=O) \cdot O \cdot CH_2 \cdot CH(NH\sim) \cdot CO \cdot NH\sim$

Blocked enzyme

Parathion is not active as such, but is metabolised by the insect to para-oxon (Gage, 1953), which is the effective insecticide. Replacement of sulphur by the more electronegative oxygen promotes nucleophilic attack on phosphorus which is also favoured by the reduction in steric hindrance due to the smaller bulk of the oxygen atom compared with sulphur. Phosphorylation occurs with the release of *p*-nitrophenol (Hartley and Kilby, 1952). The *p*-nitrophenoxy group is a particularly good leaving group because of the electron-attracting properties of the nitro group.

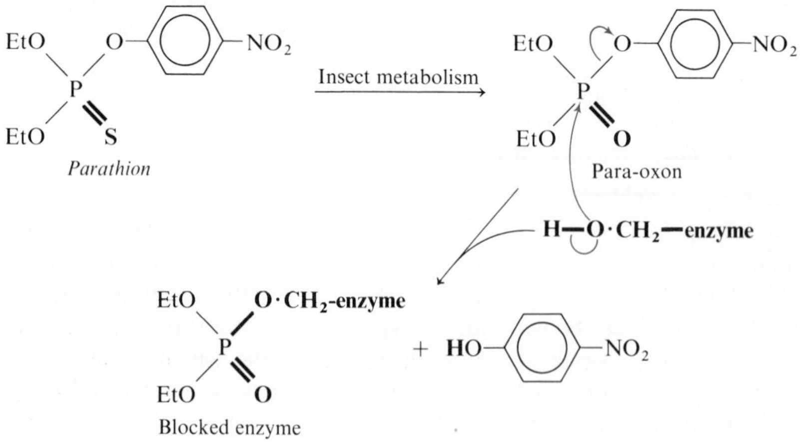

$(EtO)_2P(=S)-O-C_6H_4-NO_2$ *Parathion* $\xrightarrow{\text{Insect metabolism}}$ $(EtO)_2P(=O)-O-C_6H_4-NO_2$ Para-oxon

$H-O \cdot CH_2-$enzyme

$\longrightarrow (EtO)_2P(=O)-O \cdot CH_2$-enzyme $+ HO-C_6H_4-NO_2$

Blocked enzyme

Malathion also acts in a similar manner, but because of major differences in the metabolic pathway in insects and mammals, it is relatively non-toxic to animals (Krueger and O'Brien, 1959).

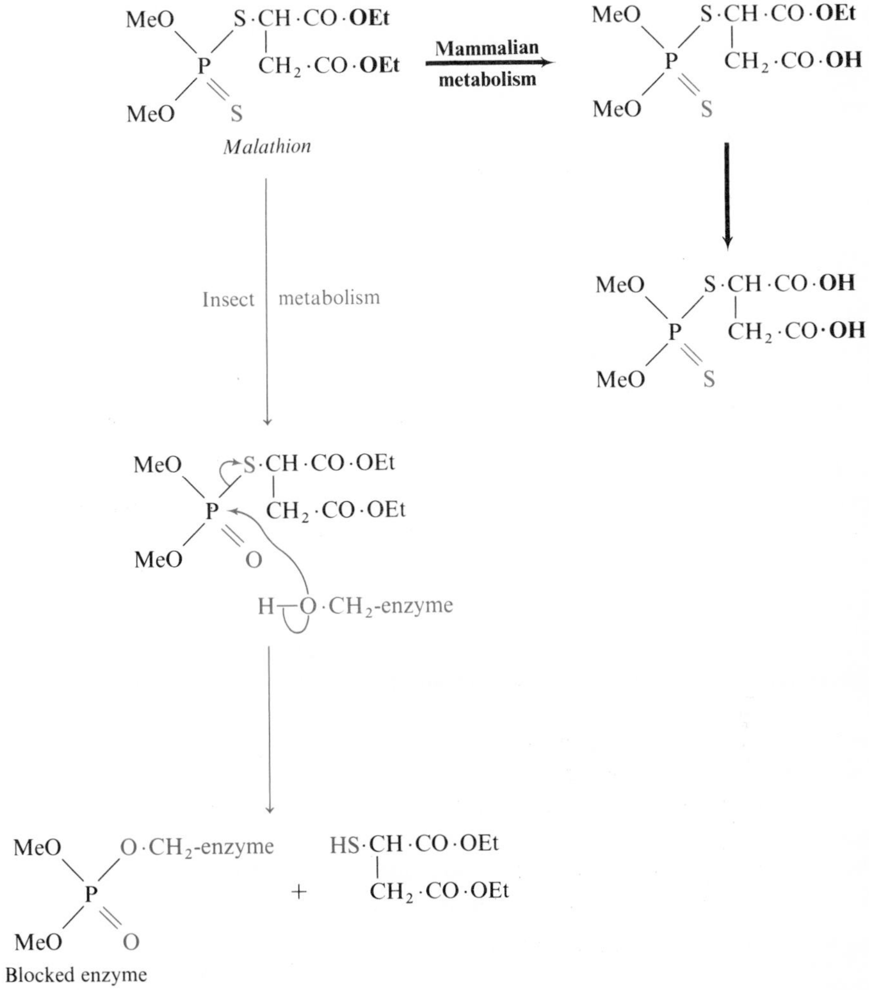

MeO S·CH·CO·**OEt**
P CH₂·CO·**OEt**
MeO S

Malathion

Mammalian metabolism →

MeO S·CH·CO·**OEt**
P CH₂·CO·**OH**
MeO S

↓

MeO S·CH·CO·**OH**
P CH₂·CO·**OH**
MeO S

Insect metabolism ↓

MeO S·CH·CO·OEt
P CH₂·CO·OEt
MeO O

H—O·CH₂-enzyme

↓

MeO O·CH₂-enzyme
P + HS·CH·CO·OEt
MeO O CH₂·CO·OEt

Blocked enzyme

Phosphorylation by phosphorus insecticides was originally thought to be irreversible. It is now established that the oximino compound, *Pralidoxime*, (pyridine-2-aldoximine methiodide) is able to reverse esterase inhibition as a result of nucleophilic attack by the powerfully nucleophilic oximino hydroxyl (**1**, 10) on the phosphoryl residue (Wilson and Meislich, 1953; Childs, Davies, Green and Rutland, 1955; Holmes and Robins, 1955). Specificity of the compound is assisted by the quaternary group, which is able to link with the cationic centre of the enzyme whilst enhancing the acidic strength of the oximino hydroxyl by electron-withdrawal.

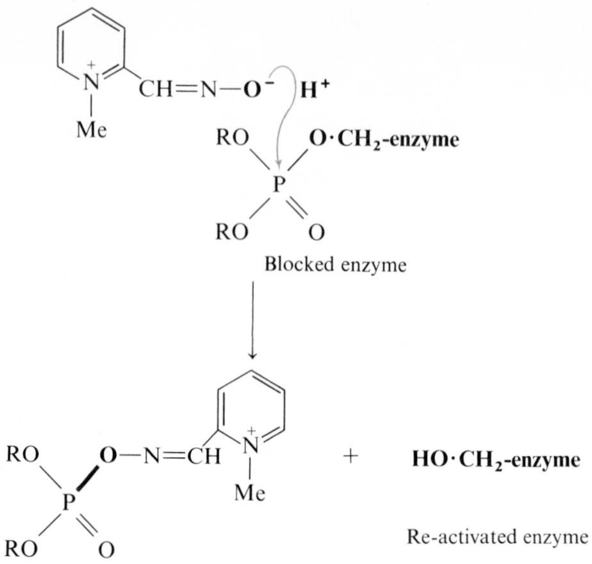

Me
$CH{=}N{-}O^-$ H^+
RO $O\cdot CH_2$-enzyme
P
RO O
Blocked enzyme

RO $O{-}N{=}CH$ + $HO\cdot CH_2$-enzyme
P Me
RO O Re-activated enzyme

Pralidoxime suffers from the disadvantage that it is itself an inhibitor of cholinesterase. A number of cyclic hydroxamic acid reactivators, which are free from this disadvantage, have been described (Bickerton, Coutts and Johnson, 1965).

Carbamate Cholinesterase Inhibitors

Certain carbamate esters, such as *Carbachol*, *Neostigmine*, *Pyridostigmine*, and *Physostigmine*, also act as anticholinesterases. Their reaction, however, is of limited duration, and for this reason they are used in human medicine not only as anticholinesterases, but also for the reversal of competitive neuromuscular blockade. *Carbachol*, being structurally similar to acetylcholine, has a parasympathomimetic action which, however, is more prolonged than that of acetylcholine owing to its resistance to hydrolysis by acetylcholinesterase. *Neostigmine*, *Pyridostigmine* and *Physostigmine* also possess some direct action at the nerve motor-end-plate in addition to their anticholinesterase activity.

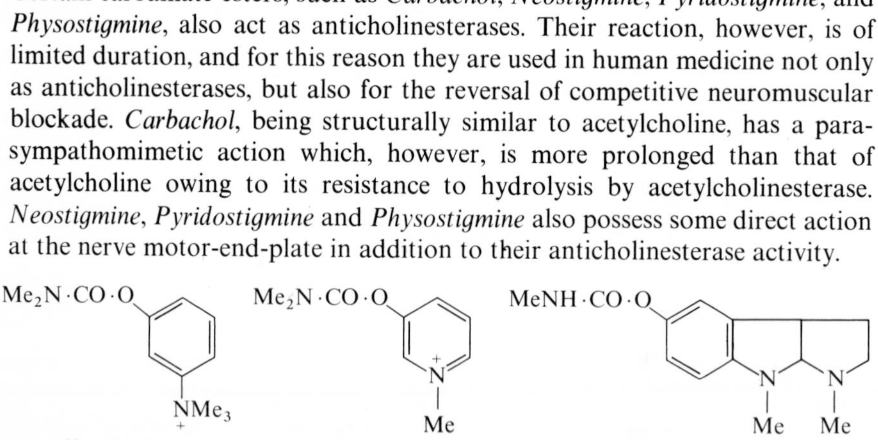

$Me_2N\cdot CO\cdot O$ — $\overset{+}{N}Me_3$ — *Neostigmine*

$Me_2N\cdot CO\cdot O$ — Me — *Pyridostigmine*

$MeNH\cdot CO\cdot O$ — N N Me Me — *Physostigmine*

The anticholinesterase activity of these carbamyl compounds appears to arise from carbamylation of the enzyme serine hydroxyls (Wilson, Harrison and Ginsburg, 1961) following ionic bonding of the quaternary ammonium group at the anionic site of the enzyme. The electron-withdrawing properties of the quaternary group promote breakdown of the intermediate facilitating the carbamylation reactions.

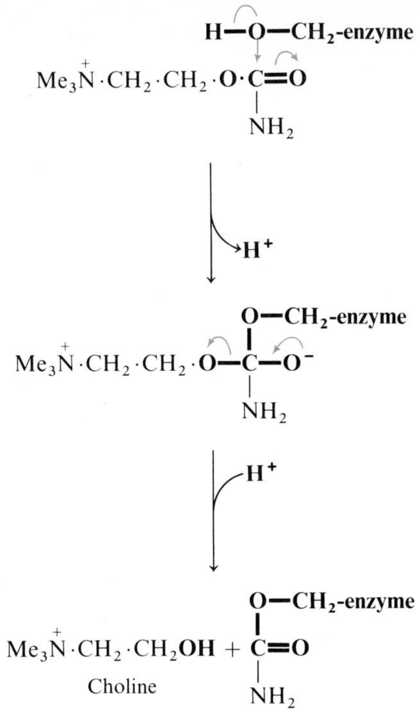

$$\mathrm{H{-}O{-}CH_2\text{-}enzyme}$$
$$\mathrm{Me_3\overset{+}{N}\cdot CH_2\cdot CH_2\cdot O\cdot C{=}O}$$
$$\mathrm{NH_2}$$

$\downarrow$ $\mathrm{H^+}$

$$\mathrm{O{-}CH_2\text{-}enzyme}$$
$$\mathrm{Me_3\overset{+}{N}\cdot CH_2\cdot CH_2\cdot O{-}C{-}O^-}$$
$$\mathrm{NH_2}$$

$\downarrow$ $\mathrm{H^+}$

$$\mathrm{O{-}CH_2\text{-}enzyme}$$
$$\mathrm{Me_3\overset{+}{N}\cdot CH_2\cdot CH_2OH + C{=}O}$$
$$\mathrm{NH_2}$$

Choline

Carbamylserine group of enzyme

Penicillins and Cephalosporins

The acylating action of penicillins (and cephalosporins) is believed to be directed towards a *β*-mercaptoethylamine ($>\mathrm{N\cdot CH_2\cdot CH_2\cdot SH}$) fragment of a transglycolase enzyme essential for cell-wall synthesis in susceptible micro-organisms (Fig. 37, **1**, 21). Exposure of sensitive strains of *Staphylococcus aureus* to sublethal

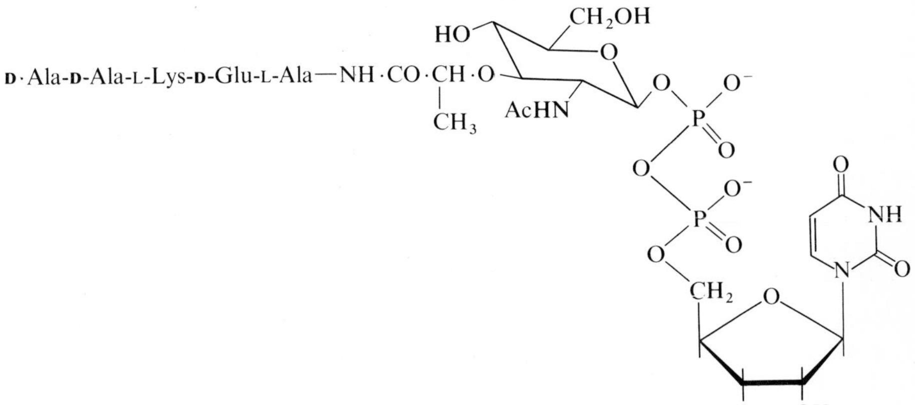

D·Ala-D-Ala-L-Lys-D-Glu-L-Ala—NH·CO·CH·O— (CH$_3$; sugar ring with HO, CH$_2$OH, O, AcHN) —O—P(O$^-$)(=O)—O—P(O$^-$)(=O)—O—CH$_2$—(ribose, HO, OH)—uracil (O, NH, N, O)

Cell-wall precursor

concentrations of *Penicillin* leads to accumulation within the cell of a cell-wall precursor (**1**, 21), a uridinediphosphorylglucopentapeptide (Park, 1952; Park and Strominger, 1957). The ratio of amino sugar to amino acid in this macromolecule is the same as that found in separated staphylococcal cell walls (Strange, 1956).

Pyrimidines are not normal cell-wall constituents, and the UDP group is found in the non-assimilated UDP-glucapentapeptide in the position normally occupied by *N*-acetylglucosamine in the intact cell wall. *Penicillin* appears to act by blocking a vital transglycolase. Attack on a β-mercaptoethylamine fragment of the enzyme is facilitated by the reactivity of the strained β-lactam ring of the penicillins. Alignment of the reacting groups is assisted by hydrogen bonding between the lactam carbonyl group and the sulphydryl hydrogen, and also by hydrogen bonding of the antibiotic carboxyl to the β-mercaptoethylamine.

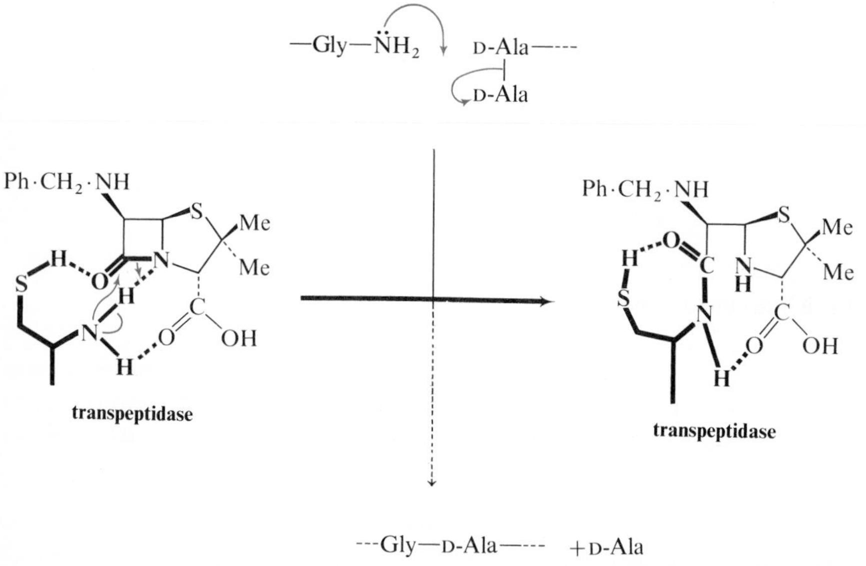

—Gly—$\ddot{N}H_2$ D-Ala—---
D-Ala

Ph·CH_2·NH
transpeptidase

Ph·CH_2·NH
transpeptidase

---Gly—D-Ala—--- + D-Ala

Biological Alkylating Agents

Most alkylating agents are toxic in some degree. The most toxic are carcinogenic; others which show a degree of selectivity between cancerous and normal tissues are carcinolytic. Nearly all biological alkylating agents are capable of esterifying carboxylic and phosphoric acids, and of etherifying phenols. The more reactive agents, *Mustine* (Mechlorethamine; Gilman and Philips, 1946), *Chlorambucil* (Ross, 1958) and *Melphalan* (Bergel, 1958) are capable of alkylating primary, secondary and tertiary amines

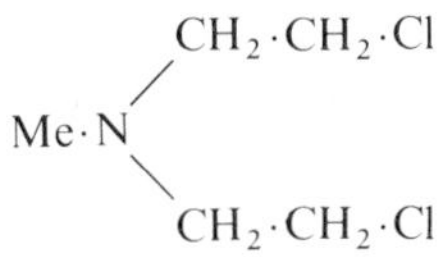

$$Me{\cdot}N(CH_2{\cdot}CH_2{\cdot}Cl)_2$$

Mustine

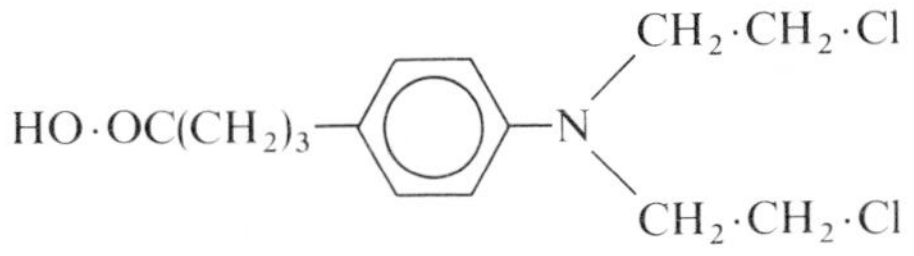

$$HO{\cdot}OC(CH_2)_3{-}C_6H_4{-}N(CH_2{\cdot}CH_2{\cdot}Cl)_2$$

Chlorambucil

$$HO{\cdot}OC{\cdot}CH(NH_2){\cdot}CH_2{-}C_6H_4{-}N(CH_2{\cdot}CH_2{\cdot}Cl)_2$$

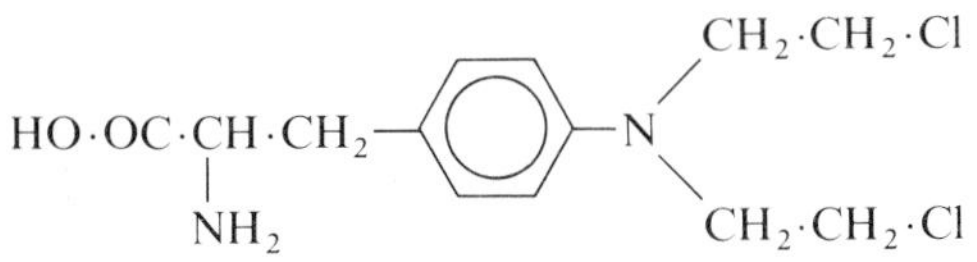

L-*Melphalan*

Such compounds appear to single out the guanine of DNA for attack in the 7-position (Lawley and Brookes, 1959). In the case of *Mustine*, the reaction is a nucleophilic attack by an S_Ni mechanism via an intermediate cyclic iminium ion. (See pp. 84–5.)

The most effective carcinolytic alkylating agents are bifunctional, and cross-linking of the DNA double helix forms an essential step in their mechanism of action (Ross, 1962). The cross-linked bis-guaninium complex is susceptible to nucleophilic attack by, for example, histidine residues of ribonuclease, so that the bridging bis-guanylethylmethylamine is deleted from the polymer. This interrupts the purine/pyrimidine base sequence of the DNA chain causing misreplication. The deleted bis-guanylethylmethylamine may also function as an antimetabolite (Timmis, 1960).

The aromatic nitrogen mustards do not form cyclic iminium ions because of the effects of electron-withdrawal on the nitrogen lone pair, so that the acyclic carbonium ion is stabilised.

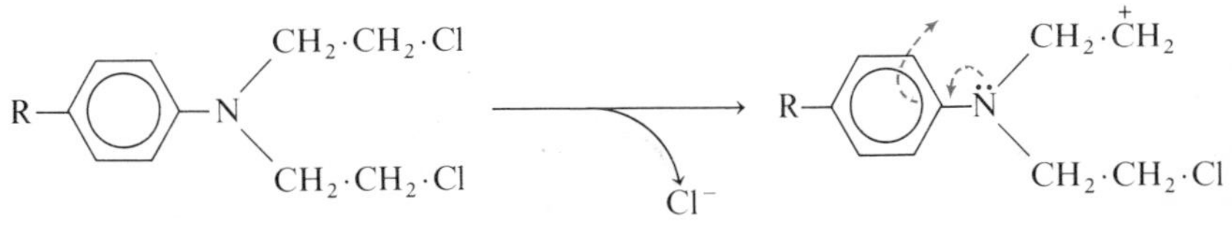

$$R{-}C_6H_4{-}N(CH_2{\cdot}CH_2{\cdot}Cl)_2 \xrightarrow{-Cl^-} R{-}C_6H_4{-}\ddot{N}(CH_2{\cdot}\overset{+}{C}H_2)(CH_2{\cdot}CH_2{\cdot}Cl)$$

Nucleophile

Cross-linked DNA

Nucleophile

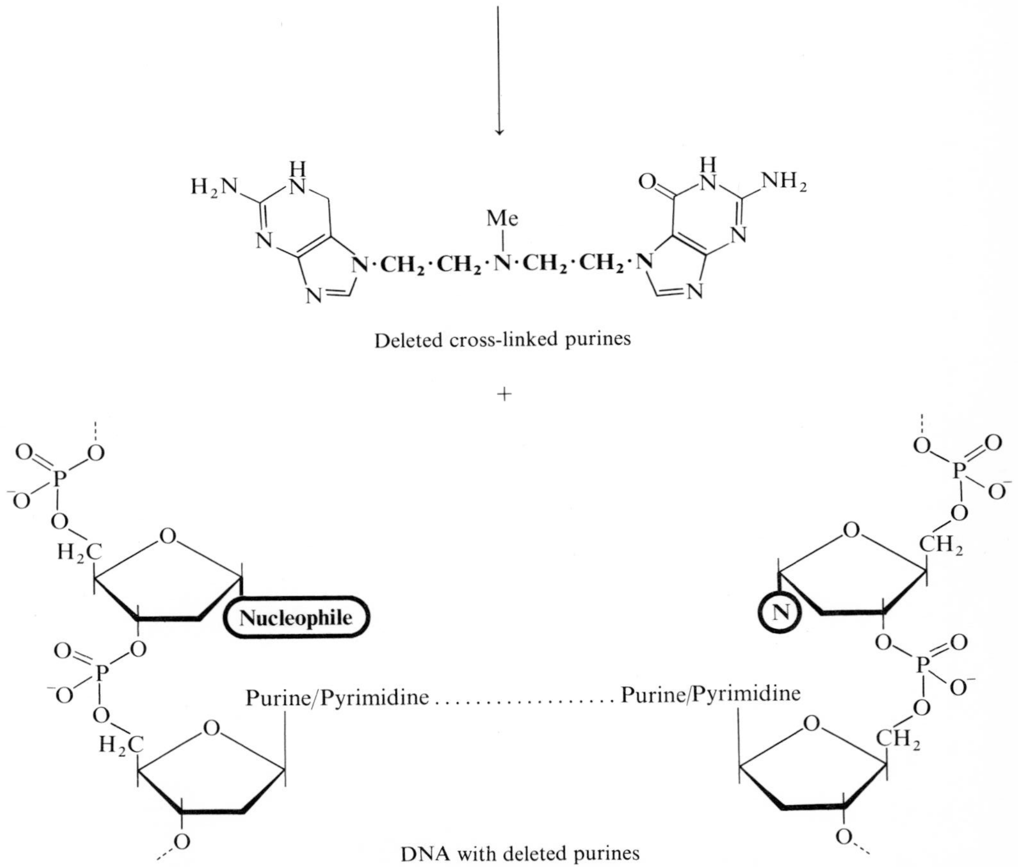

$$\longrightarrow$$

$$N\cdot CH_2\cdot CH_2\cdot N(Me)\cdot CH_2\cdot CH_2\cdot N$$

Deleted cross-linked purines

+

Nucleophile

Purine/Pyrimidine . . . . Purine/Pyrimidine

DNA with deleted purines

It is now considered that the acyclic carbonium ion is the active intermediate for all nitrogen mustards, which in the case of the aliphatic compounds is in resonance with the cyclic iminium ion.

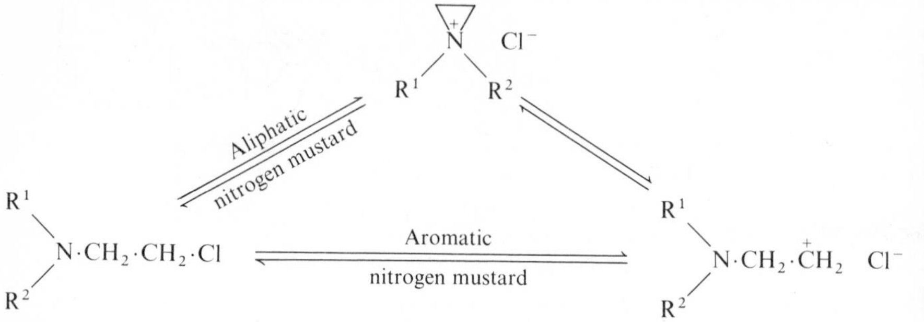

Aliphatic nitrogen mustard

Aromatic nitrogen mustard

$R^1R^2N\cdot CH_2\cdot CH_2\cdot Cl \rightleftharpoons R^1R^2N\cdot CH_2\cdot \overset{+}{C}H_2\ Cl^-$

Cyclic iminium ion

Biological active acyclic carbonium ion

In all the tri(aziridinyl) compounds, tri(aziridin-1-yl)-1,3,5-triazine, *Triaziquone* (Triazaquinone; tri(aziridin-1-yl)-1,4-benzoquinone) and *Thiotepa* (tri(aziridin-1-yl phosphine sulphide), the aziridinyl groups are linked to powerful electron-withdrawing groups, which promote facile nucleophilic attack on the ethyleneimine ring.

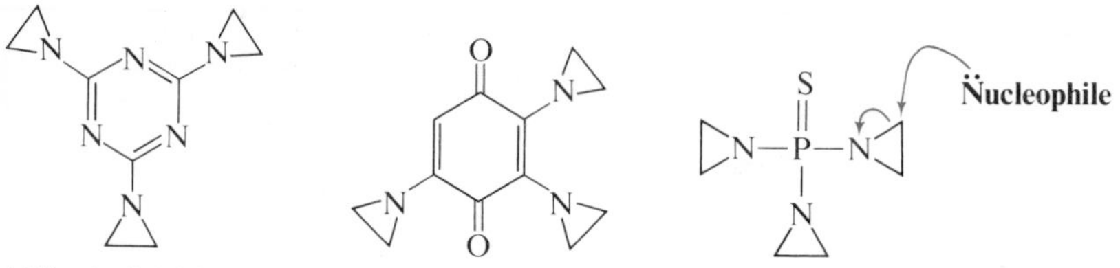

Nucleophile

Tri(aziridin -1-yl)-1,3,5-triazine

Triazaquinone

Thiotepa

METAL CHELATION

Metal ion complexes are formed from electron-donating molecules or ions (**ligands**) and a metal ion with an incomplete valency shell. Neutral molecules which function as electron donors in metal ion complexes include amines, imines, ketones and sulphides. Typical ions include carboxylate, sulphonate, phosphonate, phenolate and thiophenolate. Some typical metal ion complexes are shown in Table 15.

Complexes involving bi-, tri- or poly-dentate ligands, such as ethylenediamine, are known as **chelating** agents. Water-soluble chelating agents, which are able to confer water solubility by complexing with otherwise insoluble ions, are described as **sequestering** agents. Many metal chelates play an important rôle in a very wide variety of biological systems, such as haemoglobin, cyanocobalamin (vitamin B_{12}) and numerous enzyme systems.

Table 15. Metal Ion Complexes

Metal ion	Electronic configuration	Ligand	Complex ion	Electronic configuration of complex ion
Co^{3+}	2,8,14	$\ddot{N}H_3$	$[Co(NH_3)_6]^{3+}$	2,8,**18**,**8**
Co^{3+}	2,8,14	$H_2\ddot{N}\cdot CH_2\cdot CH_2\cdot \ddot{N}H_2$(en)	$[Co(en)_3]^{3+}$	2,8,**18**,**8**
Cu^{2+}	2,8,17	NH_3	$[Cu(NH_3)_4]^{2+}$	2,8,**17**,**8**
Cu^{+}	2,8,18	CN^-	$[Cu(CN)_4]^{3-}$	2,8,**18**,**8**
Fe^{2+}	2,8,14	CN^-	$[Fe(CN)_6]^{4-}$	2,8,**18**,**8**
Fe^{3+}	2,8,13	CN^-	$[Fe(CN)_6]^{3-}$	2,8,**17**,**8**

Ligand Selectivity

Metals, in general, show a definite order of affinity for organic ligands. Equally, different ligands show varying degrees of selectivity for a particular metal ion. The reasons for this selectivity are complex. They are, however, related to the concept of **hard** and **soft** acids and bases (HSAB; Pearson, 1963, 1966). This concept is based essentially on the principle that wherever two different atoms are bonded together, the bonding electrons are unevenly distributed, so that one atom functions as an electron acceptor, and can be considered as an acid, whilst the other plays the complementary rôle of electron donor and can be considered as a base. The significance of this concept is obvious in substances which function as acids and bases.

$$[\text{Acid}]^{+} + \ddot{\text{B}}\text{ase} \rightleftharpoons \textbf{A—B}$$

Although less readily apparent, the same concept is applicable on a much wider basis to the consideration of all chemical bonds, irrespective of whether the components can or cannot exist as discrete ionic, atomic or molecular entities. It requires the strengths of acids and bases to be determined by two parameters, intrinsic strength (S), and softness (σ), a factor which reflects the mobility of the electrons, according to the following relationship:

$$\log K = S_A S_B + \sigma_A + \sigma_B$$

It follows that if both species contributing to the bond show either high or low electron mobility, the pull of one in effect neutralises that of the other, and a stable bond results. If, however, only one of the two contributors shows high electron mobility, the bond is unstable and weak. Contributing species showing high electron mobility are described as **soft** acids or bases; those showing-low electron mobility are **hard** acids or bases. According to this nomenclature, **hard-hard** and **soft-soft** bonds are stable, whilst **hard-soft** bonds are of low stability.

Table 16. Typical Characteristics of Hard and Soft Acids and Bases

	Electron Acceptors (acids)		Electron Donors (bases)	
	Hard	Soft	Hard	Soft
Size	Small	Large	Small	Large
Charge	Large (+)	Small (+)	Large (−)	Small (−)
Electronegativity			High	Low
Electropositivity	High	Low		
Outer shell electrons				
Number	Small	Large		
Excitability	Low	High		
Unfilled orbitals				
Energy			High	Low
Accessibility			Inaccessible	Accessible
Bond type	Electrostatic	π Dative	Electrostatic	π Dative

Electron mobility (**polarisability**) is determined by a variety of factors including size, charge, electronegativity, the number and excitability of outer shell electrons and the state and accessibility of unfilled orbitals. Typical characteristics of hard and soft acids and bases are expressed in terms of these parameters in Table 16. Important examples of hard and soft acids are shown in Table 17.

Fluid Transport

Since the formation of strong rather than weak bonds is always favoured, hard acids are bound preferentially with hard bases. Thus, the hard acids, Na^+, K^+, Ca^{2+} and Mg^{2+}, not only form salts with such hard bases as HO^-, $PO_4{}^{3-}$, $CO_3{}^{2-}$, and carboxylates, but are also strongly hydrated by association with water molecules in aqueous solution. It is not surprising, therefore, that they play a major role in fluid transport, retention and excretion within and from the animal body (**1**, 14).

Drug Absorption

The tetracycline antibiotics have three ionisable groups (**1**, 22), all of which are capable of functioning as hard bases. This explains their ability to form stable salts with the hard acids, Ca^{2+}, Mg^{2+} and Al^{3+}, and why gastric absorption is delayed if tetracyclines are co-administered with calcium, magnesium or aluminium salts. The precise structure of these particular tetracycline–metal complexes has not been established, but 2:1 complexes of the softer ions, Fe^{3+}, Fe^{2+}, Ca^{2+}, Zn^{2+}, Co^{2+} and Mn^{2+}, are known.

Table 17. Some Typical Hard and Soft Acids and Bases

Hard		Soft
Acids		
Ca^{2+} Sr^{2+} Mg^{2+} Be^{2+} Mg^{2+} Fe^{3+}	Fe^{2+} Co^{2+} Zn^{2+} Ni^{2+} Cu^{2+} Sn^{2+} Pb^{2+}	Pd^{2+} Cd^{2+} Pt^{2+} Hg^{2+} CH_3Hg^+
Bases		
H_2O HO^- F^- Cl^-	Br NO_2 $SO_3{}^2$	I′ SCN^- CN^- CO NO
ROH RO^- ROR′		RSR′ RSH RS^- R_3As
$R{\cdot}CO{\cdot}O^-$ $RCO{\cdot}CH{=}CR'{\cdot}O^-$ $CO_3{}^{2-}$		
NH_3 RNH_2 $H_2N{\cdot}NH_2$	$Ph{\cdot}NH_2$ Pyridine $-N{=}N-$	$R{\cdot}CH{=}CHR'$ Benzene

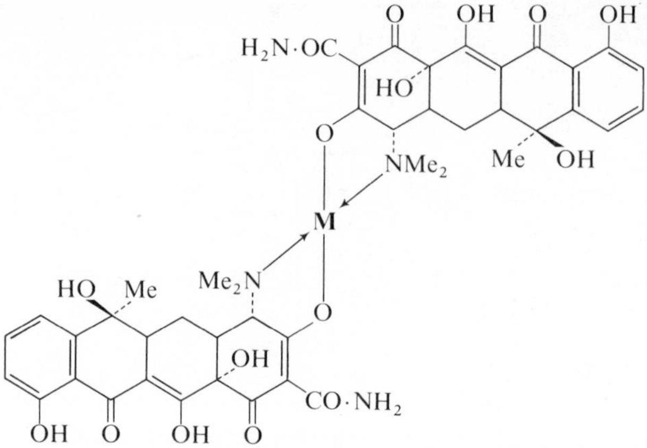

2:1 Tetracycline–transition metal complexes

In contrast, oral administration of the 2:1 *Dicoumarol*–magnesium complex gives very much higher plasma concentrations of *Dicoumarol* (Dicumarol) than administration of the drug alone, with potentiation of its anticoagulant action (Akers, Lach and Fischer, 1973). This particular hard acid–hard base complex appears, therefore, to be sufficiently stable to be absorbed intact at a rate which is significantly greater (up to 180% of controls) than that of *Dicoumarol.*

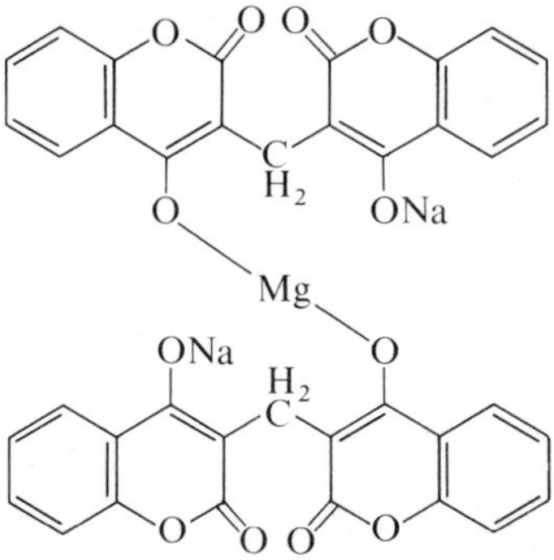

Calcium Ion Transport and Availability in Man

The rate of calcium release into the blood from cells in the bone marrow, where it is stored, is controlled by *Calcitonin*, a peptide with hard basic groups such as COO^- and NH_2, capable of forming stable complexes with Ca^{2+} (Riniker *et al.*, 1968; Neher *et al.*, 1968; Sieber *et al.*, 1968).

Calcium forms a stable complex with ligands, such as ethylenediaminetetra-acetic acid (EDTA), involving both amine and carboxylate functions, both hard bases. It is used in the form of *Trisodium Edetate Injection* in the treatment of hypercalcaemia, the stability and solubility of the calcium complex being such that it mobilises calcium from the bones and promotes its rapid excretion via the kidney.

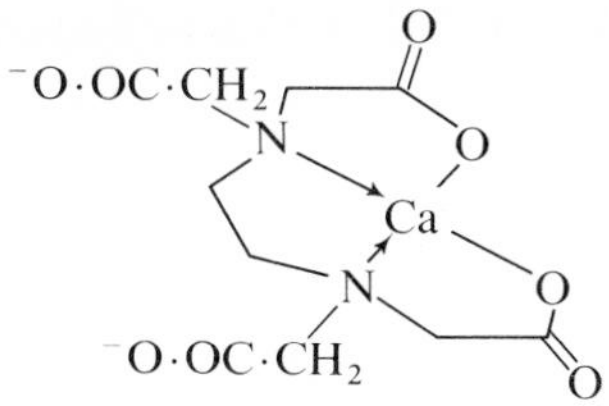

The ease with which EDTA chelates Ca^{2+} is, in itself, a complication in the treatment of metal poisoning with EDTA salts (p. 92). Thus, the complexing agent not only binds toxic metal ions, but also other hard acids such as Ca^{2+}, so that circulating levels of this physiologically-important ion are lowered, and tissues depleted. This problem, however, is partly overcome by the use of *Sodium Calcium Edetate*, $Na_2[CaEDTA]$, rather than the di- and tri-sodium salts. The Ca^{2+}-specific binding protein of the intestinal mucosa responsible for calcium absorption (**1**, 22) is presumably also a hard base.

Edetate Stabilisation of Pharmaceutical Dosage Forms

Disodium Edetate is widely used to chelate small traces of contaminating heavy metal ions as hard acid–hard base complexes, and so stabilises pharmaceutical products against metal-catalysed decomposition, particularly oxidation. It is also used in the formulation of skin antiseptics containing *Chloroxylenol* to reduce the inactivating effect of metal ions (Ca^{2+} and Mg^{2+}), present in hard water used for dilution, on the antibacterial efficiency of the phenolic antiseptic.

Metal Chelates as Antibacterials

The antibacterial action of oxine (8-hydroxyquinoline) has been ascribed to the formation of its 1:1-complex with ferric iron (Albert, Rubbo, Goldacre and Balfour, 1947) a stable hard acid–hard base complex. In support of this view, it is significant that 8-methoxyquinoline and oxine methiodide are structurally incapable of acting as metal-chelating agents, and both are without antibacterial activity in the presence of ferric iron.

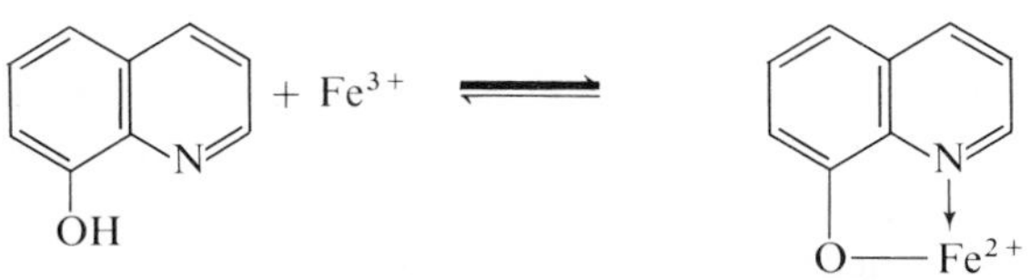

8-Hydroxyquinoline + Fe^{3+} ⇌ 1:1 Oxine–ferric iron complex

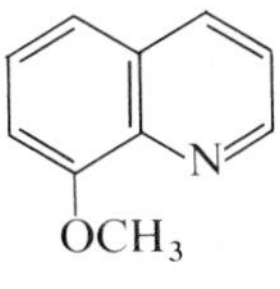

8-Methoxyquinoline

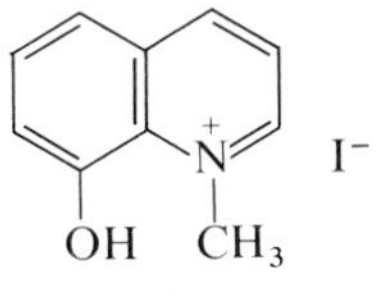

Oxine methiodide

Two explanations for the antibacterial action are possible. Either the ferric iron–oxine complex is toxic to sensitive micro-organisms in its own right, or oxine merely complexes ferric iron and/or other metal ions withholding them from vital metabolic processes. Experiments with distilled water suspensions of staphylococci, in which the organisms are viable for at least 24 hr, showed that oxine (10 μM) alone is **not** bactericidal, but that it becomes so on the addition of an equivalent amount of ferric iron. The conclusion that it is the complex which is bactericidal was confirmed in further experiments in broth media, which contained iron and other trace metals; these showed that whereas oxine at 10 μM is highly bactericidal, oxine at 625 μM is not. Albert has described this effect as **concentration-quenching** and ascribes the loss of activity to the formation of the 3:1 oxine–ferric iron complex, which in contrast to the 1:1-complex is non-toxic to the bacteria.

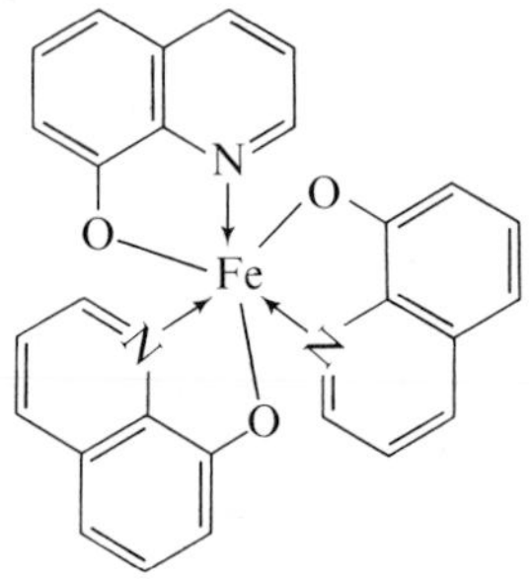

3:1 oxine–ferric iron complex

It has been suggested that *Isoniazid*, which is capable of metal chelation, owes its tuberculostatic activity to this property, since its *N*-methyl derivative, which is not a chelating agent, has little activity against *M. tuberculosis.* It is doubtful, however, whether this is an adequate explanation, as the corresponding hydrazides of nicotinic and picolinic acids, which are also metal-chelating agents, are ineffective.

Treatment of Metal Poisoning

The strong affinity of soft acids for soft bases explains the toxicity of a number of metal ions. Some of the most toxic metal ions known, including Pb^{2+}, Pd^{2+}, Cd^{2+}, Hg^{2+}, CH_3Hg^{+}, are soft acids. This toxicity arises because they combine to form strongly-bonded complexes with soft bases, such as the sulphydryl group of cysteine, which is present at the active centre of a number of enzyme proteins. Complexation once achieved is all the more difficult to reverse, because lone pairs involved in complexation are no longer available for solvation, so that the complexes are usually less soluble than their ligands.

Treatment of metal poisoning is achieved by the use of complexing agents, such as *Dimercaprol* and *Disodium Edetate*, which form complexes capable of removing the toxic metal in a readily excretable, water-soluble form (Sidbury,

Bynum and Fetz, 1953; Bessman, Rubin and Leikin, 1954). It is important to understand, however, that the antidote–metal complex does not necessarily have to be more stable than the toxic ion–enzyme complex, and indeed is often less so. This is because the function of the antidote is not the reversal of toxic ion–enzyme complex formation by competition. Instead, chelation with uncomplexed free metal ions unbalances the equilibrium between freely-circulating and enzyme-bound metal ion in favour of the former.

Disodium Edetate has been used in this way for the treatment of lead (Pb^{2+}) and vanadium (Va^{2+}) poisoning. The vanadyl–edetate complex is a hard acid–hard base complex. The corresponding Pb^{2+}–edetate chelate, however, is a soft acid–hard base complex. It is not surprising, therefore, that lead poisoning is more effectively treated with *Penicillamine*, which chelates through a combination of soft (SH) and hard (NH_2) basic groups.

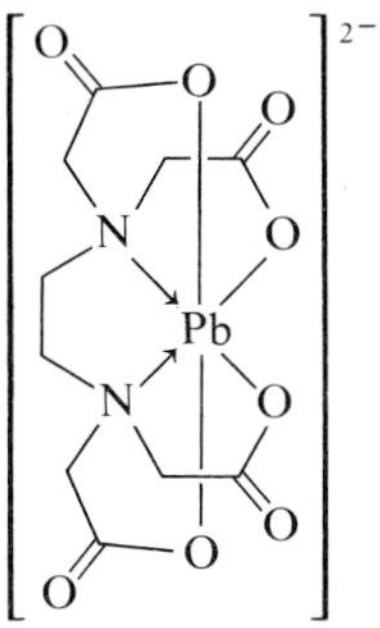

Lead Edetate

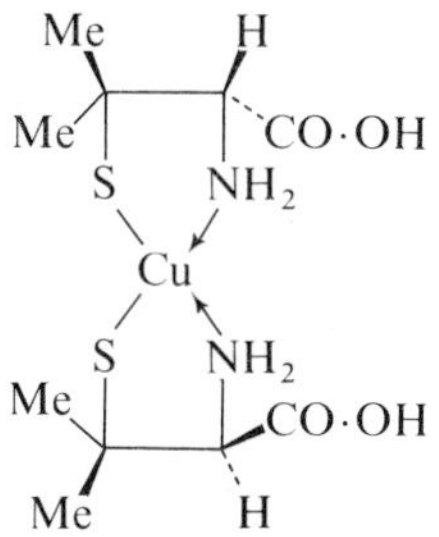

Penicillamine-Cu(II)chelate

Penicillamine is also used in the treatment of Wilson's disease (hepatolenticular degeneration), a metabolic disorder which leads to extensive deposition of Cu(II), principally in the liver, brain and kidney. *Penicillamine* is very

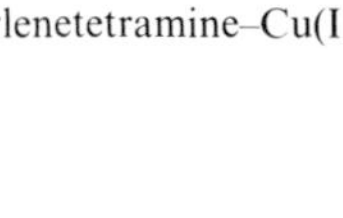

Triethylenetetramine–Cu(II) chelate

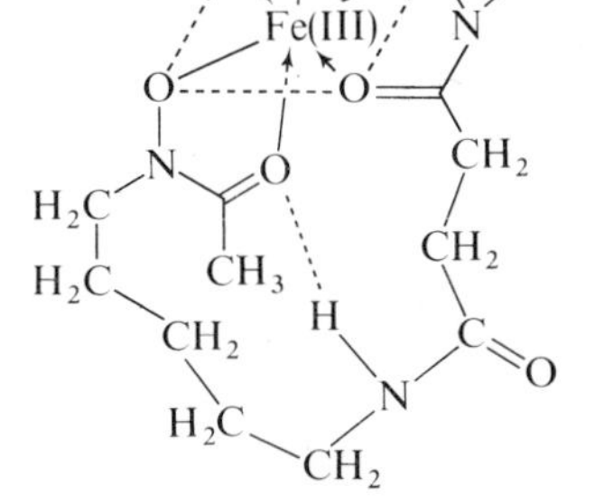

Desferrioxamine–Fe(II) chelate

effective, markedly increasing urinary excretion of copper as *Penicillamine* Cu(II) chelate. It does, however, produce serious and dangerous side-effects in some patients, and the use of triethylenetetramine (as dihydrochloride), which also forms a tetradentate complex is preferred (Walshe, 1975).

Penicillamine has also been recommended for treatment of iron poisoning. However, *Desferrioxamine*, which forms a more stable hard acid–hard base complex, is more effective and has proved of particular value in the treatment of iron overload in the Bantu of South Africa. The use of *Salicylic Acid* in the treatment of beryllium is similarly based on the formation of a stable hard acid–hard base complex.

Complex Stabilisation

The affinity of soft acids for soft bases is also an important feature of metallo-enzymes in which softer acids, such as Zn^{2+}, Cu^{2+} and Mn^{2+}, are often complexed with sulphydryl groups. In many of these complexes, however, the softness or hardness of the ion is modified by its immediate environment, thereby increasing the stability of the complex. Thus, Zn^{2+}, which is hydrated in aqueous solution, thereby becomes a hard acid, and then combines with halide ions in the order $F^- > Br^- > Cl^- > I^-$. In contrast, the Zn^{2+}-carbonic anhydrase complex, which also binds halide ions, does so in the order $I^- > Br^- > Cl^- > F^-$, indicating that the bond in this complex is a soft one, with the metal co-ordinated to soft bases. Similarly, Zn^{2+} in liver alcohol dehydrogenase and Mo^{2+} in xanthine oxidase function as soft acids and are complexed with soft sulphydryl residues.

Modification towards soft or hard behaviour is also characteristic of the cobalt–corrin and iron–porphyrin complexes. The metal ion is hardened by its presence at the centre of planar 4-co-ordinate nitrogenous structures, but the behaviour of individual complexes is modified by additional ligands present in the 6- (and 5-) co-ordinate carrier complexes of the Vitamin B_{12} (*Cyanocobalamin*) group. This has been very clearly demonstrated by ultraviolet, visible, infrared and circular dichroism spectra of cyanocobalamin complexes (Hill, Pratt and Williams, 1969; Pratt, 1972). Thus, when X^- is varied in place of CN^- from hard (Cl^-) to soft ($CH_3 \cdot CH_2^-$) donors, major spectral changes

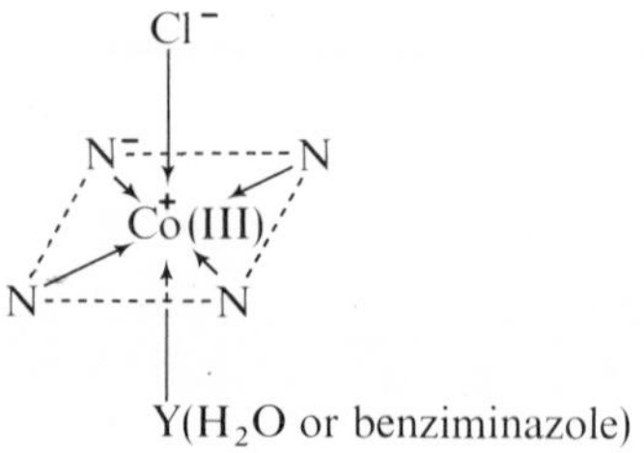

Cl^- — Co(III) — Y(H_2O or benziminazole)

6-co-ordinate corrin complex

$CH_2 \cdot CH_2^-$ — Co(III)

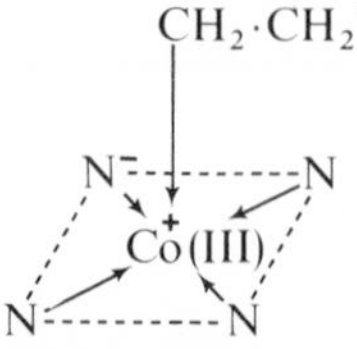

5-co-ordinate corrin complex

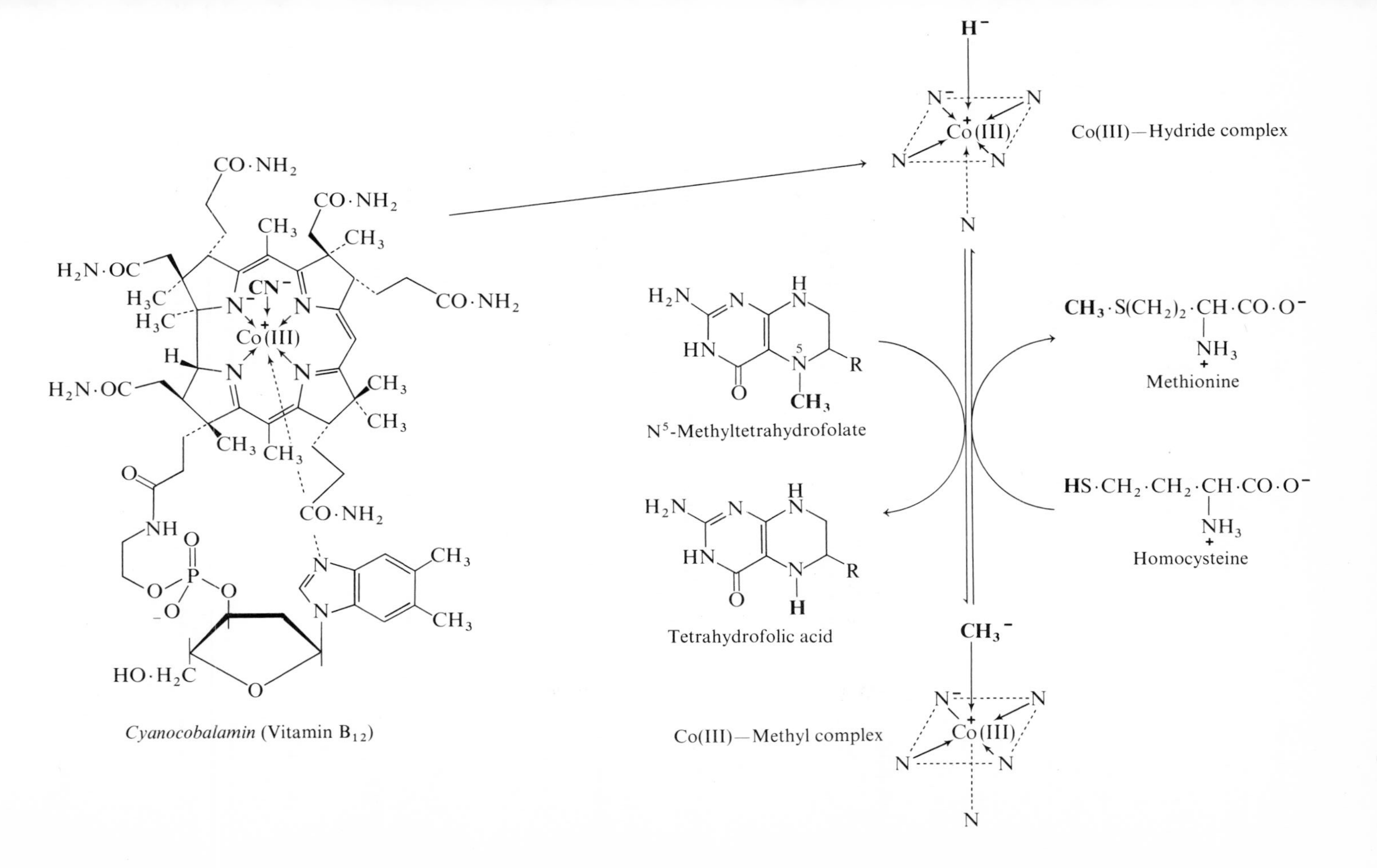

Cyanocobalamin (Vitamin B_{12})

occur which indicate a shift from 6- to 5- co-ordinate complexes, corresponding to a softening of the cobalt ion by association with the softer donors.

A marked reduction in the CN stretching is also clearly seen in the complexes (Y = CN) when X is varied from the hard donor, water (CN stretching at 2130 cm^{-1}) to the soft donor $CH_3CH_2^-$ (CN stretching at 2082 cm^{-1}; free CN^- at 2078 cm^{-1}) indicating a complete shift from 6- to 5-co-ordination.

Vitamin B_{12} (*Cyanocobalamin*), the anti-pernicious anaemia factor of liver, and related corrinoids, are six co-ordinate cobalt(III) complexes (Brink, Hodgkin, Lindsey, Pickworth, Robertson and White, 1954). *Cyanocobalamin* and some related corrinoid complexes function as coenzymes in various methyl transfer reactions as, for example, the conversion of homocysteine to methionine in human liver. The source of the methyl group is N^5-methyltetrahydrofolic acid, and transfer probably occurs via a methyl corrinoid, which is formed in nucleophilic attack by an intermediate cobalt(III) hydride complex, similar to that obtained synthetically by reduction (Johnson, Mervyn, Shaw and Smith, 1963). The valency state in such complexes, which is reflected in its colour (grey-green) and chemical behaviour, is determined by the relative electronegativities of the alternative ligands with respect to cobalt. (See structures on p. 95.)

There is some evidence that the conversion of mercury wastes by anaerobic bacteria to toxic methyl mercury and dimethyl mercury involves a corrinoid-dependent methyl transfer reaction (Wood, Kennedy and Rosen, 1968).

A second group of B_{12} enzymes, responsible for a series of isomerase reactions (Barker, 1967) which appear to involve hydride ion transfer, have a 5-deoxyadenosyl corrinoid as coenzyme (Lenhert and Hodgkin, 1961). Labelling experiments show an exchange of hydrogen between the substrate and the C'_5-adenosyl carbon, with a concomitant acyl migration, possibly through a cyclic intermediate. (See structure at the top of p. 97.)

Association with hardening or softening of the ion is also responsible for stabilisation of valency states in the iron–porphyrins. Higher oxidation states (Fe^{3+}) are normally stabilised by association with hard bases, whilst lower oxidation states are stabilised by association with soft bases. This explains the ability of carbon monoxide or cyanide (CN^-) to poison cytochrome enzymes by holding them firmly in the lower oxidation state; also, the poisoning of catalase and peroxidase by the soft bases CN^- and H_2S which favour combination with the lower, Fe(II), oxidation state rather than the normal Fe(III) state of the complexes. Similarly, *Dicobalt Edetate*, which is used as an antidote in cyanide poisoning, is a bi-(tridentate) complex of Co(II) which, being in the lower oxidation state, has a stronger affinity for the soft base cyanide (CN^-). It, therefore, depends for its action upon the ability of Co(II) to increase its co-ordination number to four, or, possibly, six by association with CN^- ions. (See structure at the bottom of p. 97.)

Sodium nitroprusside, which is used by injection in the control of hypertensive crises, is a further example of a complex which is formed when iron in the

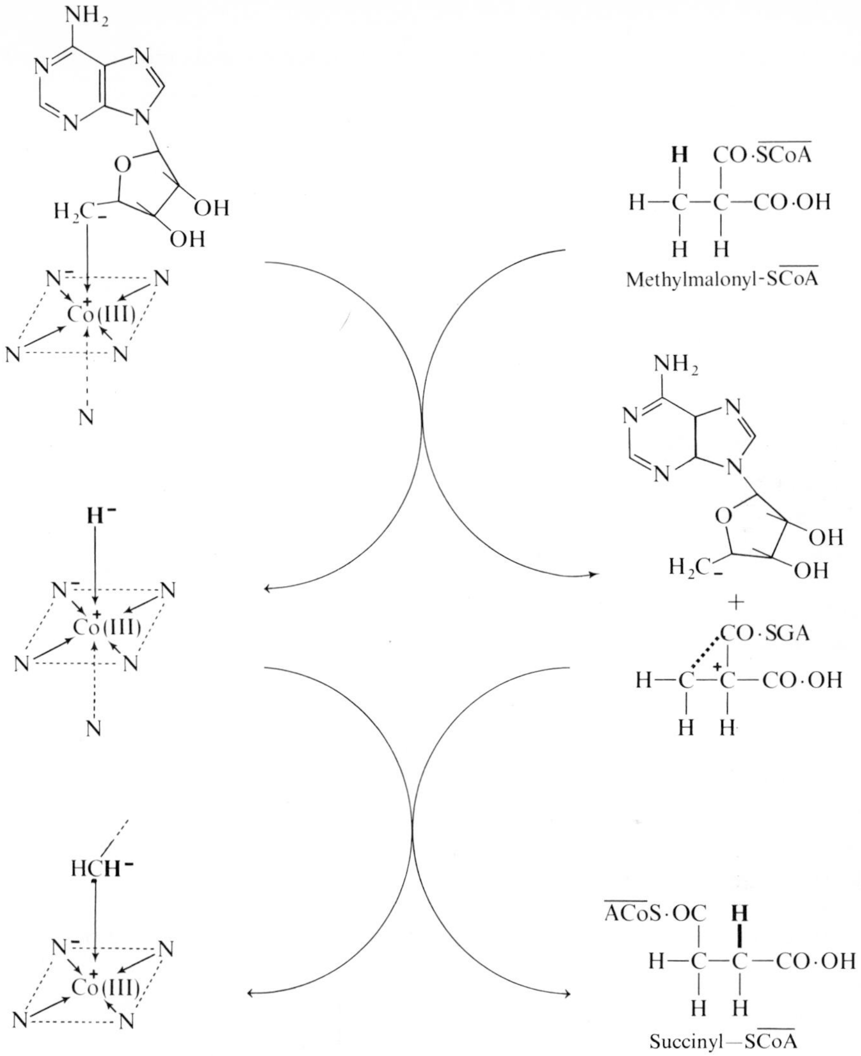

NH_2

OH

OH

H_2C^-

Co(III)

H CO·SCoA

H—C—C—CO·OH

H H

Methylmalonyl-SCoA

H^-

Co(III)

NH_2

OH

OH

H_2C^-

+

CO·SGA

H—C—C—CO·OH

H H

HCH$^-$

Co(III)

ACoS·OC H

H—C—C—CO·OH

H H

Succinyl—SCoA

CN^-

Co Co

CN^-

CN^-

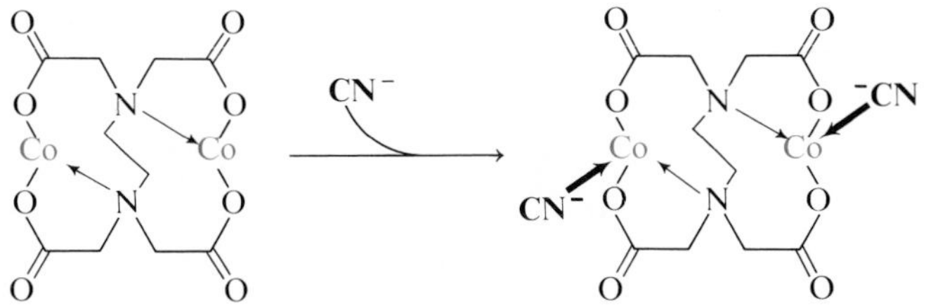

hard higher oxidation state Fe(III) already softened by association with soft CN^- ions, is still further stabilised by association with the soft base, nitric oxide.

$$[Fe(III)(CN)_6]^{3-}Na_3^+ + NO \longrightarrow [Fe(III)(CN)_5NO]^{2-}Na_2^+ + NaCN$$

Light Stability of Complex Ions

A number of Fe(III) and Co(III) complexes show instability to light. Thus, sodium nitroprusside is extremely sensitive to light and both solid and injection solutions must be protected from its effects. The decomposition, which represents a hardening of the Fe(III) environment by displacement of CN^- and its replacement by water, can be followed by observing accompanying changes in absorption intensity at its ultraviolet absorption maximum.

$$Na_2{}^+[Fe(CN)_5NO]^{2-} + \mathbf{H_2O} \xrightarrow{h\nu} Na^+[Fe(CN)_4, NO, \mathbf{H_2O}]^- + \mathbf{NaCN}$$

Photo-aquation with hardening of the Co(III) complex and expulsion of CN^- is also the characteristic of photodecomposition of *Cyanocobalamin* in aqueous solution (Pratt, 1964). The corresponding expulsion of $CH_3{}^-$, $CH_3CH_2{}^-$ and similar carbanions, however, is not favoured, and alkyl cobalamins undergo an alternative photoreduction on exposure to light in aqueous solution, with expulsion of an alkyl radical and reduction of the cobalt to the lower valency Co(II) state. The reaction is completely reversible under nitrogen, but becomes

$$H_3C^-\text{–Co(III)} + H_2O \xrightarrow{h\nu} H_2O\text{–Co(II)} + CH_3\cdot$$

$$CH_3\cdot + \cdot O\text{—}O\cdot \longrightarrow H_3C\text{—}O\text{—}O\cdot \longrightarrow H_2C\text{=}O + HO\cdot$$

irreversible with the formation of aquocobalamin in the presence of air. The rate of the reaction is also enormously increased in air by a factor of about 1200 over that under nitrogen, $t_{\frac{1}{2}}$ being about 1 min in air. The expelled methyl radical is oxidised via the methylperoxy radical to formaldehyde and formic acid (Dolphin, Johnson and Rodrigo, 1964). Experiments with ^{14}C-labelled methyl cobalamin also provide evidence of methane formation by abstraction of hydrogen from the corrin ring system, and methyl substitution in the corrin ring (Hogenkamp, 1966).

3 Stereochemical Factors in Biological Action

INTRODUCTION

Stereospecificity in Drug Action

The idea of stereochemical specificity as a key factor in biological action has its origins in the observations of Biot (1815) and Pasteur (1848). It was Biot who first observed in 1815 that organic substances, such as tartaric acid, sugar and turpentine, were capable of rotating plane-polarised light. Such substances are said to be optically active (**1**, 13) and are designated **dexrotatory** if the rotation is clockwise (positive rotation; (+)) or **laevorotatory** if the rotation is anti-clockwise (negative rotation; (−)). Pasteur, however, not only conceived and demonstrated the fundamentals of optical asymmetry in the tartaric acids, but was quick to note the abundance of many naturally occurring optically active compounds. More important still, Pasteur (1860) observed that moulds and yeast could differentiate the optically active (+)- and (−)-tartrates by selective utilisation of one isomer in preference to the other in the first experimental demonstration of the importance of stereochemistry in biological action. A few years later, Lewkowitsch (1883) showed that *Penicillium glaucum* selectively oxidises (+)-lactic, (+)-mandelic and (−)-glyceric acids in preference to their enantiomorphs, and in 1902 McKenzie reported that dogs utilised (+)-β-hydroxybutyric acid more readily than the corresponding (−)-isomer. Also in that year, Neuberg and Wahlgemuth (1902) found that less (−)-arabinose is excreted in the urine of rabbits than the (+)-isomer after oral or subcutaneous administration.

These, and other examples of stereospecificity in biological action, were discussed by Hirsch (1918) and later by Cushny (1926) in his study of biological relations of optically isomeric substances. Cushny's appreciation of the incidence of stereochemical factors in biological reactions provides the basis for current awareness of their importance. He was also the first to advocate biological examination of enantiomorphic molecules to ascertain what he described as 'the configuration of the tissues'. Some thirty years later, Pauling (1956), after examining the basis of antigen—antibody reactions came to the general conclusion that 'complementariness in molecular structure of some sort is responsible for biological specificity in general', and went on to specify 'many enzymes, perhaps all enzymes except those involved in the transfer of electrons during oxidation—reduction reactions, are effective vecause of complementariness of structure of the reacting molecules'.

This idea of stereochemical complementariness in the structure of reacting molecules is merely an extension and refinement of the concept of 'molecular fit' between drug and receptor, which was first enunciated by Emil Fischer (1894) in his lock and key analogy. The development of this and similar ideas over the intervening years has concentrated attention on the structural parameters of drug molecules for maximum pharmacological activity or, where clinical studies are involved, for optimum therapeutic response. The concentration of effort on the structure of the drug has resulted naturally from the rapid growth and development of organic chemistry, and has brought large numbers of closely-related compounds with similar physical and chemical properties into the hands of the pharmacologist. The comparative ease with which such relatively small molecules can be studied by modern chemical and physico-chemical techniques in parallel with the examination of their biological properties has led to many advances in the field of medicine. Knowledge of their basic interactions with bio-receptors at molecular level, however, is superficial in the extreme. This is partly due to the paucity of detailed stereochemical studies only now being seriously tackled, but even more to the innumerable gaps in knowledge of particular biochemical systems with which individual drugs react.

The extent of the problem is readily perceptible if one considers the complexity of drug action on the intact animal. The desired pharmacological response, which ideally should be a specific response in a particular tissue, is in reality modified by innumerable additional factors which may favour or mitigate against the required response. These include the processes of drug absorption, distribution, metabolism and excretion. These in turn embrace the factors affecting the penetration of membranes, the temporary deposition of drug in storage tissues, and other more permanent effects of metabolic involvement, which are discussed more fully elsewhere (**2**, 4 and 5). Even where these complications can be temporarily set aside, effective understanding of the prime pharmacological response requires a knowledge of bio-receptors, which needs to be as precise and extensive as our present-day knowledge of the drugs which react with them.

Receptor Conformation

Many enzyme substrates and coenzymes appear to have structural and stereochemical parameters which are either similar to or complementary to those of the drugs which react with them. The enzymes themselves, the peptides, hormones, and many of the structural and protoplasmic components of living tissue, however, are macromolecular, being either protein, lipoprotein, phospholipoprotein, polysaccharide or nucleic acid. Such components possess a complexity, and, frequently, a flexibility of structure of a totally different order from that of the drug. These factors contribute equally with the drug to the chemical and stereochemical parameters of drug–receptor interactions, yet in many cases the extent to which they intervene is not known and cannot be measured.

Whilst the complementary rôle of receptor geometry was sensed by Cushny (1926), Pauling (1956), Nachmansohn (1952), Koshland (1958) and others, firm experimental evidence of its contribution and importance in drug—macromolecular interactions is only now appearing. Koshland (1963) has, however, shown that the serine phosphoryl group on phosphoglucomutase is activated in the presence of its substrate, glucose-6-phosphate, by unfolding of the enzyme protein, and Nachmansohn proposed, as far back as 1952, that acetylcholine induced a conformational change in the molecule of acetylcholinesterase. It is now established that a serine primary hydroxyl group is the primary nucleophile in the latter reaction, and inactivation of the enzyme by selective phosphorylation with [^{32}P]-di-isopropylphosphofluoridate (Cohen, Oosterbaan, Jansz and Berends, 1959; Sanger, 1963) and tetraethyl pyrophosphate (TEPP) has been shown by ORD studies to be accompanied by conformational changes (Fig. 20; Kitz and Kremzner, 1968).

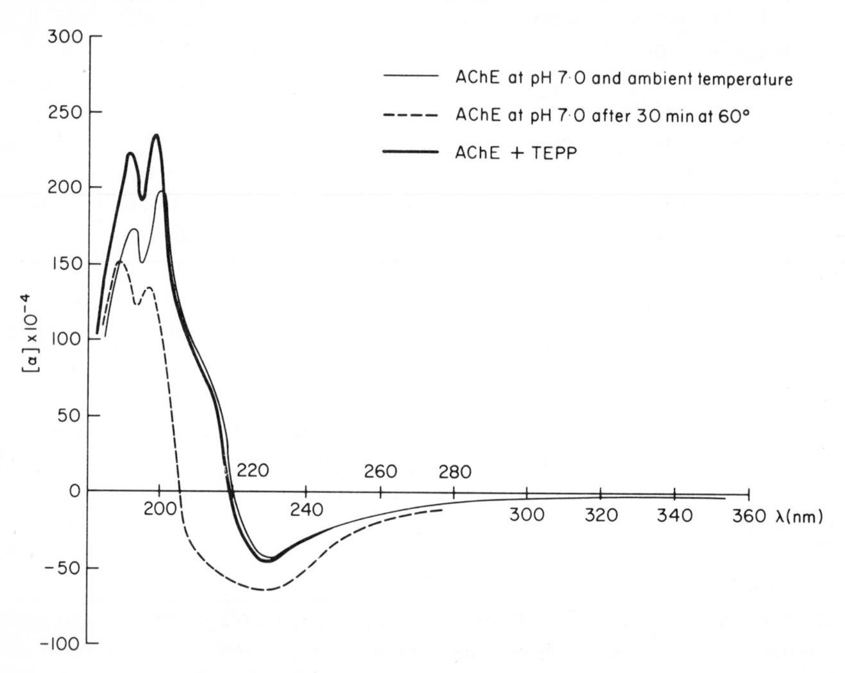

Fig. 20 Optical rotatory dispersion of acetylcholinesterase showing the effects of temperature and of interaction with TEPP

Krupka (1966, 1967) has shown by pH dependence studies of the hydrolysis of both cationic and neutral ester substrates using bovine erythrocyte cholinesterase that two basic groups of pK_a 5.5 and 6.3 are present at the active centre of the enzyme, which are catalytically active when unprotonated. The group of pK_a 5.5 functions in acetylation and is apparently located at least 9Å from the anionic site; the second group of pK_a 6.3 functions in de-acetylation and is situated within 5Å of the anionic site. The catalytic ability of this latter group is neutralised on protonation, the positive charge so generated apparently repelling cationic substrates or inhibitors from the active centre. The anionic site itself is protonated at pH 4.0–4.5 and is probably the side-chain carboxyl group of an aspartate or glutamate residue. A fourth group of pK_a 9.2 functions catalytically in the protonated form, and appears to be more than 10Å from the anionic site. The active centre is represented diagrammatically by Krupka as in Fig. 21.

The basic groups of pK_a 5.5 and 6.3 are considered to be histidine residues (Krupka, 1967) and probably function as follows:

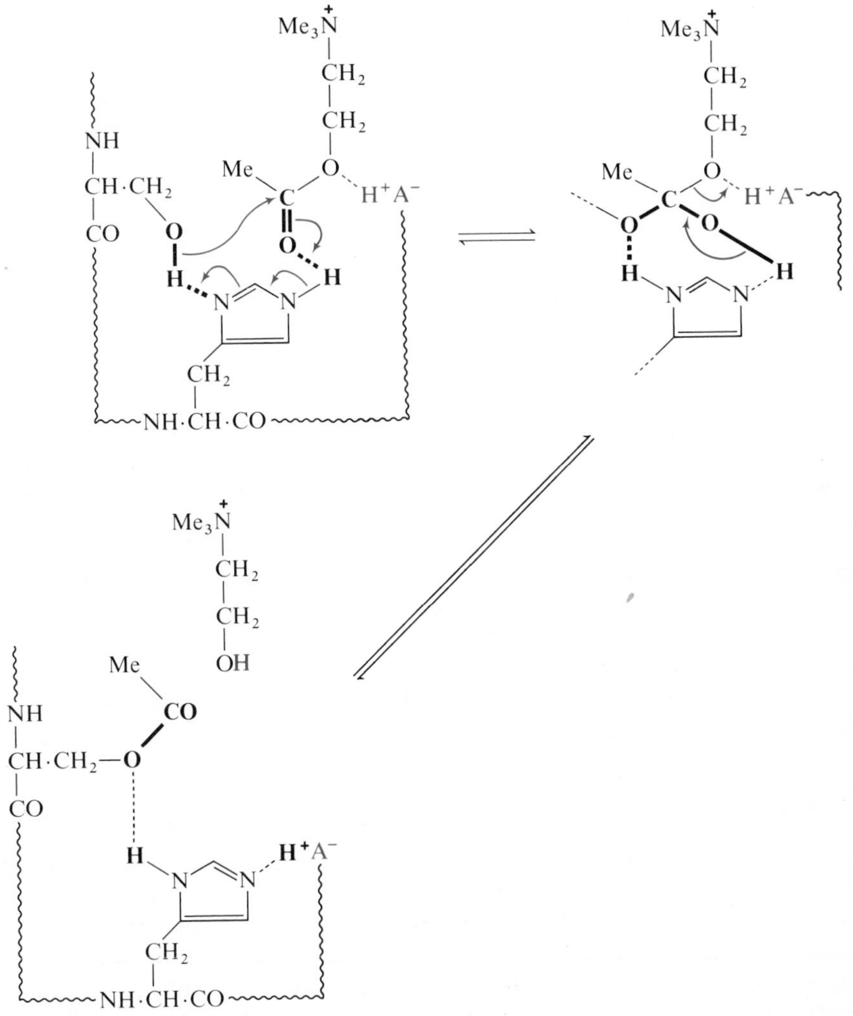

Few other drug—macromolecular studies have been carried out in such detail, though it is clearly evident that similar studies are essential to a complete understanding of any drug–receptor interaction. It is, however, readily apparent from this one study that such reactions are greatly influenced by stereochemical

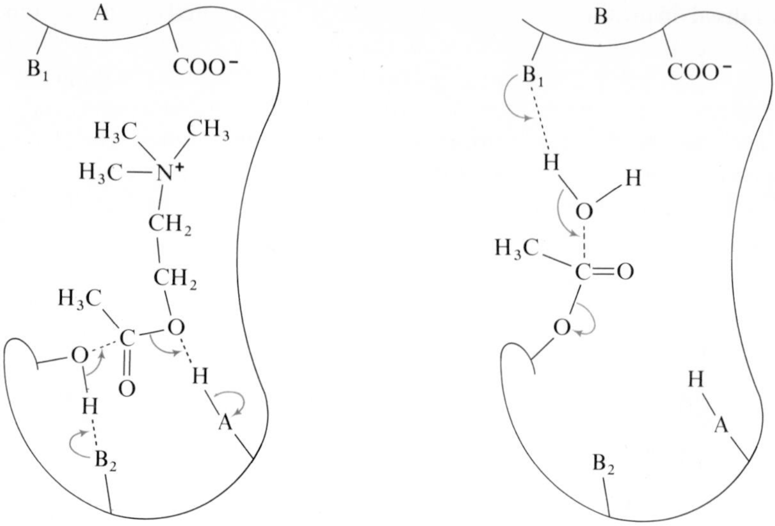

B_1 Basic gp pK_a 6.3 COO$^-$ Anionic site
B_2 Basic gp pK_a 5.5 HA Acidic gp pK_a 9.2

A E–S complex. Ammonium ion held at anionic site
B_2 and AH catalyse transfer of acetyl group to serine (acetylation)

B Acetyl–enzyme (EA) after conformational change bringing the acetyl residue near B_1, which then catalyses hydrolysis of EA

Fig. 21 Diagrammatic representation of acetylcholinesterase hydrolysis of acetylcholine

considerations in both the drug molecule and its bio-receptor. It follows, therefore, that even in the absence of precise information on the nature, structure and conformation of the receptor molecule, drug structure and stereochemistry are not only important *per se*, but also indirectly informative about essential chemical and stereochemical features of the receptor site, and the course of drug–receptor interaction.

The Shapes and Significance of some Molecular Sub-Units

In order to appreciate the full significance of stereochemical factors in drug–receptor interactions, it is essential to visualise the shapes of the interacting species both in the ground state, and in their activated states if these form an important part of the reaction process. One must begin, therefore, by examining the stereochemical parameters of simple organic structural units.

The shapes of drug molecules depend upon their bonding characteristics, and in particular, the resulting bond lengths and bond angles of the structural units from which they are composed. Bond lengths are determined by the form and extent of orbital overlap (**1**, 1). Bond angles reflect primarily the geometry of bonding orbitals, but may be modified by steric, electronic and strain effects. The geometry of individual functional groups may, therefore, be determined from a knowledge of orbital geometry (**1**, 1), the co-ordination number of the central atom of the group, and the equivalence or non-equivalence of the attached (ligand) atoms. The co-ordination number of any structural unit is defined as the number of ligand atoms bonded covalently or co-ordinately (but not electrovalently) to the central atom. Thus, carbon is capable of exhibiting co-ordination numbers from one to four, depending on orbital hybridisation and its state of combination. Other elements of group five and six of the Periodic Table, which are commonly incorporated in the structure of medicinal agents, may have correspondingly higher co-ordination numbers when combined in their highest valency state.

MONOGONAL AND DIGONAL ARRANGEMENTS

Monogonal Arrangements

There are relatively few examples of structural units with a co-ordination number of one. Hydrogen chloride and the cyanide ion are typical. Both possess conical symmetry due to differences in size of the ligand atoms, with a single axis of symmetry passing through the apex and the centre of the base of the cone. The cyanide ion, which is highly toxic, is extremely small relative to the metabolic systems with which it reacts. There is no evidence that its shape is in any way relevant to its mode of action, which rests solely on its ability to form stable complexes capable of holding vital oxidation–reduction enzymes in one of their essential oxidation states (**2**, 2).

H—Cl —C≡N axis of symmetry

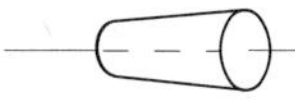

Digonal Arrangements

Stable carbon compounds with a co-ordination number of two arise as a result of *sp* hybridisation in carbon (**1**, 1), so that the nuclei of the constituent atoms lie in a straight line with a dihedral bond angle of 180°. In carbon dioxide, the ligands (oxygen) are equivalent, and hence the C=O bond lengths are equal. In contrast, each digonal acetylenic carbon has non-equivalent ligands (carbon and hydrogen) with C—C and C—H bonds formed by *sp*–*sp* and *sp*–*s* overlap respectively. Both carbon dioxide and acetylene, however, have cylindrical symmetry analogous to that of the hydrogen molecule.

O=C=O (1.16, 1.16 Å) — axis of symmetry

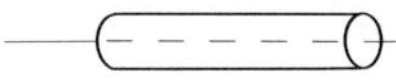

H—C≡C—H (1.06, 1.20 Å) — axis of symmetry

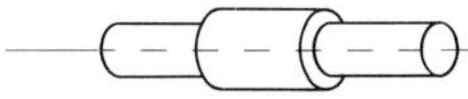

H—C≡N (1.06, 1.16 Å) — axis of symmetry

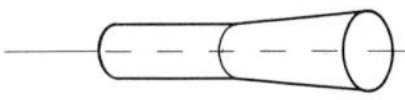

Structure Shape

Other triatomic molecules and functional groups of biological interest include water, the molecule of which is non-linear, but planar with symmetrically-disposed equivalent bonds. Aliphatic ether groups, as in the anaesthetics, *Diethyl Ether*, *Vinyl Ether* and *Methoxyfluorane*, and the thio-ether link, as in the phenothiazines (p. 157), are similarly non-linear and planar. The symmetry of these compounds, however, is dependent upon the nature of the substituents. Alcohol, phenol, thiol and thiophenol links are similarly non-linear and planar, but with their non-equivalent ligands, they give rise to non-symmetrical molecules.

Symmetrical digonal groups

Water

Dimethyl Ether

Dimethyl sulphide

Non-symmetrical digonal groups

Methoxyfluorane

Methanol

Phenol

PLANAR TRIGONAL ARRANGEMENTS

Carbonium Ions

Both planar and non-planar trigonal arrangements are possible. Carbonium ions of the type $R_3\overset{+}{C}$(R = H or alkyl) in which the ligands are equivalent, are symmetrical and planar, with bond angles of 120°.

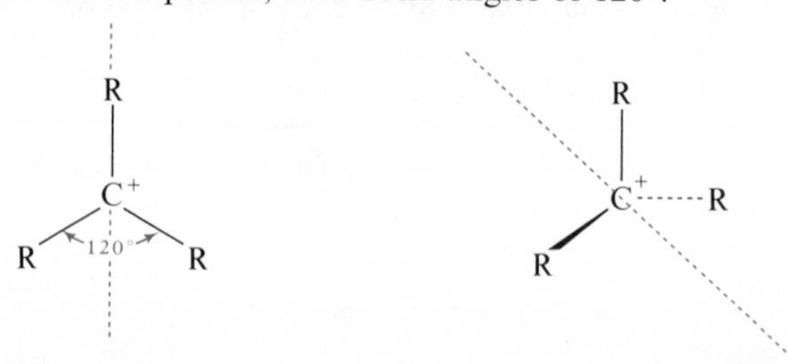

Two-fold axis of symmetry

Three-fold axis of symmetry at right angles to the plane of the ion

Such symmetrical arrangements are not only planar, but possess a three-fold axis of symmetry passing through the central atom at right angles to the plane of the ion. By this, we mean that each rotation about the axis through an angle of 120° (360°/3) results in a three-dimensional structure which is indistinguishable from the original. Additionally, the ion possesses three two-fold axes of symmetry along each of the axes joining the ligand and the central atom. The orientation of the vacant *p*-orbital is such that it ensures attack from either side of the planar structure along the three-fold axis of symmetry.

Non-symmetrical or irregular trigonal carbonium ions are encountered more frequently. Such compounds which are still **planar** may be of two types. Either only two of the three ligands are equivalent or all three ligands are different. In both cases, electronic and steric differences in the ligands lead to some departure from regular trigonal symmetry. In consequence there is divergence from the typical trigonal bond angle of 120°, and the possibility that steric effects may influence the direction from which attacking species may approach. Typical examples of such irregular trigonal arrangements occur in the following carbonium ions:

R^1, R^2, R^2 (angles a, a, b) — R^1, R^2, R_3 (angles a, c, b)

The planar structure of carbonium ions plays an important part in the racemisation of certain optically active compounds. Thus, the acid-catalysed racemisation of optically active secondary and tertiary alcohols is due to re-hydration of the planar trigonal carbonium ion intermediate, which can occur equally readily by attack from either side of its plane of symmetry.

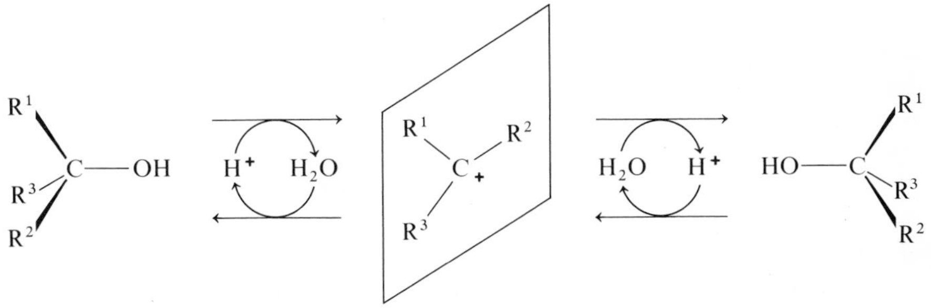

$R^1R^2R^3$C—OH ⇌ (H^+, H_2O) $R^1R^2R^3C^+$ ⇌ (H_2O, H^+) HO—C$R^1R^2R^3$

Deformation of Bond Angles

Divergence in bond angles from that of the symmetrical trigonal angle (120°) is unlikely to be large, unless there is considerable chain-branching close to the central carbon atom, e.g. if one of the alkyl substituents is t-butyl. More significant divergences are found in the non-symmetrical trigonal arrangements of formaldehyde, ethylene and acetaldehyde.

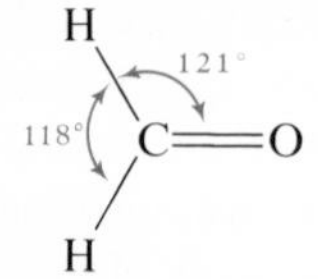

Formaldehyde 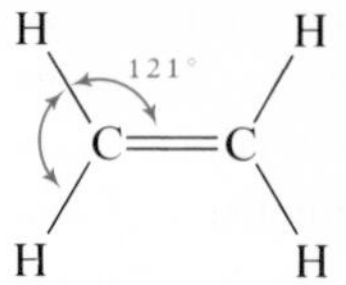 Ethylene 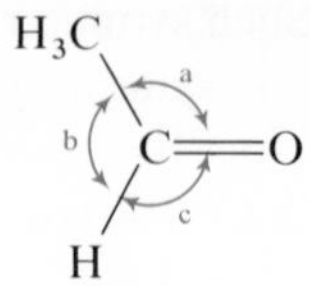 Acetaldehyde

Aromatic and Heteroaromatic Compounds

Molecular Planarity

In certain special cases,. deformation of non-symmetrical trigonal arrangements is opposed by other structural parameters, and the structural unit is constrained into retaining the regular trigonal bond angle of 120°. This occurs in the benzene molecule as a result of the regular hexagonal symmetry of the planar ring structure.

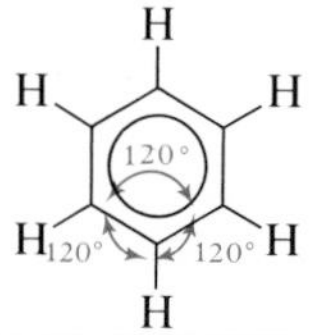

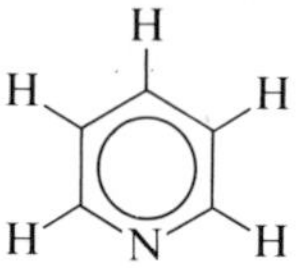

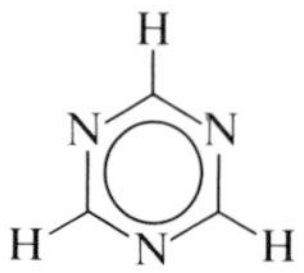

Six-membered heteroaromatic compounds, such as pyridine and pyrimidine, will similarly be planar and virtually hexagonal, though, because they are lacking in the symmetry elements inherent in the carbocyclic benzene ring, some bond deformation is inevitable. This planarity of aromatic and heteroaromatic rings is a significant factor in a number of physiological and pharmacological interactions where face-to-face alignment of co-planar structures is essential.

Base-pairing in RNA and DNA

Co-planarity of purine and pyrimidine bases in complementary strands is itself an essential requirement for effective hydrogen bonding and base-pairing in the double helix of nucleic acids. The angle at which the bases lie with respect to the polymer backbone is fixed by the co-planarity of the planar heterocyclics and the C_1 atom of the pentose units to which they are linked. As a result, the base pairs are stacked parallel to each other and approximately at right angles to the long axis of the helix. The double and triple hydrogen bonds linking the individual bases in each pair ensure their co-linear edge-to-edge alignment (**1**, 23).

The parallel stacking of co-planar base pairs in DNA gives rise to structures in which attack on these strand-stabilising heterocyclics is hindered in non-planar molecules. Thus, intercalation of quinoline and acridine antimalarials and the anti-cancer antibiotic, *Actinomycin D*, between the purine–pyrimidine base pairs of DNA (**2**, 2) is only possible because of the co-planarity of the opposing units. This not only ensures insertion of the drug, but also assists stabilisation of the resulting complexes by van der Waals forces and charge-transfer complexing.

Antibacterial Acridines

There is evidence that acridines owe their antibacterial action to their ability to block RNA-polymerase, and that this action is correlated with their flat molecular structure. The antibacterial action of acridines is pH- and p*K*-dependent in a way that establishes that this activity is due to the ionised form (Albert, Rubbo and Goldacre, 1941; Albert, Rubbo, Goldacre, Davey and Stone, 1945; Albert and Goldacre, 1948; **2**, 2).

Despite the fact that some acridines are able to penetrate the cell wall of certain organisms, there is indirect evidence that their bacteriostatic action is on the bacterial cell surface. Thus, 10-methylacridinium compounds, which are fully ionised irrespective of pH and therefore unlikely to penetrate the cell wall readily, are just as efficient bacteriostats as acridines which are subject to pH-dependent ionisation. It has also been shown that 9-aminoacridine, which is strongly antibacterial against *E. coli* and *B. lactis aerogenes*, does not penetrate into the cytoplasm of these organisms.

The action of acridines at the bacterial cell surface seems to be centred on the inhibition of RNA-polymerase, an enzyme responsible for the synthesis of RNA, which is 85% inhibited by 3,6-diaminoacridine (*Proflavine*) at concentrations as low as 30 μM (Hurwitz, Furth, Malamy and Alexander, 1962). It has been shown that the target molecule with which 9-aminoacridine (*Aminacrine*) similarly combines in *E. coli* has pK_a9 or higher and it has been suggested that this may be the phenolic hydroxyl of a tyrosine residue (Albert, 1965). An alternative point of attack, however, could be the —CO—NH— group of uridylic acid which has pK_a 9.4 (Jordon, 1960). The importance of molecular planarity for optimum activity of the acridine antiseptics supports this view. The requirement of a minimum molecular flat surface area for activity amongst acridines and analogous bases also suggests that charge-transfer and van der Waals bonding forces contribute. This is evident in the lack of antibacterial activity in 4-aminopyrimidine and 4-aminoquinoline, both of which, like the active acridines, are almost completely ionised at pH 7.3. Increasing the flat surface area of these molecules by the introduction of co-planar styryl substituents restores the antibacterial activity. Similarly, destruction of the completely flat surface of the acridine molecule, as in 9-amino-1,2,3,4-tetrahydroacridine, reduces antibacterial activity to negligible proportions.

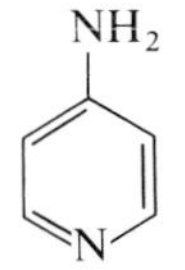

4-Aminopyridine
Min. bact. activity <1 in 5000

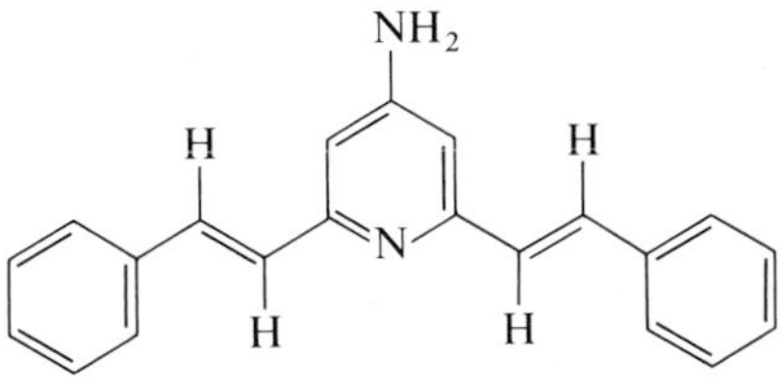

4-Amino-2,6-distyrylpyridine
1 in 80 000

4-Aminoquinoline
Min. bact. activity <1 in 5000

4-Amino-2-styrylquinoline
1 in 80 000

9-Aminoacridine

9-Amino-1,2,3,4-tetrahydroacridine

Geometrical Isomerism about the Ethylenic Bond

Rotational Stability

The linking of two trigonal sp^2 hybridised carbon atoms, in which the three σ-bonding atomic orbitals are planar and mutually inclined trigonally at an angle of 120°, gives rise to the planar ethylenic bond (**1**, 1), and the possibility of geometrical isomerism. The π-bond is formed by sideways overlap of the remaining unhybridised *p*-orbital from each of the carbon atoms comprising the bond. These atomic orbitals lie at right angles to the plane of the three sp^2-hybridised orbitals and in consequence the area of overlap is greatest above and below this plane. A degree of stability is achieved by this overlap which creates an energy barrier opposing the otherwise free rotation of the molecule about the C—C sigma bond. As a result, two stable forms of the molecule are possible, in both of which all six elements lie in a single plane, interchangeable only by the input of sufficient energy to overcome the barrier to rotation created by orbital overlap. The two forms differ only in the relative orientation of the substituents X and H and may be considered to be related by the physical rotation through 180° about the C—C bond.

cis *trans*

Each isomer of a geometric pair shows distinctly different physical properties from the other, and depending on the nature of the substituents, sometimes exhibits distinctly different chemical properties. Thus, maleic and fumaric acids (*q.v.*) show differences of melting point, solubility and dissociation constant

(pK_a), and, whilst maleic acid (*cis*-isomer) readily forms an anhydride when heated, its *trans*-isomer, fumaric acid, does not.

Absolute Configuration about Double Bonds

The use of the terms *cis* and *trans* is not always unambiguous. Thus, it is not possible to identify the geometrical isomers of the compound, CHCl=CBrI, without at the same time specifying which of the four substituents H, Cl, Br and I are *cis* and which are *trans*. To overcome this difficulty, a system of nomenclature known as the **E** and **Z** convention has been devised (Blackwood, Gladys, Loening, Petrarca and Rush, 1968) which permits the unambiguous identification of the individual isomers. This system makes use of the Cahn, Ingold, Prelog (1956) sequence rules (**1**, 13) for the determination of absolute configuration about asymmetric carbon atoms in the following way.

(a) Determine which of the two groups attached to each of the doubly bound atoms has the higher priority according to the sequence rules.

(b) That configuration in which the two groups of higher priority lie on the **same side** of the reference plane is designated the **Z** isomer (from the German Zusammen, meaning together).

(c) That configuration in which the two groups of higher priority lie on **opposite sides** of the reference plane is designated the **E** isomer (from the German, Entgegen, meaning against).

The following examples showing group priorites demonstrate the application of the rule.

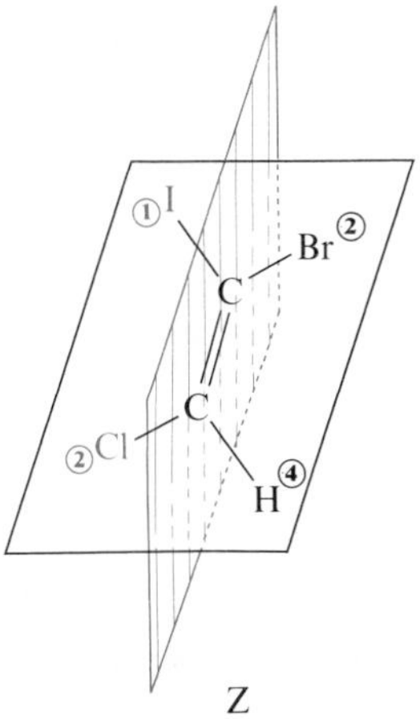

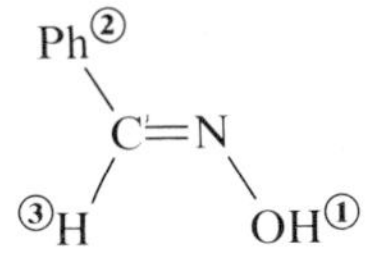

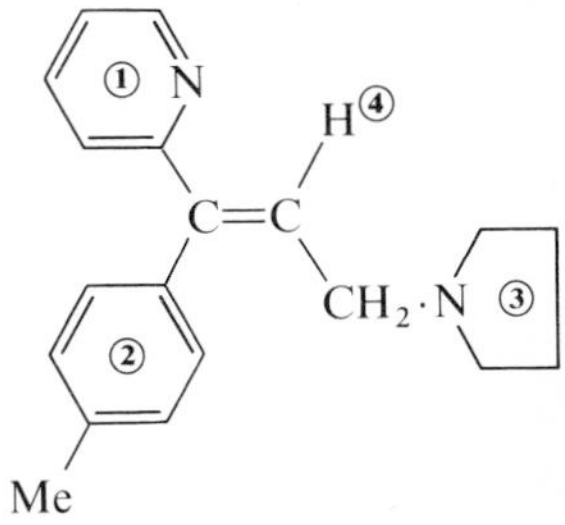

Stereospecific Activity of Geometrical Isomers

Cis-trans isomerism about ethylenic bonds undoubtedly plays an important part in biological reactions of both physiologically important compounds and a number of drug substances whose shape is fixed by the geometry of the double bond system. Thus, ethylenic dibasic acids, fumarate and *cis*-aconitate, have specific rôles as intermediates in the citric acid cycle (**2**, 1), which because of their specific double bond geometry, and the stereospecificity of the enzymic acid-catalysed hydrations involved, give rise to products (malate and citrate respectively) of precisely defined stereochemical configuration (**1**, 4). Similarly, the isomerisation of 4-maleylacetoacetate to 4-fumarylacetoacetate in the catabolism of tyrosine, and the functioning of the visual pigments in the retina provide examples of metabolically vital reactions, which are dependent upon cis-trans transitions. Likewise, the activity of synthetic oestrogenic hormones resides almost exclusively in the *trans*-isomers, the molecules of which, in contrast to those of the *cis*-isomers, closely resemble, sterically, those of the natural hormones (**1**, 22). A growing number of pharmacologically and therapeutically active compounds, similarly, have requirements of precisely defined geometry about the double bond system, which are essential for activity.

Isomerisation of 4-Maleylacetoacetate

The isomerism of 4-maleylacetoacetate to 4-fumarylacetoacetate is an important step in the metabolic oxidation of tyrosine, which normally proceeds either by ring hydroxylation via 3,4-dihydroxyphenylalanine (DOPA; **1**, 19), or by transamination and decarboxylation via *p*-hydroxyphenylpyruvic acid and homogentisic acid as follows:

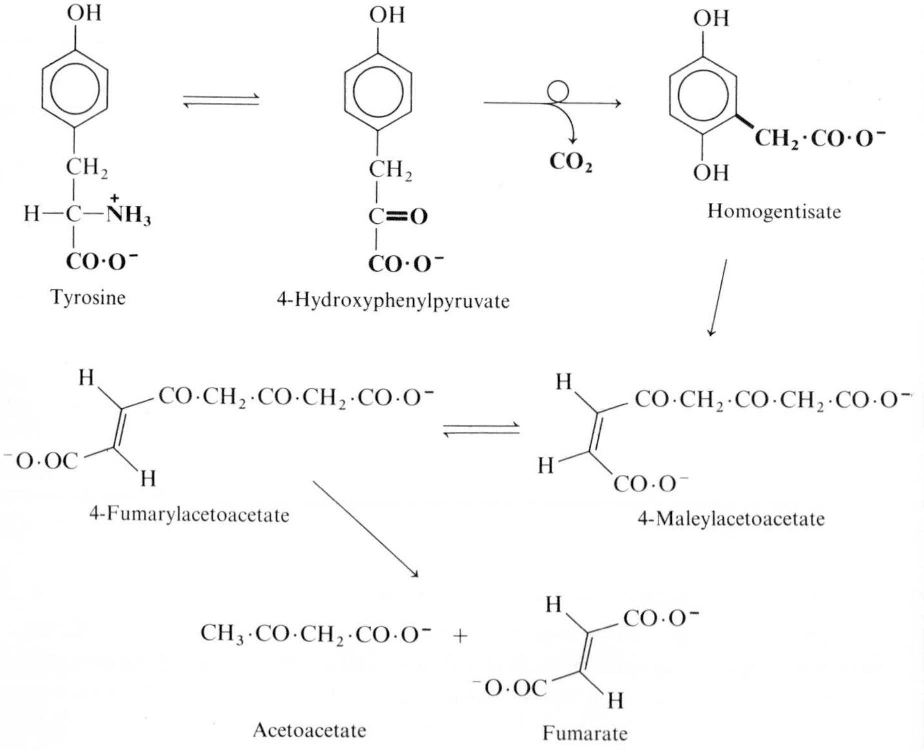

Tyrosine ⇌ 4-Hydroxyphenylpyruvate → (CO_2) Homogentisate

$CH_2 \cdot CO \cdot O^-$

Homogentisate → 4-Maleylacetoacetate ⇌ 4-Fumarylacetoacetate

$CO \cdot CH_2 \cdot CO \cdot CH_2 \cdot CO \cdot O^-$

4-Fumarylacetoacetate → $CH_3 \cdot CO \cdot CH_2 \cdot CO \cdot O^-$ + $^-O \cdot OC\text{–}CH{=}CH\text{–}CO \cdot O^-$

Acetoacetate   Fumarate

The isomerism is catalysed by maleylacetoacetate isomerase, which requires glutathione as coenzyme (Knox and Edwards, 1955; Edwards and Knox, 1956), and quite probably is achieved by a radical mechanism.

H, $CO \cdot CH_2 \cdot CO \cdot CH_2 \cdot CO \cdot O^-$
H, $CO \cdot O^-$
4-Maleylacetoacetate

H, $CO \cdot CH_2 \cdot CO \cdot CH_2 \cdot CO \cdot O^-$
$^-O \cdot OC$, H
4-Fumarylacetoacetate

RS•
Maleylacetoacetate isomerase

RS•

H, $CO \cdot CH_2 \cdot CO \cdot CH_2 \cdot CO \cdot O^-$
H, $CO \cdot O^-$
SR

$^-O \cdot CO \cdot CH_2 \cdot CO \cdot CH_2 \cdot CO$, H
H, $CO \cdot O^-$
SR

A block in the oxidation pathway sometimes occurs as an inborn error of metabolism at the oxidative step between homogentisic acid and 4-maleylacetoacetate, due to a deficiency of homogentisic oxidase. This condition results in urinary excretion of homogentisic acid, which rapidly becomes oxidised with darkening of the urine due to pigment formation (alkaptonuria).

Visual Pigments

The key substance in the visual process is retinene (Vitamin A aldehyde; *all-trans* retinal). This is derived from *Retinol* (Vitamin A) by metabolic oxidation with alcohol dehydrogenase, an enzyme requiring NAD as coenzyme.

The visual pigment, rhodopsin, is formed by combination of a geometric isomer of retinene, 11-*cis*-retinal, in the form of its Schiff's base, 11-*cis*-retinylidenephosphatidylethanolamine (Morton and Pitt, 1955; Krinsky, 1958; Bonting and Bangham, 1967), with the protein, opsin, which is present in the retina. The high stereospecificity of this reaction is evident from the fact that only the 9-*cis*-and 11-*cis*-retinal Schiff's bases are capable of complexing with opsin, whilst the continuity of the visual process depends on the inability of the corresponding *all-trans*-Schiff's base to form a stable complex (Hubbard and Wald, 1952; Brown and Wald, 1956). The primary conversion of *all-trans*-retinal to 11-*cis*-retinal is accomplished by retinenic isomerase, which requires light as the energy source to form the energetically less favourable *cis*-isomer. (See reaction pathway on pp. 116–7.)

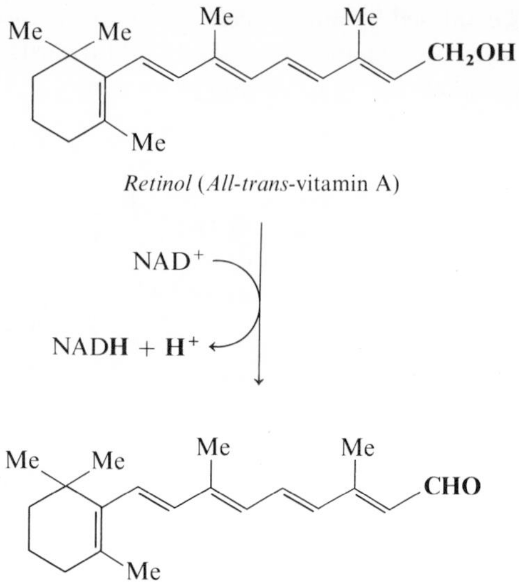

Retinol (*All-trans*-vitamin A)

NAD^+

$NADH + H^+$

All-trans-retinal

Rhodopsin is stable only in the dark. On exposure to light, it is isomerised to metarhodopsin I, an opsin–*all-trans*-retinylidenephosphatidylethanolamine complex. Within about 100μs of its formation, this complex undergoes a further transformation to metarhodopsin II, in which the *all-trans*-retinal moiety is displaced from its phosphatidylethanolamine Schiff's base to form a new retinylidene–lysine Schiff's base (Poincelot *et al.*, 1969; Daemen and Bonting, 1969). (See pp. 116–17.) The long wavelength absorption maximum of rhodopsin (λ_{max} 498 nm) has been ascribed to protonation of the Schiff's base by the phosphatidylethanolamine phosphate anion, since no such maximum occurs when this group is blocked in the corresponding benzyl ester, which has similar light absorption characteristics as metarhodopsin II (λ_{max} 380 nm; Daemen and Bonting, 1969). The small wavelength shift of about 20 nm in the isomerism of rhodopsin to metarhodopsin I, runs counter to that expected for the lengthening of the chromophore in the cis → trans isomerisation and is probably due to conformational changes in the protein structure, since this step is accompanied by changes in the ORD spectra (Takagi, 1963; Kito and Takezaki, 1966).

It is well-known that protonated Schiff's bases are highly reactive to nucleophilic attack. According to Daemen and Bonting (1969), therefore, conformational changes in the isomerisation of rhodopsin to metarhodopsin I, bring the lysine residue ε-amino group of opsin into position for immediate nucleophilic attack on the *all-trans* retinylidenephosphatidylethanolamine leading to the formation of metarhodopsin II. The shift of retinaldehyde to the ε-amino group of lysine could be responsible for visual excitation by causing cation exchange across the rod sac membrane (Bonting, 1969). (See structure pages 116 and 117.)

Structural Specificity of Oestrogens

At the molecular level, oestrogens are bound to a specific binding protein in the uterus, which is the target organ. Here, they bring about an increased synthesis of RNA, which in turn stimulates protein synthesis (Hechter and Halkerston, 1965).

The natural oestrogens, oestradiol and oestriol, combine the rigid three-dimensional features of rings B, C and D of the steroidal skeleton with the planar aromatic ring A. The essential structural features for oestrogenic activity are the phenolic hydroxyl group of ring A and the 17β-hydroxyl of ring D, or an analogous oxygen function. The importance of the phenolic hydroxyl is demonstrated in recent work, which has led to the isolation of a specific binding protein from calf uterus (Puca and Bresciani, 1969). Furthermore, the binding of 6,7-^{3}[H]-17β-Oestradiol is inhibited only by other analogous phenolic oestrogens, such as *Oestradiol* (Estradiol) itself or *Stilboestrol* (Stilbestrol; Table 18; Duncan, Lyster, Clark and Lednicer, 1963).

Table 18. Inhibition of Binding of 6,7-^{3}[H]-Oestradiol-17β by Oestrogenic, and Non-oestrogenic Compounds *in vitro* (Duncan *et al*., 1963)

Compound	Bound 6,7-^{3}[H]-oestradiol-17β (% of control)
17β-Oestradiol	14
17α-Oestradiol	109
Stilboestrol	2

It has long been realised that the geometry of the *trans*-stilbene skeleton of *Stilboestrol* confers a close stereochemical similarity to the molecule of *Oestradiol*, and it is accepted that this provides an adequate explanation of the observation that *trans*-stilboestrol is more active than the *cis*-isomer (Dodds, Goldberg, Lawson and Robinson, 1939; Dodds, Goldberg, Grünfeld, Lawson and Robinson, 1944). Crystallographic studies of *Stilboestrol* (Hospital, Busetta, Bucourt, Weintraub and Baulieu, 1972) have also shown that whereas it crystallises, solvent-free and with a symmetrical structure from non-polar solvents, crystallisation from aqueous methanol gives a solvated product, with one hydroxyl group solvated by water and the other by methanol. Significantly, this non-symmetrically solvated molecule adopts an asymmetrical conformation in which one phenolic ring is twisted in a plane at right angles to the other, imparting a degree of thickness to the molecule comparable to that which results from ring D and C_{18} in *Oestradiol*. In this connection, it is relevant that 17-noroestradiol is 2×10^{-5} times less active than *Oestradiol* (Edgren and Johns, 1960). (See structure at the top of p. 118.)

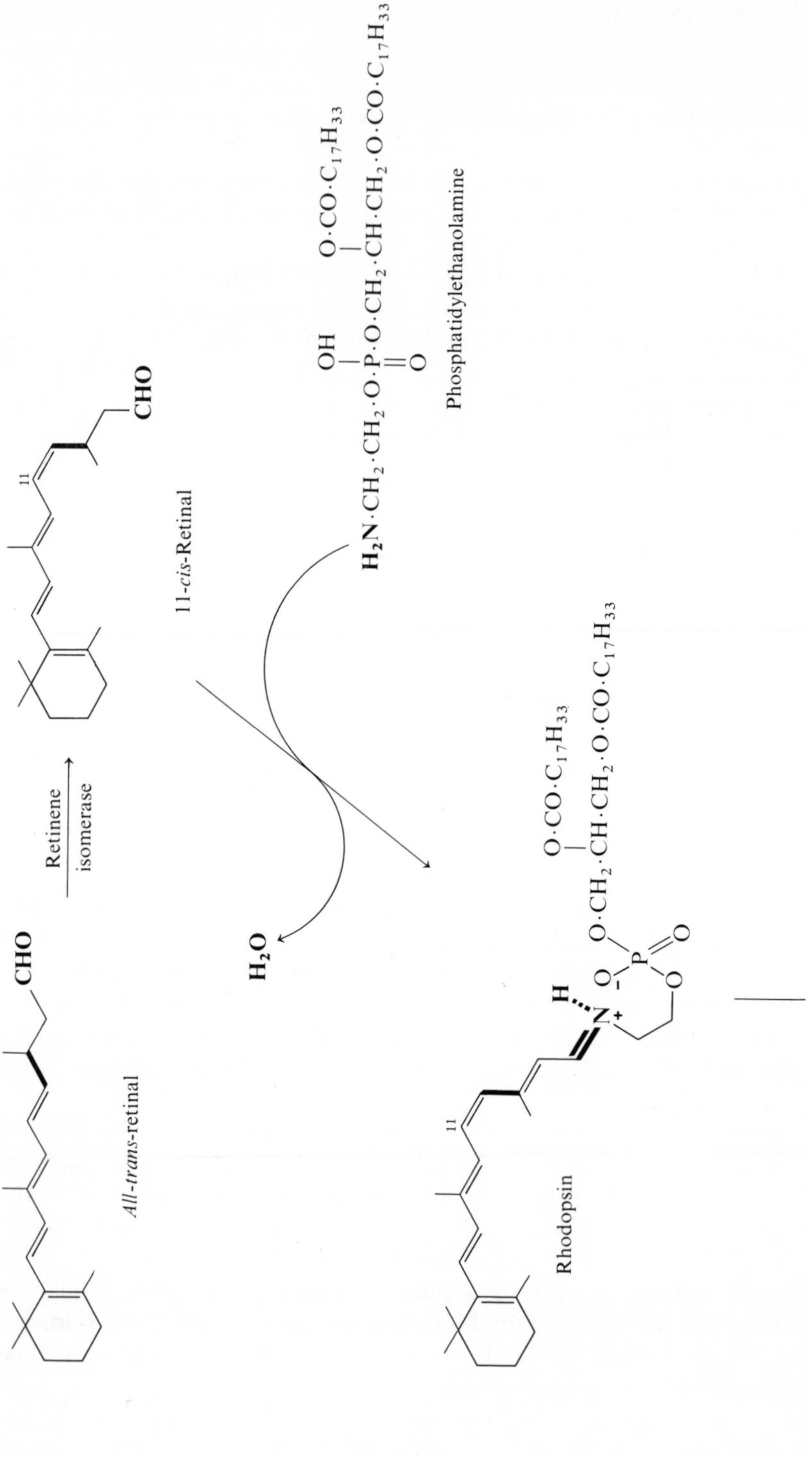

All-trans-retinal

Retinene isomerase

11-*cis*-Retinal

$H_2N \cdot CH_2 \cdot CH_2 \cdot O \cdot P(OH)(=O) \cdot O \cdot CH_2 \cdot CH(O \cdot CO \cdot C_{17}H_{33}) \cdot CH_2 \cdot O \cdot CO \cdot C_{17}H_{33}$

Phosphatidylethanolamine

H_2O

Rhodopsin

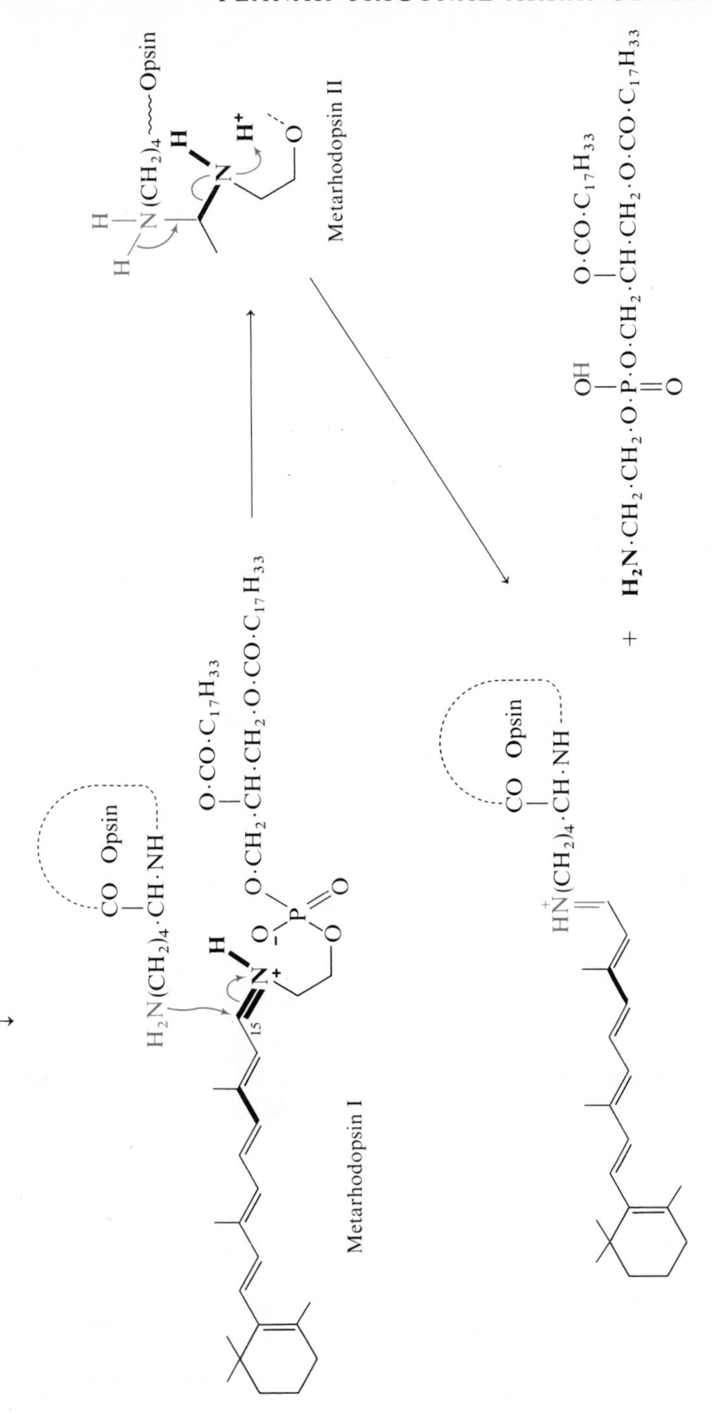
hν
Metarhodopsin I
Metarhodopsin II
Opsin
CO Opsin
CH·NH
O·CO·C17H33
O·CH2·CH·CH2·O·CO·C17H33
H2N·CH2·CH2·O·P·O·CH2·CH·CH2·O·CO·C17H33
+

Oestradiol-17β (R^1 = H; R^2 = H)
Oestriol (R^1 = H; R^2 = OH)
17α-Ethinyloestradiol (R^1 = C≡H; R^2 = H)

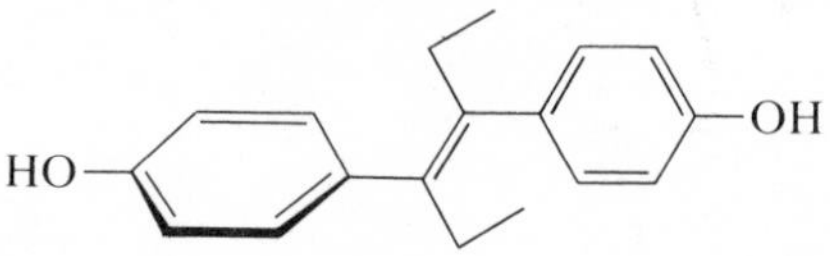

Stilboestrol

It is interesting to observe that meso-*Hexoestrol* (Hexestrol), which is a much more potent oestrogen than (±)-*Hexoestrol* (Wessely and Welleba, 1941), is capable of adopting an identical conformation, whereas the racemate is not. The (3*R*,4*R*)-(+)-isomer, however, has about then times the potency of the (3*S*, 4*S*)-(−)-isomer and it is apparent that the former does more closely resemble the key conformation than the other.

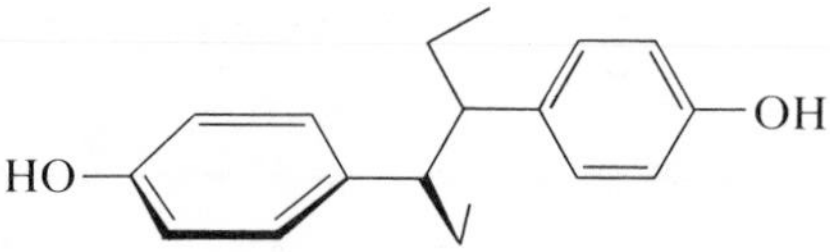

meso-*Hexoestrol*

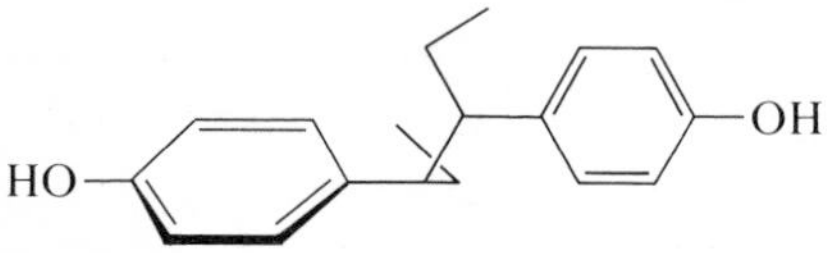

(3*S*, 4*S*)-isomer

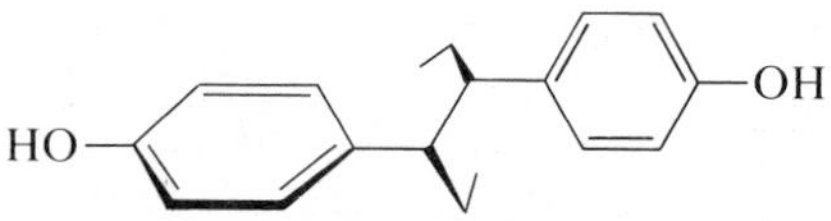

(3*R*, 4*R*)-isomer

Hexoestrol optical isomers

Similar differences in oestrogenic potency have also been shown in the following pair of triphenylethylene compounds, of which the *cis*-compound is much more active than the *trans*, in accord with the same stereochemical requirement

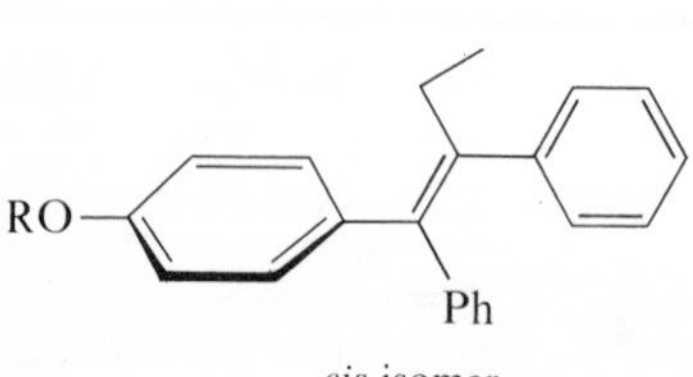

cis-isomer

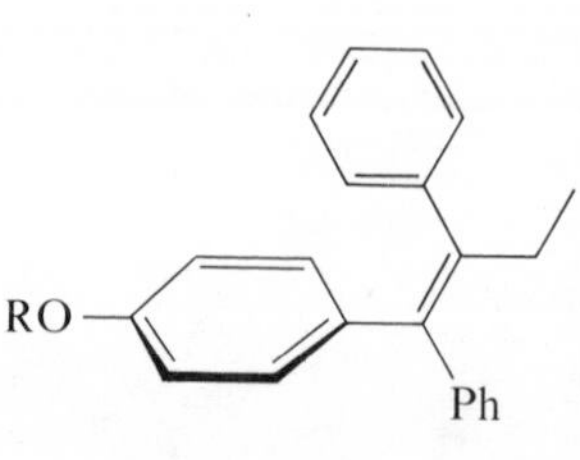

R = $Me_2N \cdot CH_2 \cdot CH_2$—

trans-isomer

(Saunders, 1966; Vitali, Gardi, Falconi and Ercoli, 1966). Not only is the *trans*-compound a weak oestrogen, but it antagonises the effect of concomitantly-administered oestrogen.

The activity of *Dienoestrol* and *Chlortrianisene* can be accounted for on a similar stereochemical basis. In contrast to earlier reports, *Dienoestrol* (Dienestrol) has been shown by spectroscopic and X-ray diffraction studies to have the structure in which the *p*-hydroxyphenyl and methyl groups are cis about each double bond (Doyle, Stewart, Filipescu and Benson, 1975).

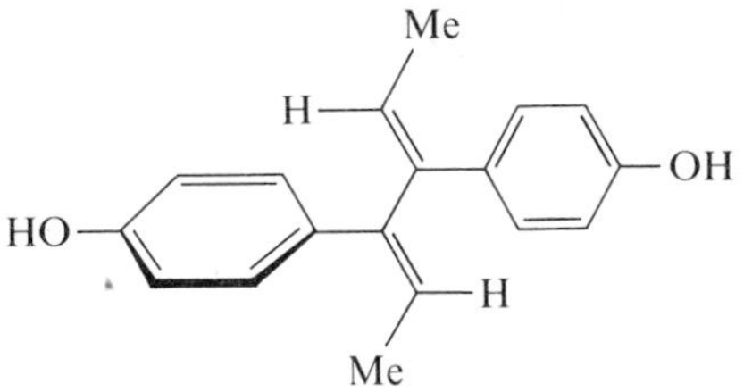
Me
H
OH
HO
H
Me

Dienoestrol

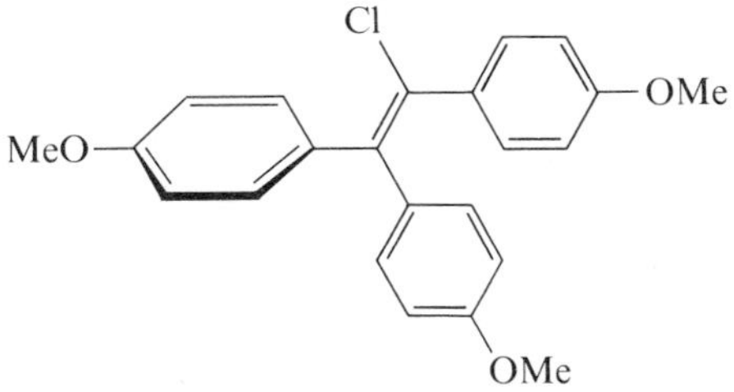
Cl
OMe
MeO
OMe

Chlortrianisene

Busulphan

The action of *Busulphan* as a tumour growth inhibitor has been clarified by comparison of its properties with those of unsaturated analogues of known stereochemistry. Thus, the corresponding butyne and both *cis* and *trans*-bismethanesulphonoxybutenes are active as tumour initiators on mouse skin (Roe, 1957), but whereas both the butyne and the *trans*-butene are inactive as tumour growth inhibitors, the *cis*-butene derivative has growth inhibitory activity (Haddow and Timmis, 1951). According to Timmis (1951), this suggests that the activity of the *cis*-compound and of the saturated, and hence flexible, *Busulphan*, depends on their ability to form a heterocyclic derivative by 1,4-bis-alkylation of some suitably disposed and reactive amino group.

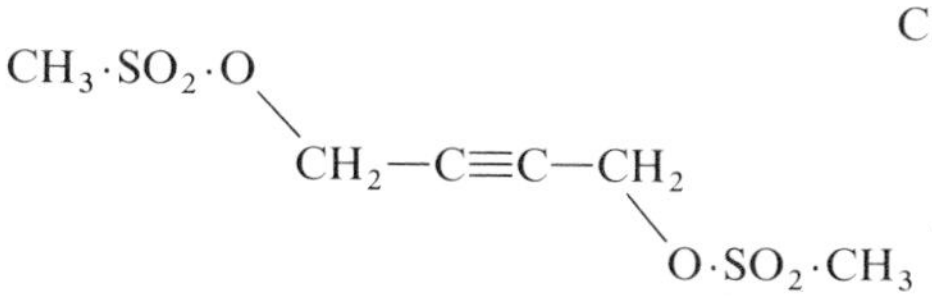
CH3·SO2·O
CH2—C≡C—CH2
O·SO2·CH3

1,4-Bismethanesulphonoxy-2,3-butyne

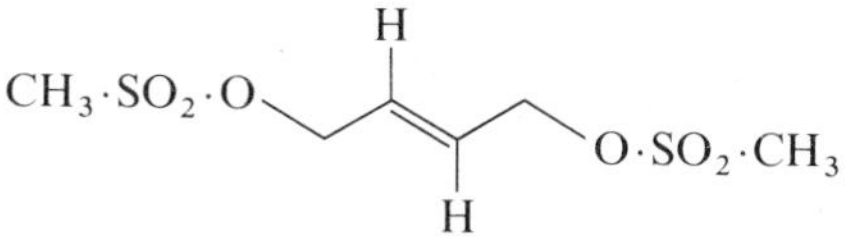
H
CH3·SO2·O
O·SO2·CH3
H

trans-1,4-Bismethanesulphonoxy-2,3-butene

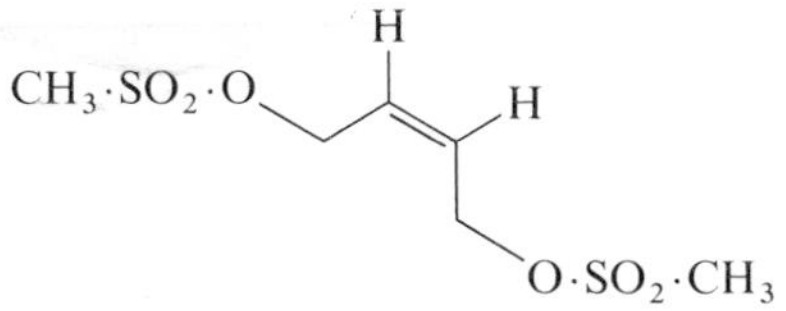
H
CH3·SO2·O
H
O·SO2·CH3

cis-1,4-Bismethanesulphonoxy-2,3-butene

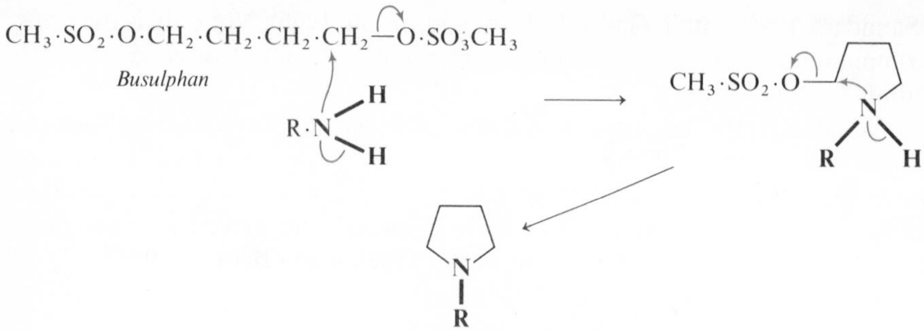

$CH_3 \cdot SO_2 \cdot O \cdot CH_2 \cdot CH_2 \cdot CH_2 \cdot CH_2 - O \cdot SO_3CH_3$

Busulphan

Steric Parameters in Esters and Amides

Rotation about the C—O—C bonds of esters is severely restricted by resonance.

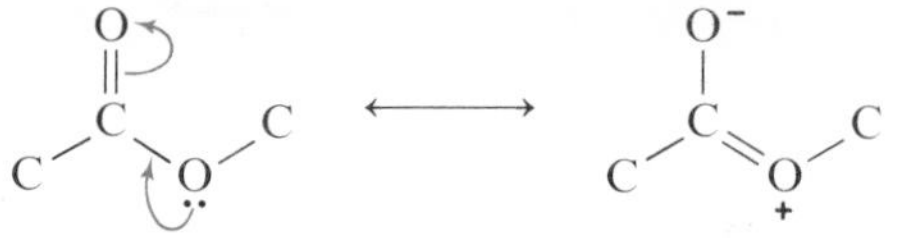

Electron diffraction studies of aliphatic esters show that the *cis*-configuration predominates with only a small contribution from the *trans*-form (O'Gorman, Shand and Schomaker, 1950).

cis *trans*

Similarly, studies of *N*-methylacetamide and other amides have shown that due to resonance, the molecule is also essentially planar (Mizushima, Simanouti, Nagakura, Kuratani, Tsuboi, Baba and Fujioka, 1950), whilst spectroscopic evidence indicates that the *trans* H—N—C=O structures are some 8 kJ more stable than the corresponding *cis*-forms.

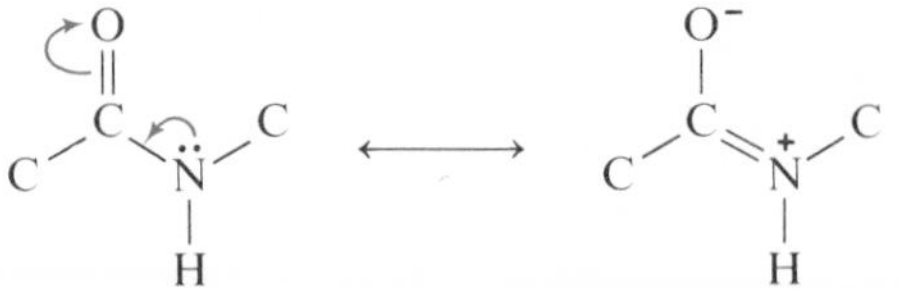

As a result, amides are essentially isosteric with esters, a fact of no little consequence in explaining the like ability of structurally-related esters and amides to function as local anaesthetics (**2**, 2).

Isosterism of Esters and Amides

Puromycin

Puromycin inhibits ribosomal synthesis of proteins by interference with the formation of transfer-RNA's. The latter, of which there are some twenty odd, esterify the various amino acids through the terminal adenosine units of tRNA before being aligned for protein synthesis by combination with mRNA. Once correctly aligned, the amino acids combine to form a protein with the required amino acid sequence. The terminal unit of all tRNA's consists of adenosine. *Puromycin* is an amidic analogue of the terminal unit of tyrosinyl-tRNA, and is taken up by mRNA in place of the latter (**1**, 23), thus blocking the reaction step responsible for the formation of the protein peptide chain (Nathans, 1964). *Puromycin* has been used to treat amoebiasis (Hutchings, 1957).

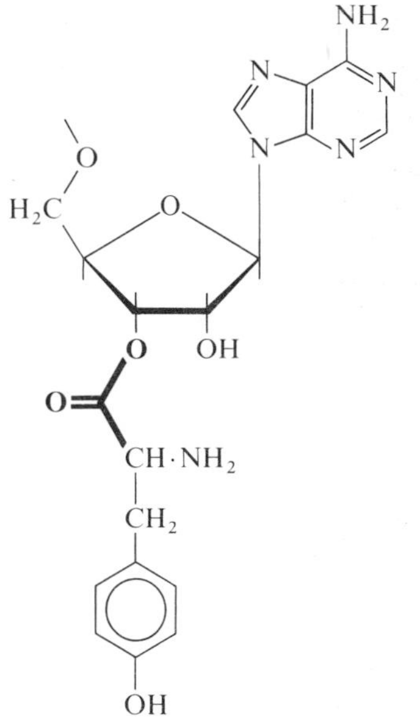

Terminal tyrosinyladenosyl unit of tRNA

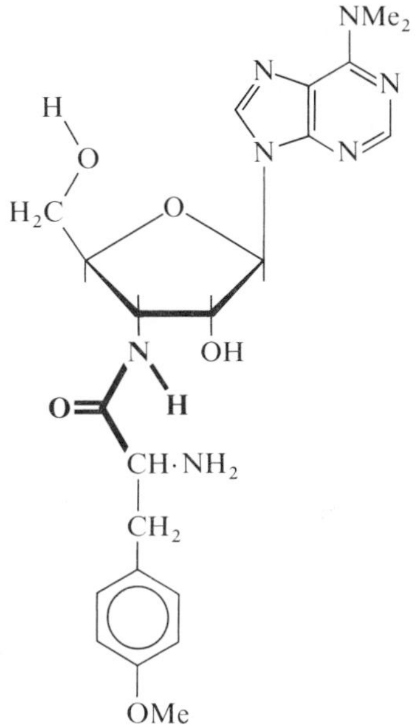

Puromycin

NON-PLANAR TRIGONAL ARRANGEMENTS

Carbanions

Non-planar trigonal arrangements occur in carbanions (R_3C^-) and radicals ($R_3C\cdot$) where the non-bonding electrons occupy one (unfilled) of the four tetrahedrally-disposed orbitals. The three bonds are thus effectively disposed pyramidally in relation to the central atom. Similar pyramidal arrangements occur in the hydrated proton (H_3O^+), in oxonium ions ($R_2\overset{+}{O}H$) and in protonated alcohols ($R\overset{+}{O}H_2$).

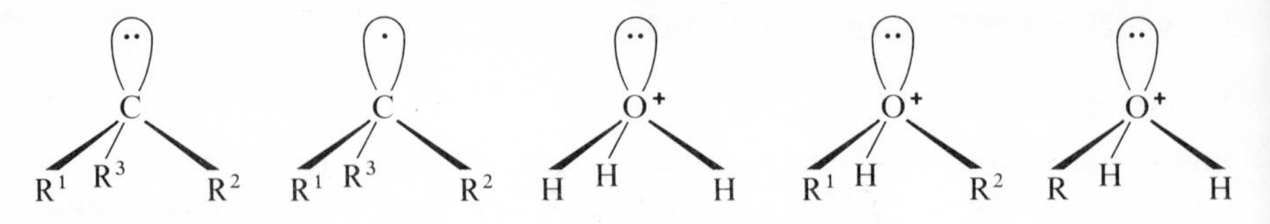

The pyramidal structure of the carbanion has important consequences in certain stereospecific reactions. Thus, it is evident that the enzyme-catalysed condensations of dihydroxyacetone phosphate and D-glyceraldehyde to form D-fructose in presence of aldolase and triosephosphate isomerase are stereospecific. The condensations involve an enzyme-linked carbanion intermediate formed by displacement of one particular hydrogen atom from dihydroxyacetone phosphate, depending on the enzyme system involved (Rose, 1958). This is demonstrated in enzyme-catalysed exchange of dihydroxyacetone phosphate with tritiated water in which the tritium atom is placed in a sterically-equivalent position to carbon-4 of the condensation product by aldolase, and in the alternative position by triosphosphate isomerase.

$CH_2{\cdot}O{\cdot}PO_3H_2$
CO
HO—C—H
H

Dihydroxyacetone phosphate (DHAP)

H_2O → $H_3\overset{+}{O}$

Aldolase

$CH_2{\cdot}OPO_3H_2$
HO CO H

$^3H_3\overset{+}{O}$ → H_2O

$CH_2{\cdot}O{\cdot}PO_3H_2$
CO
HO—C—H
T

H_2O

Triose phosphate isomerase

$H_3\overset{+}{O}$

$CH_2{\cdot}OPO_3H_2$
H CO OH

$^3H_3\overset{+}{O}$ → H_2O

$CH_2{\cdot}O{\cdot}PO_3H_2$
CO
HO—C—T
H

The stereospecificity of these enzyme-catalysed condensations and the involvement of the two possible enantiomeric carbanions (Bloom and Topper, 1958) is demonstrated in the reaction of each of the specifically-tritiated dihydroxyacetone phosphates with D-glyceraldehyde in the presence of aldolase, giving rise to D-fructose with and without retention of tritium respectively.

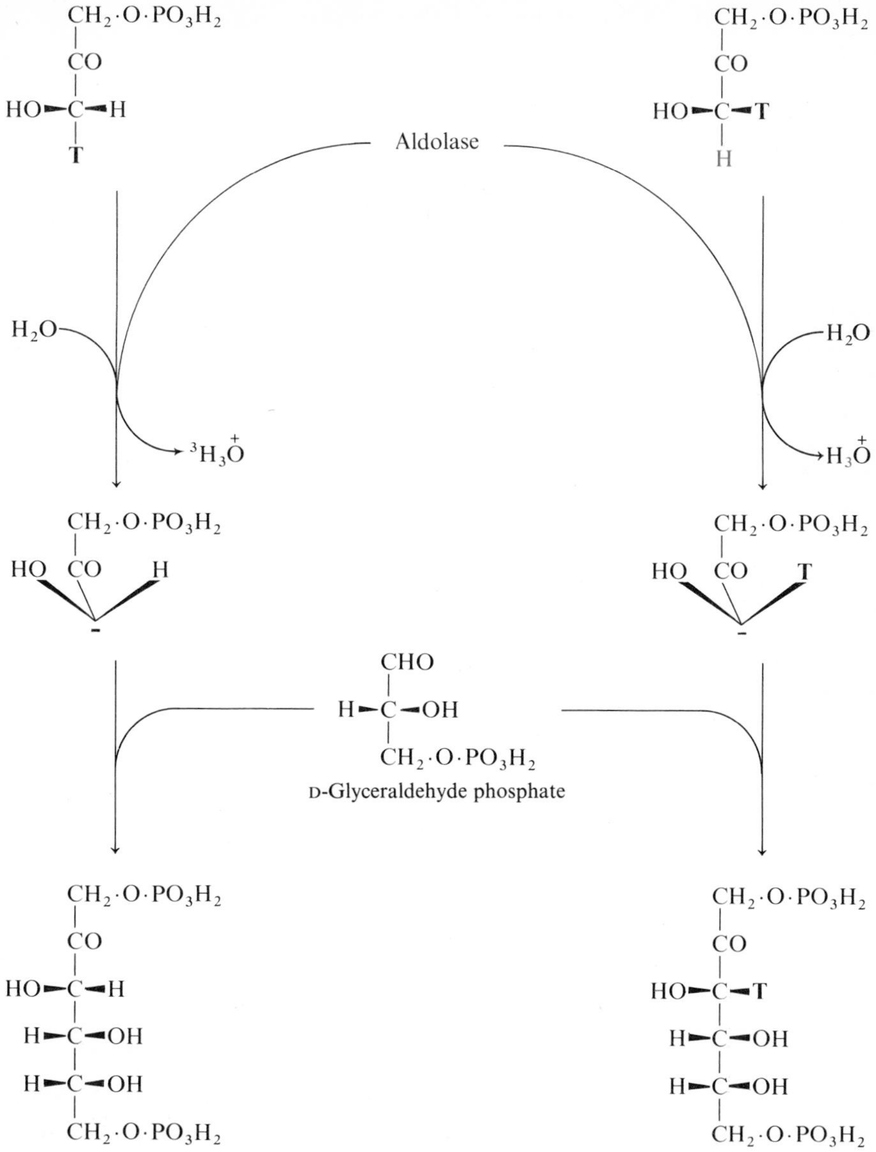

$CH_2{\cdot}O{\cdot}PO_3H_2$ — CO — HO–C–H — T

Aldolase

$CH_2{\cdot}O{\cdot}PO_3H_2$ — CO — HO–C–T — H

H_2O → $^3H_3\overset{+}{O}$

H_2O → $H_3\overset{+}{O}$

$CH_2{\cdot}O{\cdot}PO_3H_2$ — HO CO H (carbanion, −)

$CH_2{\cdot}O{\cdot}PO_3H_2$ — HO CO T (carbanion, −)

CHO — H–C–OH — $CH_2{\cdot}O{\cdot}PO_3H_2$

D-Glyceraldehyde phosphate

$CH_2{\cdot}O{\cdot}PO_3H_2$ — CO — HO–C–H — H–C–OH — H–C–OH — $CH_2{\cdot}O{\cdot}PO_3H_2$

$CH_2{\cdot}O{\cdot}PO_3H_2$ — CO — HO–C–T — H–C–OH — H–C–OH — $CH_2{\cdot}O{\cdot}PO_3H_2$

Amines and Sulphoxides

A similar pyramidal arrangement also occurs in primary, secondary and tertiary amines. The arrangement is unstable and normally undergoes a continuous and rapid **inversion** of configuration like an umbrella being turned inside out and back again. The speed of inversion is such that the molecule is effectively planar and the apparent asymmetry of tertiary nitrogen compounds is lost. As a result, the great majority of such amines are not optically active, since the two interconvertible forms, although mirror images, are superimposable.

The resultant configuration of secondary amines (R^1R^2NH) is isosteric with that of the corresponding ethers ($R^1 \cdot O \cdot R^2$). A few examples of stabilised forms are known in which the nitrogen is locked in a rigid polycyclic structure, which is resolvable into optically active components, but these examples are the exception rather than the rule.

Sulphoxides also have a pyramidal arrangement. In contrast to amines, however, the arrangement is stable, so that asymmetrically-substituted sulphoxides give rise to non-superimposable mirror image forms, which are therefore resolvable into optically active components.

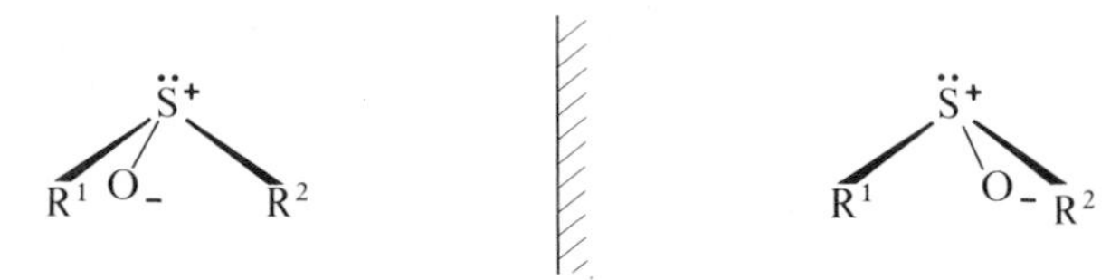

This type of structure occurs in the uricosuric agent, *Sulphinpyrazone* (Sulfinpyrazone), which has been obtained in its (+)- and (−)-forms. The former is obtained by metabolic oxidation of the corresponding sulphide.

Sulphinpyrazone

The oxidation of penicillins to their *S*-oxides is stereospecific. The stability and rearrangement of penicillinic *S*-oxides to the corresponding cephalosporin derivatives is also stereospecific (**1**, 18).

Sulphonium Compounds

Sulphonium compounds, like the analogous quaternary ammonium compounds, are capable of acting as ganglion and neuromuscular blocking agents. They have the same pyramidal structure as sulphoxides and are capable of being resolved. The optical isomers are, therefore, theoretically capable of providing information on the significance of geometry about the central atom upon the activity of the compound.

TETRAHEDRAL ARRANGEMENTS

Optical Isomerism

Four-centred co-ordinate structural units may be either planar or tetrahedral. Certain four-centred co-ordinate metal complexes of nickel, copper, and platinum are four square planar, as in the *Penicillamine* Cu(II) complex, and the platinum diammono dichloride complex, $Pt(NH_3)_2Cl_2$, now being used experimentally in the treatment of cancer. Analogous carbon compounds and the quaternary ammonium ion, on the other hand, are tetrahedral.

Asymmetric substitution of such units gives rise to optical isomerism, and often to a high degree of specificity in biological interactions. This is usually evident in marked differences in body distribution, activity, and excretion, which may or may not be inter-related. Thus, the (+)-isomer of *Hexobarbitone* (Hexobarbital) is an active hypnotic in both the rat and man, in contrast to its (−)-enantiomer, which has little or no hypnotic effect (Breimer and van Rossum, 1973). It is not surprising, therefore, that the (−)-isomer is eliminated much faster than the more active (+)-isomer in man ($t_{1/2}(+)/(-) = 3.2/1$), but anomalously, the relative elimination rates of the isomers is in the reverse order in the rat.

Stereospecificity of optical isomers in biological action is considered to arise in one isomer because of its ability to achieve a three-point attachment with its receptor molecule, while its enantiomer will only be able to achieve a two-point attachment with the same receptor.

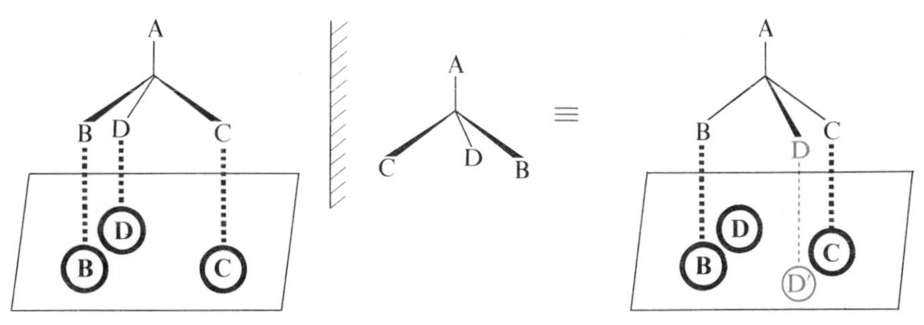

The nature of the interaction may be different at all three centres, i.e. ionic, van der Waals attraction, charge-transfer complex formation, or H-bond. Additionally, now that it is established that receptor molecules can undergo major conformational changes to accept a drug, it is clear that the fourth substituent may, on occasion, also play a vital rôle, so that the difference then becomes one involving either a three- or four-point contact with the receptor.

Catecholamines

There are many examples of drugs which have high specificity in only one of a pair of enantiomers. One of the most fully documented is that of the sympathomimetic amines, D-(−)-noradrenaline, D-(−)-adrenaline and D-(−)-isopropylnoradrenaline, which are some 50 to 1000 times more potent than their enantiomers, depending on which biological property is examined (Table 19). The absolute stereochemistry of the laevorotatory isomers has been established as **R** (**1**, 13).

Table 19. Bronchodilator Activity of some Enantiomeric Amines

Amine	Relative Bronchodilator Activity
D-(−)-Noradrenaline	1
L-(+)-Noradrenaline	0.014
D-(−)-Adrenaline	58
L-(+)-Adrenaline	1.3
D-(−)-Isopropylnoradrenaline	270
L-(+)-Isopropylnoradrenaline	0.33

(Luduena, von Euler, Tullar and Lands, 1957)

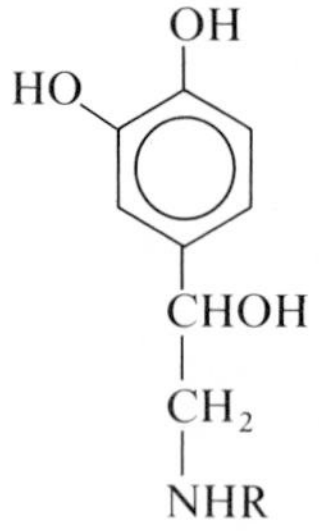

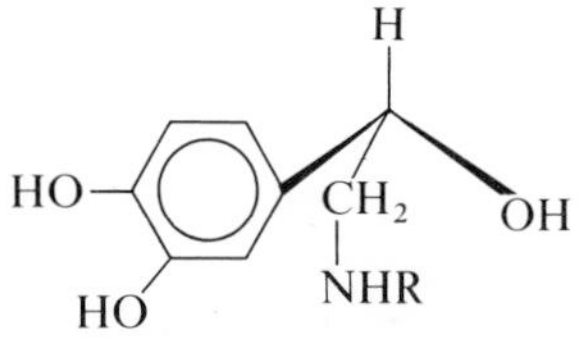

Noradrenaline (R = H)
Adrenaline (R = CH_3)
Isopropylnoradrenaline (R = $CHMe_2$)

The dissociation constant of amino groups is such that these compounds are completely ionised at physiological pH. As far back as 1933, Easson and Stedman explained the different activities of the adrenaline (Epinephrine) enantiomers in terms of a three-point attachment to a receptor molecule involving the basic cationic group, the dihydroxyaromatic structure and the alcoholic hydroxyl

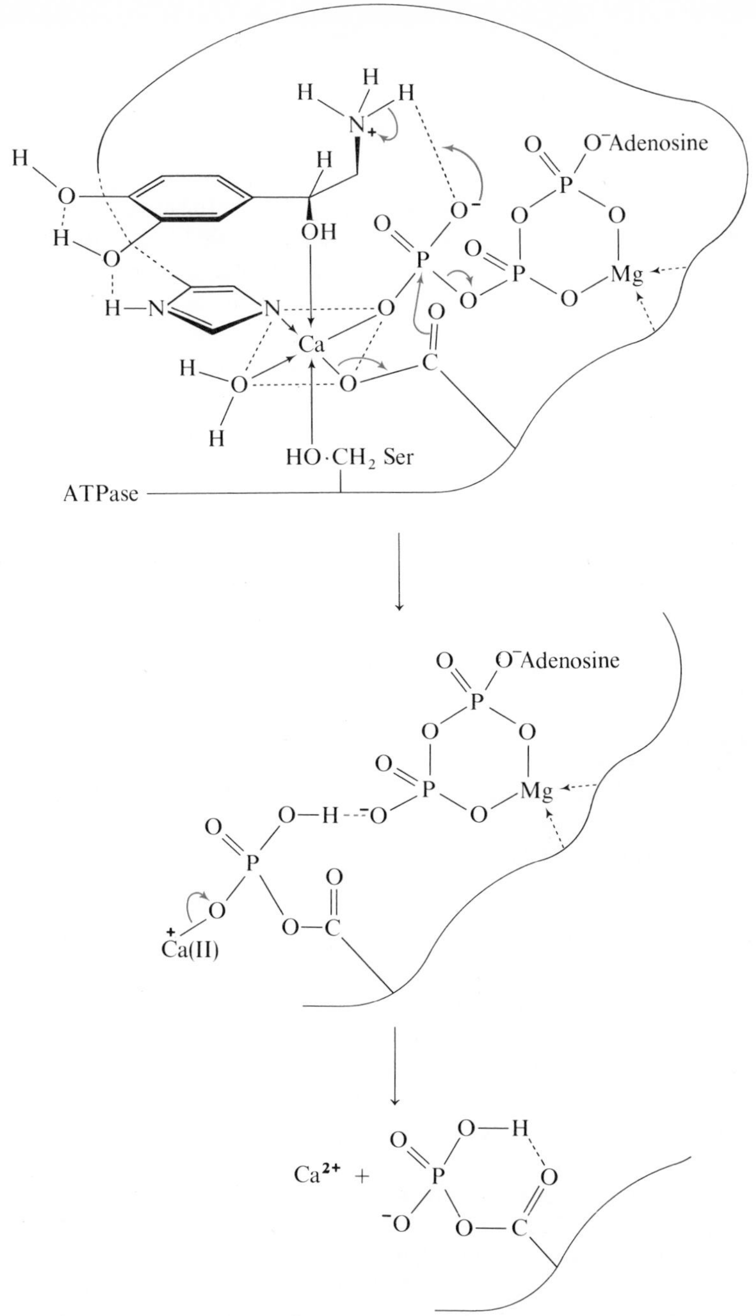

Fig. 22 Noradrenaline-assisted catalysis of ATP–ADP conversion by ATPase with release of Ca^{2+} ions

group, all of which are essential for maximum activity. If, as considered by Belleau, the catechol group and the cation are key binding centres, then the hydroxyl group will be correctly aligned only in the (−)-isomer. This conclusion is supported by Blaschko (1950), who observed that 2′-deoxyadrenaline has approximately the same effect on arterial blood pressure as (+)-adrenaline.

The importance of the cationic head has been emphasised by Belleau (1958), who points out that the ethyleneiminium ion formed by the adrenergic blocking agent, *Dibenamine*, is isosteric with the ionised phenylethylamine component of the sympathomimetic amines. The positively-charged ethyleneiminium ion acts as a biological alkylating agent forming a covalent bond with a nucleophilic centre, which is by deduction presumably also the anionic receptor for the sympathomimetic amines. The relative spatial configuration of the ethyleneiminium ion and the associated aromatic ring correlates sufficiently well with the geometry of the cationic head and catechol ring of the sympathomimetic amines for reaction to occur at the same biological site.

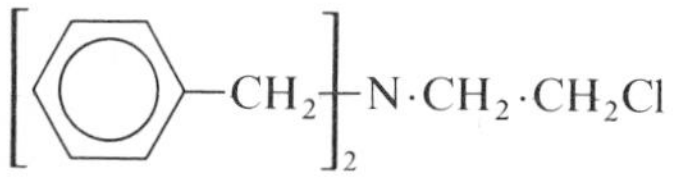

$$[C_6H_5-CH_2]_2N\cdot CH_2\cdot CH_2Cl$$

Dibenamine

$$(PhCH_2)(CH_2Ph)\overset{+}{N}(CH_2CH_2)\quad Cl^-$$

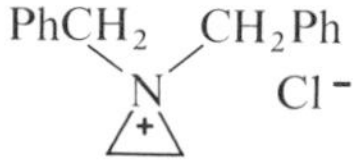

Ethyleneiminium ion derived from *Dibenamine*

The agonistic α-response (Ahlquist, 1968) seems to be mediated by a breakdown of the ATPase–ATP enzyme substrate complex to ADP and H_3PO_4 which is catalysed in the presence of catecholamine (Belleau, 1960, 1963). Essentially, the protonated amino group promotes a nucleophilic attack by the enzyme at the terminal phosphorus atom of ATP. Kier (1969) has shown that the calculated lowest energy conformation of catecholamines occurs when the rotation angle about the —CH(OH)—CH_2NHR bond is 180° between the catechol and $\overset{+}{N}H_3$ groups, and that this is consistent with *his* concept of the α-receptor. Catecholamine stimulation of the sodium–potassium activated membrane ATPase, which is concerned with ion transport, is considered to result in release of membrane-bound calcium, and this initiates muscle contraction (Fig. 22).

The effectiveness of *Adrenaline* at the α-receptor site is only marginally impaired by the bulk of the *N*-methyl substituent. Larger groups, however, such as the isopropyl group in *Isoprenaline*, inhibit ionic interaction to the detriment of α-stimulation. In consequence, *Isoprenaline* exhibits only typical β-receptor agonistic activity, i.e. inhibition of smooth muscle tonus and myocardial stimulation (Ahlquist, 1968). Since *Isoprenaline* also is the most effective of the three sympathomimetic amines in stimulating the formation of cyclic-AMP, Belleau (1966, 1967) has suggested that the β-adrenergic agonistic response is implicated in the supply of cyclic-AMP. The cationic ammonium group is essential for formation of the receptor complex with the cell-wall enzyme adenyl cyclase. This activates the enzyme allostericly for conversion of ATP to cyclic-AMP on the inside of the cell.

The alcoholic hydroxyl group appears to be relatively unimportant in the β-agonistic response. The catechol group, on the other hand, plays a significant rôle mediated by a hydrated Mg^{2+} ion in complex formation at the enzyme surface. Calculations (Pedersen, Hoskins and Cable, 1971) relating to the preferred conformation of noradrenaline show that whereas there are two conformations with minimal energy (Fig. 23; $\theta = 60^\circ$ and 180°), formation of the $Li(H_2O)_2$ complex as in co-ordination with a hydrated Mg^{2+} ion at the β-receptor leads to disappearance of the rotational minimum at 180°, so only that at 60° remains.

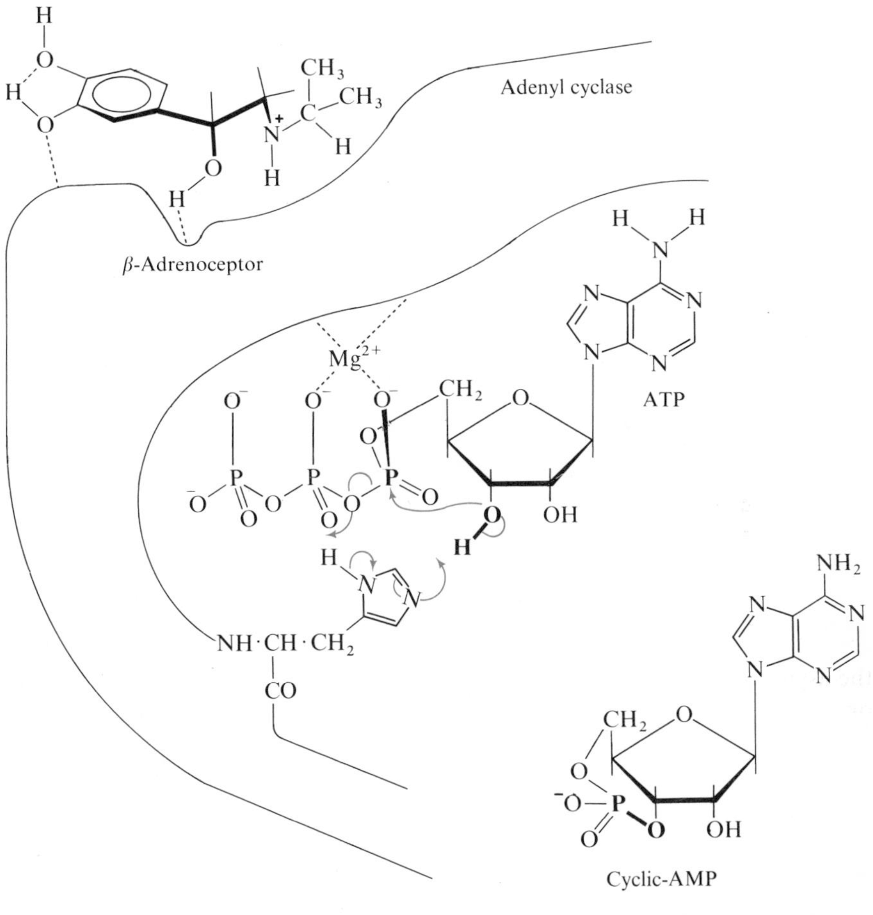

Pedersen, Hoskins and Cable (1971) point out that Li^+ and Mg^{2+} have almost identical ionic radii, but as Li^+ has a maximum co-ordination number of four (compared to six for magnesium), it may weaken the bonding of noradrenaline in the noradrenaline storage regions, when used clinically in the treatment of manic depression.

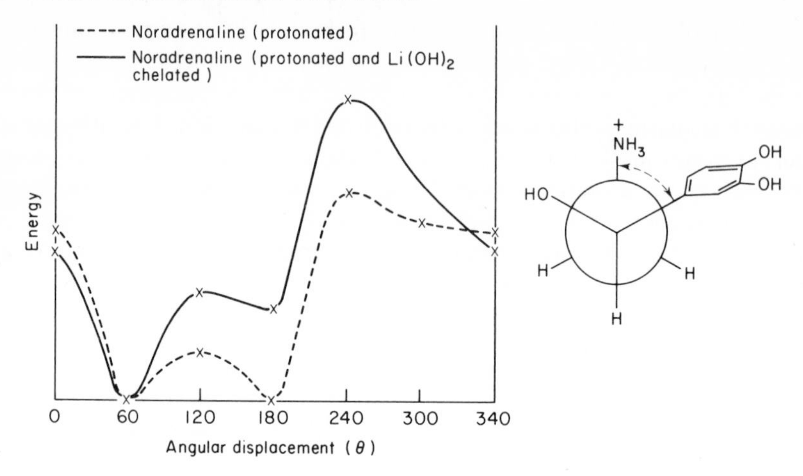

Fig. 23 Calculated energy for various angular displacements about the $H_3\overset{+}{N}\cdot H_2C$——CH(OH) bond of *Noradrenaline*

Optical Activity and Drug Antagonism

The presence of an optical or other asymmetric centre does not necessarily lead to differences in the potency of enantiomers. This is only apparent where the asymmetric centre plays a key rôle in determining the orientation of three or more biologically active groups in the drug molecule. In such a case, the enantiomer is able to provide for the correct alignment of only two of the required groups, and in consequence is generally of considerably lower potency, as in the case of D-(−)- and L-(+)-*Noradrenaline* (p. 126). It follows, too, that the racemate has the average potency of the two contributing enantiomers. Considerable care is essential to ensure complete resolution of optical isomers required for biological testing, where these are obtained by resolution of enantiomers rather than by stereospecific syntheses (Casy, 1975). This becomes significant where there are large differences (> 100 fold) in the potency of two enantiomers, since whilst small amounts of the weaker isomer will not substantially affect the potency of the stronger isomer, even relatively small amounts of the stronger isomer can very materially alter the apparent potency of the weaker enantiomer.

Compounds with dual pharmacological effects can exhibit either the same or opposite stereospecificity for the two receptor targets. Thus, *Dextropropoxyphene*, the (2*S*, 3*R*)-(+)-enantiomer, is a potent analgesic with an ED_{50} in the mouse hotplate test about six times that of *Levopropoxyphene* (Casy and Myers, 1964; Eddy, 1959). In contrast, however, *Levopropoxyphene* has powerful antitussive properties, whilst those of *Dextropropoxyphene* are negligible. In a few rare cases, other factors may intrude and one enantiomer may actually antagonise the agonistic effect of the other. Thus, D-(−)-isoprenaline is a strong β-adrenergic agonist, and its enantiomer, L-(+)-, a weak β-agonist (Lands,

Luduena and Tullar, 1954), but whilst the former also has weak α-receptor agonist properties (at high drug concentration), the L-(+)-isomer is actually a weak α-receptor antagonist (Luduena, 1962). L-(+)-Isoprenaline, therefore, blocks the relaxation of rabbit uterine muscle by its enantiomer. Similarly, the ED_{50} in albino rats is 35.5 mg/kg for the D-(−)-isomer, 107 mg/kg for the L-(+)-isomer and 136 mg/kg for the racemate (±) respectively, showing that the (+)-isomer actually reduces the toxicity of the (−)-isomer. A further example of this type of antagonism is seen in the microbiological field. Both L- and DL-dihydro-orotic acids support the growth of *Lactobacillus bulgaricus* 09. The D-isomer, however, not only lacks activity, but also reversibly inhibits the growth-promoting properties of its enantiomer and the racemate (Miller, Gordon and Engelhardt, 1953).

Conformational Isomerism

Non-bonded Interactions

When two or more tetrahedral carbon atoms are linked to each other, the single bonds joining them may permit free rotation about the bond. Non-bonded repulsive interactions between the substituents attached to one such carbon and those of the adjacent atoms influence their relative orientation in space. These interactions are least when the substituents are fully staggered, greatest when they are eclipsed, and of intermediate effect when they are partially staggered, as illustrated by the accompanying potential energy diagram for a four-carbon unit (Fig. 24).

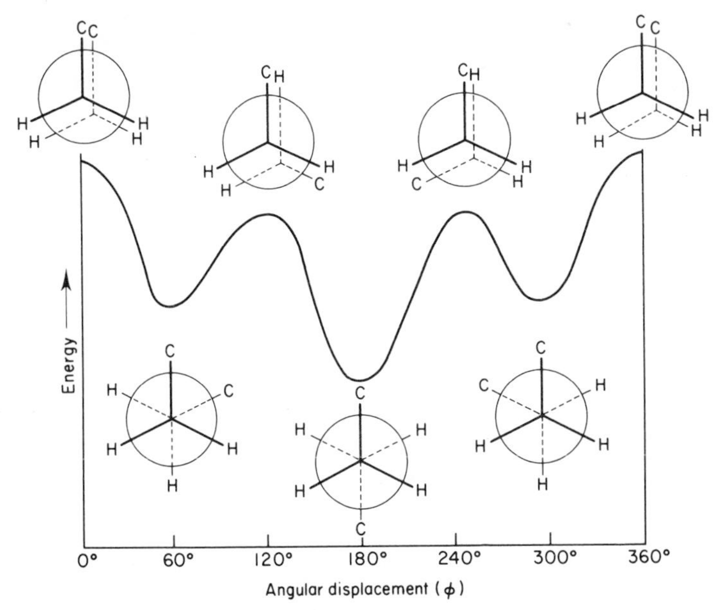

Fig. 24 Variation of interaction energy with conformation in saturated hydrocarbons

The conformation shown at 180° is fully staggered; those at 60° and 300° partially staggered (skew); conformations at 0° and 360° are fully eclipsed and those at 120° and 240° partially eclipsed. The interaction energy between the two terminal methyl groups varies from 0–17 kJ depending on the angular displacement (Taylor, 1948; Szasz, Sheppard and Rank, 1948). Similarly, the interaction energy between staggered and eclipsed forms of ethane has been calculated as about 16 kJ mol^{-1} (Hoffmann, 1963), and found by microwave determination to lie between 12 and 14 kJ mol^{-1} (Kier, 1967, 1968).

Hydrocarbon Chain Lengths

The low energy fully-staggered conformation is characteristic of hydrocarbon chains (**1**, 3), and, in the absence of other opposing forces, leads to the adoption of the zig-zag conformation of the fully-extended carbon chain. Molecular dimensions of aliphatic drug molecules are usually calculated on the basis of the fully staggered chain. However, Gill (1959, 1965) has calculated that the relative probabilities at 37°C for the fully staggered and the two partially staggered conformations in *n*-butane are 1.00, 0.272 and 0.272 respectively. So that, allowing for the equivalence of the two skew conformations, approximately two-thirds of the molecules in a sample of butane are fully extended at any one moment in time and one-third have a skew conformation. This effect reduces the extended length of hydrocarbon chain.

This probability factor is important in considering the overall intramolecular dimensions of pharmacologically active molecules. Thus, Elworthy (1963a and b) has shown experimentally that there is considerable interfacial energy between hydrocarbon and solvent in aqueous solution which leads to a contraction of the aliphatic chain, as predicted by Gill. In particular, conductimetric measurements show that the inter-nitrogen distance in polymethylene-bis-quaternary ammonium compounds is less than that calculated for a fully extended hydrocarbon chain, an observation of considerable relevance to the mechanism of action of ganglion blocking and neuromuscular blocking agents. Thus, the inter-nitrogen distance of decamethonium (Table 20), which has the optimum neuromuscular blocking activity of a series of polymethylene-α,ω-bis-trimethylammonium compounds, correlates well with estimates of the inter-nitrogen distance in tubocurarine and related compounds (Edwards, Stenlake, Carey and Lewis, 1959).

Table 20. Inter-nitrogen Distances in Polymethylene-bis-onium Compounds

$R^3\overset{+}{N}(CH_2)_n\overset{+}{N}R_3$

Compound	R	n	Inter-nitrogen distances (Å) Found	Calculated (fully extended)
Hexamethonium	Me	6	6.3	9.5
Decamethonium	Me	10	9.5	13.5
Deca-ethonium	Et	10	10.2–13.1	13.5

Effects of Rotamer Energy on Oxidative Stability of Catecholamines

The oxidative stability of catecholamines to Fenton's reagent (H_2O_2 and $Fe(NO_3)_3$) is in the order *Adrenaline* (R = Me) > *Isoprenaline* (R = iPr) > *Noradrenaline* (R = H) (Clauder, Radics, Szabo and Varga, 1968). The +I effect of the *N*-alkyl substituents in the secondary amines increases the rate of cyclisation of the quinone intermediate over that from *Noradrenaline* (Norephinephrine). The enthalpy of activation for adrenochrome and *N*-isopropylnoradrenochrome formation is also greater than that for noradrenochrome formation. This is due to greater energy being required for rotation about the C—C bond in the secondary amines in order to achieve the correct orientation for nucleophilic attack on the ring.

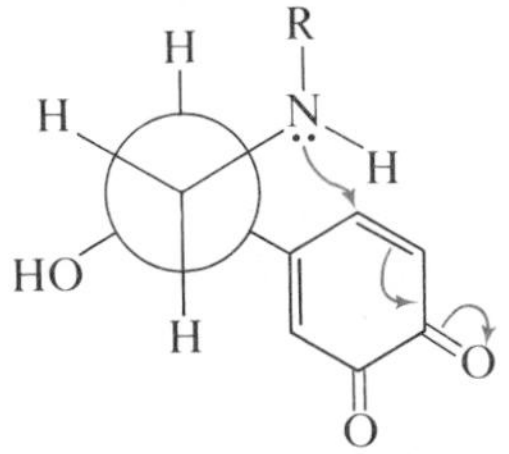

Adrenochrome (R = Me)

Conformational Factors in the Anti-Parkinsonism Effects of Dopamine

Extended Hückel calculations (Rekker, Engel and Nys, 1972) and evidence based on potential energy functions and PMR spectroscopy (Bustard and Egan, 1971) lead to the conclusion that the *trans* conformation of *Dopamine* is preferred, marginally, to the two *gauche* conformations.

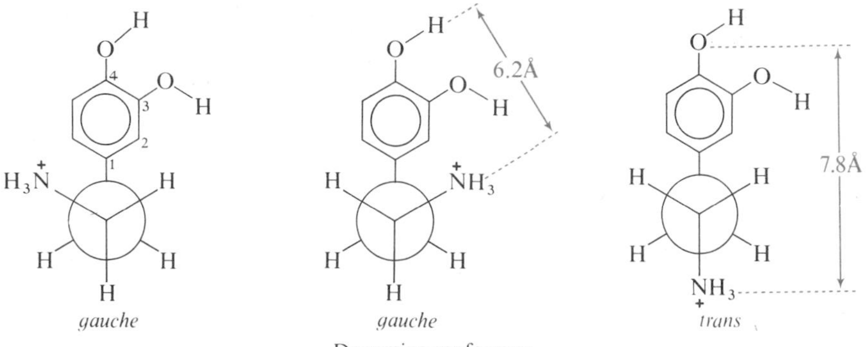

gauche *gauche* *trans*

Dopamine conformers

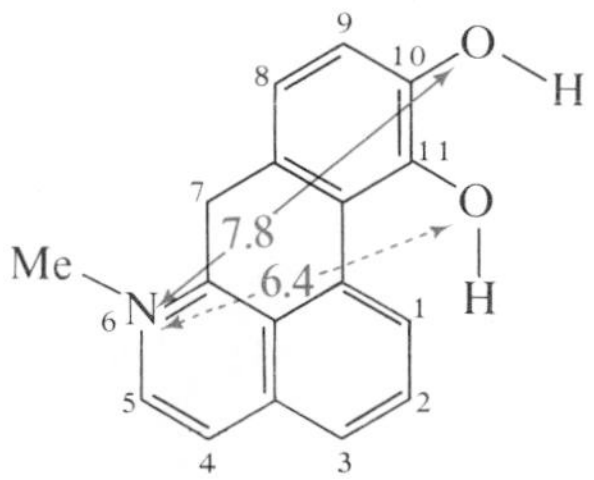

Apomorphine

The significance of the *trans*-conformation in dopaminergic activity is supported by comparison with *Apomorphine* (10,11-dihydroxyaporphine), which also acts at dopamine receptors. Both 10- and 11-hydroxyl groups are essential to activity, since neither aporphine (Pinder, Buxton and Green, 1971) nor the 10-methoxy-,11-methoxy- nor the 10,11-dimethoxyaporphines (Cannon, Smith, Modiri, Sood, Borgman, Aleem and Long, 1972) are active, whilst 11-hydroxyaporphine is only weakly active (Granchelli, Neumeyer, Fuxe, Ungerstedt and Corrodi, 1971). It is, therefore, relevant that the N—O^{10} distance in *Apomorphine* (7.8 Å) is the same as the corresponding N—O^4 distance in the *trans-Dopamine* conformer, whilst the shorter N—O^{11} distance in *Apomorphine* (6.4 Å) corresponds to that of the N—O^4 distance in the two *gauche*-dopamine conformers.

Conformational Isomerism in Benzothiadiazine-1,1-dioxides

Wohl (1970) has used extended Hückel calculations to determine the dihedral angle between the plane of the nucleus and that of the 3-substituents in a series of benzothiadiazine-1,1-dioxides. Although this correlates with the relative 'antihypertensive' potencies (Table 21), it is established that the rotamer conformation also affects the net charge on the 3-substituent, and that only in the case of the Δ^3-cyclopentenyl derivative does the preferred rotamer (dihedral angle = 90°) for minimum rotational energy coincide with that required for the maximum net charge on the 3-position. High charge density at both 3- and 5-positions of the benzothiadiazine ring is shown to be important, with the

Table 21. Preferred Rotamers of 3-Substituted Benzothiadiazine-1,1-dioxides

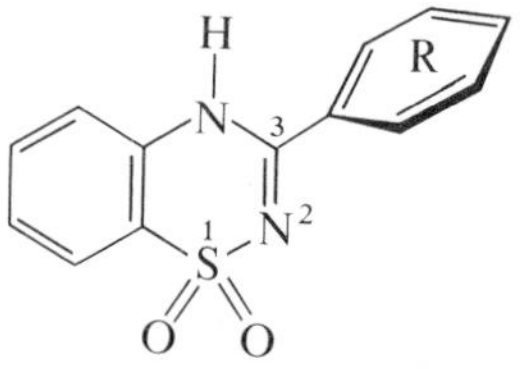

Substituent (R)	Rotational Energy Barrier [1]	Preferred Rotamer [2]	Rotamer with Maximum Net Charge on 3-position	Relative Order of Increasing Anti-hypertensive Potency
2-Furyl	7.8	0		1
Δ^1-Cyclopentenyl	3.6	0	90°	2
Phenyl		45		3
Δ^2-Cyclopentenyl	3.9	45	120°	4
Δ^3-Cyclopentenyl	4.1	90	90°	5

[1] Rotational Energy Barrier ≡ energy difference in kJ mol^{-1} between rotamers of highest and lowest energy

[2] Dihedral angle between benzothiadiazine ring and plane of substituent ring

latter acting as a centre for receptor–drug electron transfer. Steric effects affecting the 5-position, as with the introduction of a 4-methyl substituent, and electron-withdrawal at the 3-position with, for example, a trifluoromethyl substituent, significantly lower activity.

Acetylcholine Conformation

The molecule of acetylcholine is flexible and capable of adopting a number of different conformations. Proton magnetic resonance spectroscopy of acetylcholine in deuterium oxide (Culvenor and Ham, 1966) has shown that the $\overset{+}{N}—C—C—O$ grouping is in a *gauche* arrangement in aqueous systems, and that the $CH_2 \cdot O \cdot CO \cdot Me$ group has the normal conformer population distribution of a primary ester.

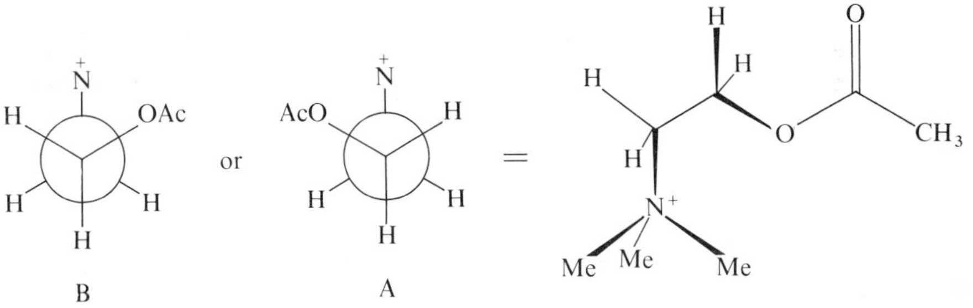

B or A =

It is particularly important to note that whilst X-ray diffraction (Canepa, Pauling and Sörum, 1966) and neutron diffraction (Brennan, Ross, Hamilton and Shefter, 1970) of acetylcholine bromide and erytho-(±)-α,β-dimethylacetylcholine iodide confirm that the $\overset{+}{N}—C—C—O$ group is *gauche* in the solid state, the ester conformation is markedly different from that found by PMR spectroscopy, and not relevant to the situation in aqueous solution. Studies on the crystal structure of β-methylacetylcholine (Chothia and Pauling, 1969a and b) also establish a *gauche* conformation capable of assuming virtually an identical arrangement of the quaternary head and acetoxy group to that of acetylcholine in the (*S*)-(+)-enantiomer only. It is significant, too, that a similar orientation cannot be maintained in the (*R*)-(−)-enantiomer due to steric repulsion between the β-methyl and trimethylammonium groups (Belleau

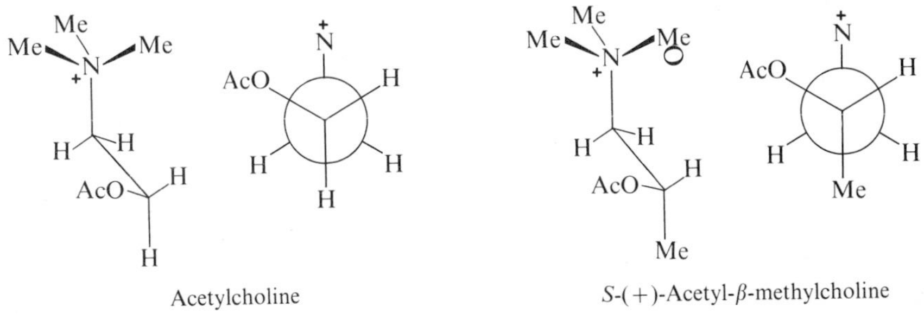

Acetylcholine

S-(+)-Acetyl-β-methylcholine

and Pauling, 1970). Since, also, only the (*S*)-(+)-enantiomer is capable of acting as a substrate for acetylcholinesterase and possesses muscarinic activity equivalent to that of acetylcholine, the ability to adopt the *gauche* conformation of the (*S*)-(+)-enantiomer or a close approximation to it seems to be essential for activity.

Despite numerous attempts, no simple correlation has been established between the predominant conformation and the potency of action of cholinergic drugs at either the nicotinic or muscarinic receptor (Partington, Feeney and Bergen, 1972). Nonetheless, certain cyclic analogues (pp. 140 and 141) show a high degree of stereospecificity (Chiou, Long, Cannon and Armstrong, 1969).

Conformation and Isomerism in three- and four-membered Saturated Ring Systems

The influence of bond angle and dihedral angle strain on the shape and reactivity of three- and four-membered rings has been discussed elsewhere (**1**, 3). Not only do such systems show enhanced reactivity due to ring strain compared with corresponding five- and six-membered ring compounds, but the distortion of carbon–carbon bonds creates a planar ring system. As a result, 1,2-disubstituted cyclopropanes and cyclobutanes are able to exist as *cis*- and *trans*-isomers. Where the two substituents are identical, the *cis*-compound has a plane of symmetry, and only the *trans*-isomer is capable of resolution into its two enantiomorphs.

plane of symmetry

HO·OC CO·OH

cis-Cyclopropane-1,2-dicarboxylic acids

CO·OH HO·OC

HO·OC CO·OH

trans-Cyclopropane-1,2-dicarboxylic acids

When, however, the two substituents are different, as for example in *Tranylcypromine Sulphate*, both *cis*- and *trans*-isomers are capable of resolution into enantiomorphs. The unresolved *trans*-racemate is used as an antidepressant, on account of its powerful monoamine oxidase inhibitory properties; it is, for example, some 55 times more effective than *Iproniazid* in potentiating the convulsant effects of tryptamine *in vivo*. The *cis*-isomer is also effective as a monoamine oxidase inhibitor, but only some 24 times as potent as *Iproniazid*.

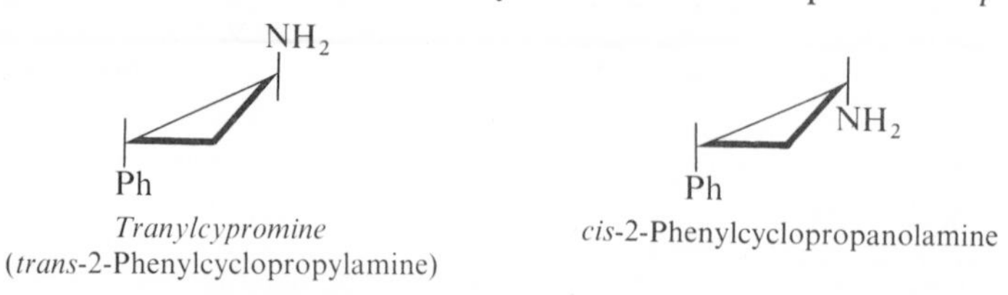

Tranylcypromine
(*trans*-2-Phenylcyclopropylamine)

cis-2-Phenylcyclopropanolamine

Much greater stereochemical specificity is shown by *cis*- and *trans*-2-acetylcyclopropyltrimethylammonium iodides (Table 22). Thus, whereas both the

Table 22. Muscarinic and Nicotinic Activities of Isomeric 2-Acetylcyclopropyltrimethyl ammonium (ACTM) Iodides

$O{\cdot}CO{\cdot}CH_3$ / $Me_3\overset{+}{N}$ — *trans*-isomer

$Me_3\overset{+}{N}$ $O{\cdot}CO{\cdot}CH_3$ — *cis*-isomer

Compound	Muscarinic Activity		Neostigmine Potentiation (×-fold)	Nicotinic Activity Frog Rectus Abdominis
	Guinea pig ileum	Dog blood pressure		
Acetylcholine	1.0	1.0	41	1.0
(+)-*trans*-ACTM	1.13	4.7	23	0.013
(−)-*trans*-ACTM	0.0022	0.023	2.8	0.0028
(±)-*cis*-ACTM	0.0001	—	—	0.0042

(±)-*cis*- and the (−)-*trans*-isomers have negligible activity at both muscarinic and nicotinic receptors, the (+)-*trans*-isomer is almost equipotent with acetylcholine at muscarinic receptors, though with little nicotinic activity. The (+)-*trans*-isomer is hydrolysed by acetylcholinesterase at about 96% of the rate for acetylcholine, compared with 59% for the (−)-*trans*-isomer. This high specificity as a substrate for acetylcholinesterase also correlates well with the relative potentiation of the isomers' muscarinic effects on dog blood pressure by *Neostigmine* (Table 22; Armstrong, Cannon and Long, 1968; Chiou, Long, Cannon and Armstrong, 1969).

Conformation and Isomerism in five-membered Saturated Ring Systems

Ring Conformation

In contrast to the three- and four-membered carbocyclic compounds, cyclopentane and analogous five-membered saturated ring structures adopt strain-free puckered conformations (**1**, 3). Two such conformations are possible, known as **envelope** and **twist** respectively, the particular conformation adopted being dependent on the substituent pattern, which dictates key factors such as non-bonded interactions. However, these ring conformations appear to exercise only a minor influence on biological activity, their major function being to place substituents and hetero-atoms in geometrical and optical arrangements with high bioactive specificity. These latter effects are seen in highly specific metabolic transformations of particular furanose sugars (**1**, 21) and in muscarine, which mimics certain of the actions of the neurotransmitter, acetylcholine.

Envelope Twist

Muscarinic Activity

The minimal influence of ring conformation is shown by comparison of (±)-muscarone with (±)-4,5-dehydromuscarone, and of (±)-muscarine with (±)-4,5-dehydromuscarine in which the twist conformation of the first member in each pair is restricted by inclusion of the planar double bond to an envelope conformation in the two unsaturated compounds, without any appreciable effect on potency (Waser, 1961; Table 23).

Table 23. Activity of Muscarinic Derivatives Relative to Acetylcholine (1.0) in Lowering Cat Blood Pressure (Waser, 1961)

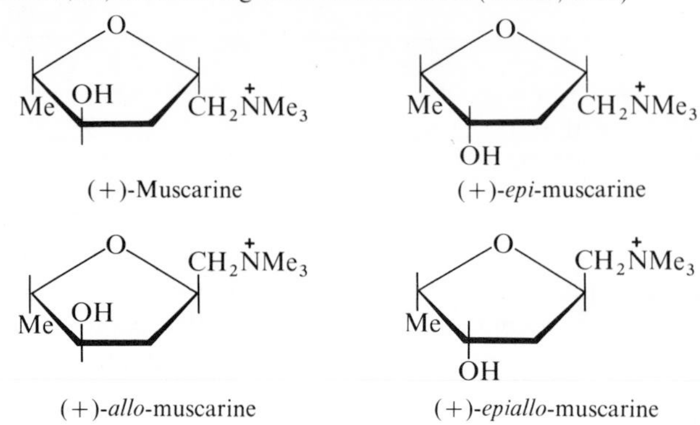

Compound	Potency
(+)-Muscarine	2.7
(−)-Muscarine	0.003
(±)-Muscarine	0.8
(±)-*epi*-Muscarine	0.003
(±)-*allo*-Muscarine	0.005
(±)-*epiallo*-Muscarine	0.008
(±)-4,5-Dehydromuscarine	0.8
(±)-2-Desmethylmuscarine	0.008

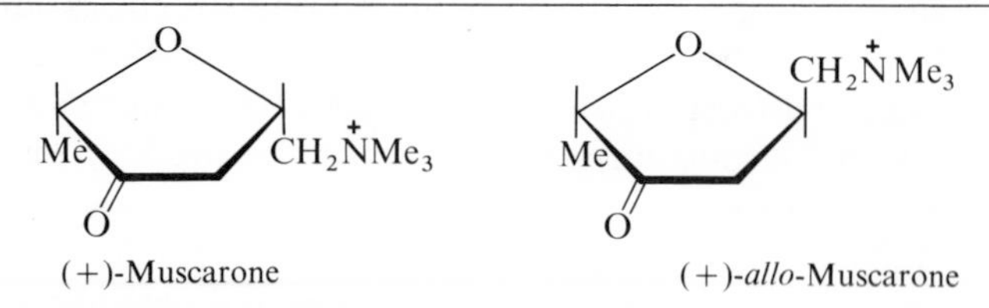

Compound	Potency
(+)-Muscarone	2.7
(−)-Muscarone	8.0
(±)-Muscarone	5.3
(±)-*allo*-Muscarone	2.7
(±)-4,5-Dehydromuscarone	5.3
(±)-2-Desmethylmuscarone	0.13

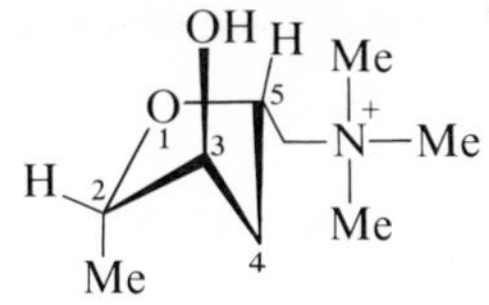

(2*S*, 3*R*, 5*S*)-(+)-Muscarine

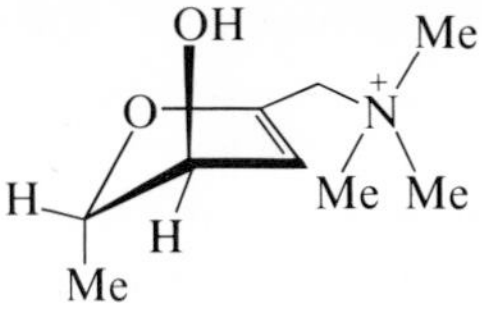

(2*S*, 3*R*)-(+)-4,5-Dehydromuscarine

The importance of the 2-methyl substituent is evident from the low potency of the 2-desmethyl analogues in both muscarine and muscarone series. That its rôle is steric is shown by the fit of the (+)- and (−)-muscarine conformations with the two *gauche* acetylcholine conformations. It is significant, however,

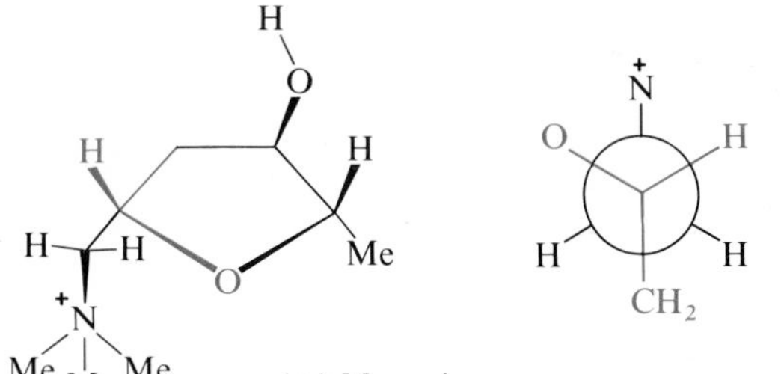

(+)-Muscarine

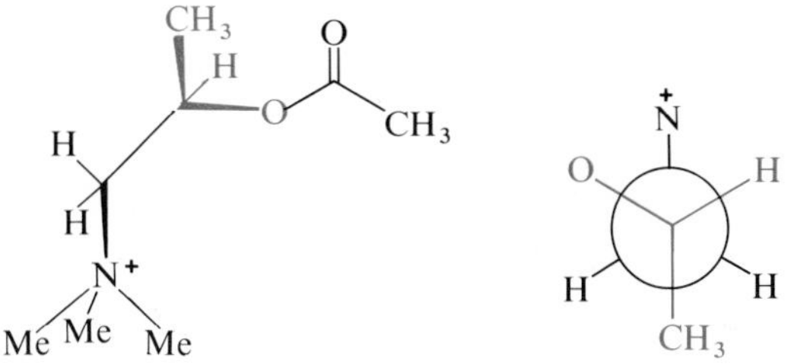

(*S*)-(+)-β-Methylacetylcholine

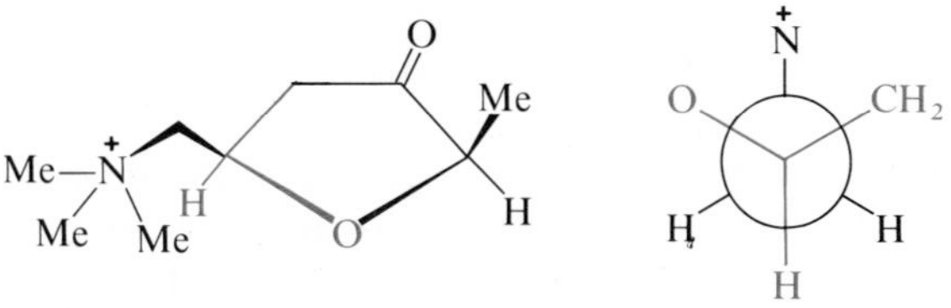

(−)-Muscarone

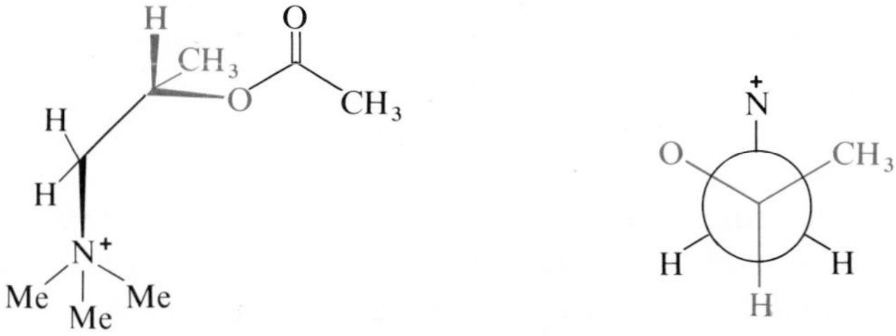

(*R*)-(−)-β-Methylacetylcholine

that (+)-muscarine is 2.7 times more potent than acetylcholine and about 1000 times more active than (−)-muscarine in depressing blood pressure. In contrast, both muscarone enantiomers are more potent than acetylcholine, with (−)-muscarone almost three times the potency of its enantiomer. It would appear, therefore, that both *gauche* acetylcholine conformations are capable of effecting a muscarinic response. It is, however, of considerable interest that (+)-muscarine has the same *gauche* conformation as the active (*S*)-(+)-*β*-methylacetylcholine, whilst (−)-muscarone adopts the alternative unfavourable conformation of inactive (*R*)-(−)-*β*-methylacetylcholine.

In this respect, (−)-muscarone also contrasts with the following dioxolane isomers. The more active enantiomer is not only 6 times more potent than acetylcholine and about 100 times more active than its enantiomer, but also like (+)-muscarine is capable of adopting the same *gauche* conformation as (*S*)-(+)-*β*-methylacetylcholine (Belleau and Puranen, 1963).

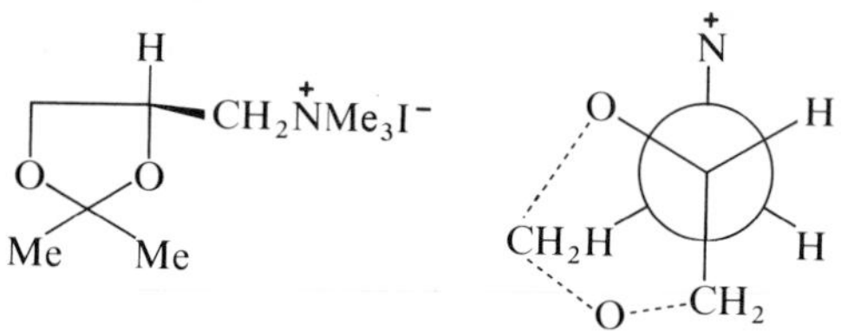

Muscarinic activity 6.0 (ACh = 1.0)

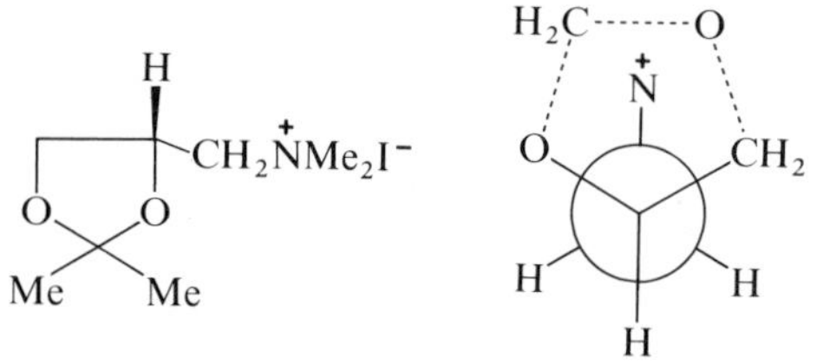

Muscarinic activity 0.06 (ACh = 1.0)

The apparently anomalous stereospecificity of the muscarinic receptor for (−)-muscarone has been explained by postulating the presence of an accessory nucleophilic site at the active centre (Belleau and Puranen, 1963), which is specific for the (*R*)-(−)-*β*-methylacetylcholine conformation. A further example of this apparent inversion of optical specificity is evident in the 1,4-bridged six-membered ring compound, (*R*)-(−)-3-acetoxyquinuclidine methiodide. Only the (*R*)-enantiomer of this compound is active at the muscarinic receptor and at the acetylcholinesterase binding site, yet this enantiomer has a rigid structure which fixes the receptor and esteratic functional groups in an eclipsed conformation, which approximates more closely to the *gauche* conformation of the relatively inactive (*R*)-(−)-*β*-methylacetylcholine (Belleau and Pauling, 1970), than to that of the more active (*S*)-(+)-*β*-methylacetylcholine.

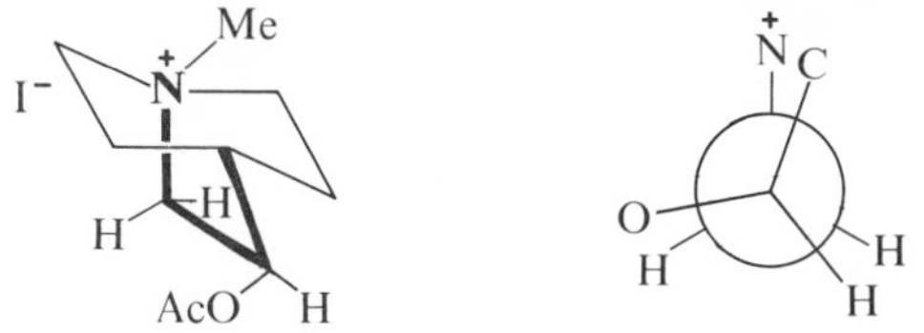

(*R*)-(−)-3-Acetoxyquinuclidine methiodide

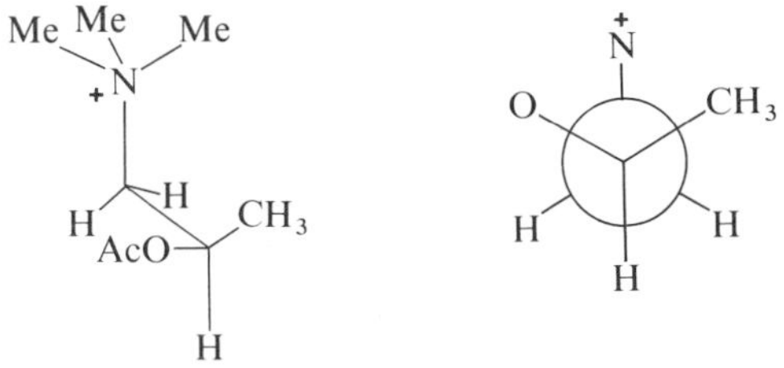

(*R*)-(−)-β-methylacetylcholine

Tetramisole

Tetramisole is a highly active anthelmintic against practically all gastrointestinal and pulmonary nematodes, but almost inactive against cestodes and trematodes (Van den Bossche and Janssen, 1967). It is, however, inactive against bacteria, fungi or protozoa, and relatively non-toxic to humans. The anthelmintic activity of *Tetramisole* resides largely in the (*S*)-(−)-enantiomer, which *in vitro* and *in vivo* is about twice as active against various nematodes as the racemate, and several times more active than (*R*)-(+)-tetramisole.

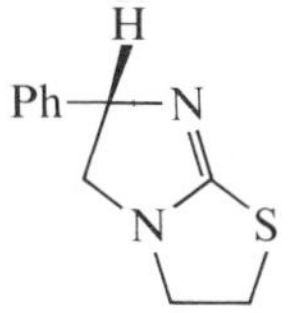

(*S*)-(−)-Tetramisole

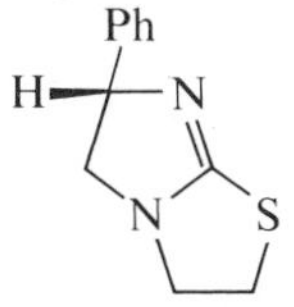

(*R*)-(+)-Tetramisole

According to Van den Bossche and Janssen (1967), (*S*)-(−)-tetramisole acts as a stereospecific inhibitor of succinic dehydrogenase in *Ascaris* muscle. This enzyme is essential to the formation of succinate which plays a key rôle in the formation of α-methylbutyrate and α-methylvalerate, the major end products of *Ascaris* muscle metabolism. *Pyruvate*, which is formed by glycolysis (**1**, 21) in man, is largely converted in *Ascaris* muscle by carbon dioxide fixation to succinate via malate and fumarate or, alternatively, to acetate by oxidative decarboxylation. Succinate is further decarboxylated to propionate, and α-methylbutyrate and α-methylvalerate are formed by the condensation of succinate with acetate and propionate respectively.

Tetramisole isomers inhibit the enzymic activity of purified succinic dehydrogenase from *Ascaris* muscle, and hence decrease the succinate content of muscle suspensions in order of effectiveness laevo $>$ ($\pm$) $>$ dextro. Because

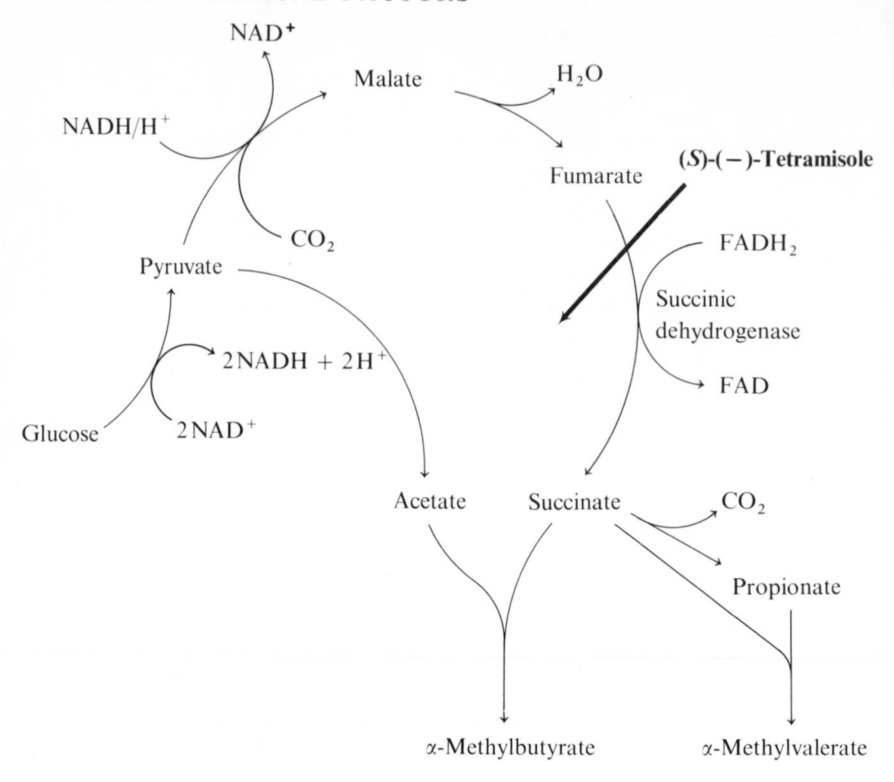

fumarate is the terminal electron acceptor (in place of oxygen) in the anaerobic metabolism of *Ascaris* muscle, the reduction of fumarate to succinate is essential for the reoxidation of NADH formed in the glycolysis of glucose to pyruvate; the reduction also provides the muscle with a source of ATP (Bueding, 1962). In accord with this, the isomers in the same order of effectiveness (laevo > $(\pm)$ > dextro) also decrease NAD^+ and increase inorganic phosphate in *Ascaris* muscle suspensions.

Pilocarpine

The miotic activity of *Pilocarpine*, which has the (2*S*, 3*R*)-configuration (Hill and Barcza, 1966), is not shown by isopilocarpine ((2*R*, 3*R*)-configuration). The latter is thermodynamically more stable than *Pilocarpine* (Nunes and Brockmann–Hanssen, 1974). Solutions of *Pilocarpine* must, therefore, be protected from the effects of heat or alkali to prevent formation of the inactive isomer.

Et, O, O, N·Me, N — *Pilocarpine*

Et, O, O, N·Me, N — Isopilocarpine

Conformation and Isomerism in Six-membered Saturated Ring Systems

Strainless Rings

Cyclohexane normally adopts a non-planar chair conformation, which is strainless, as a result of each of its six carbon atoms being out of plane in respect of its immediate neighbours (**1**, 3). There is also considerable evidence that the majority of six-membered heterocyclic compounds adopt a similar conformation.

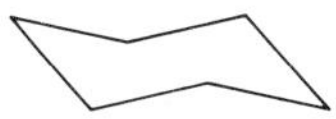

Cyclohexane 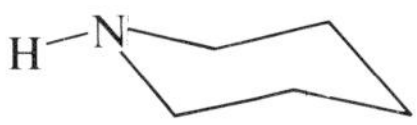 Piperidine 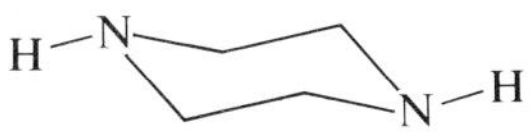 Piperazine

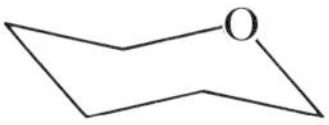

Tetrahydropyran

The distinction between axial and equatorial substituents is important in relation to the geometry, reactivity, and specificity of pharmacologically active molecules based on these systems. As already discussed (**1**, 3), these monocyclic ring structures are flexible, and the effect of monosubstitution is such that the more stable conformation is generally that in which the substituent is equatorial as, for example, in methylcyclohexane and in cyclamic acid.

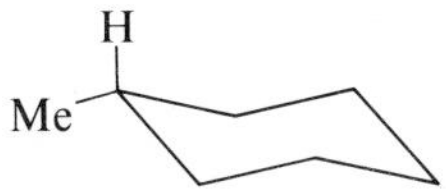

Methylcyclohexane

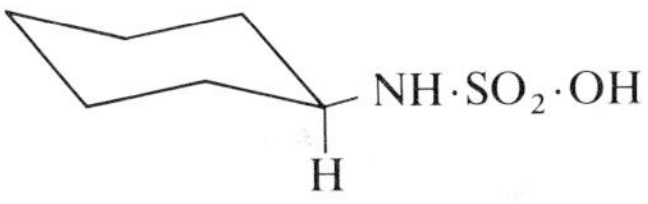

Cyclamic acid

The single diphenylhydroxymethyl substituent in the central nervous stimulant, *Pipradol*, is similarly equatorially-oriented. Rotation about the equatorial bond is constrained by hydrogen bonding between the side-chain hydroxyl and the nitrogen lone pair. It is of interest that the CNS receptor is completely stereospecific for the (*R*)-enantiomer, since the racemate has only half its activity.

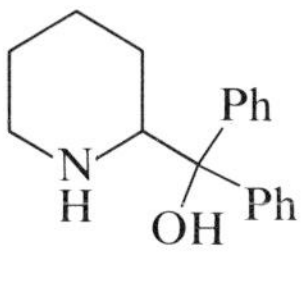

Pipradol 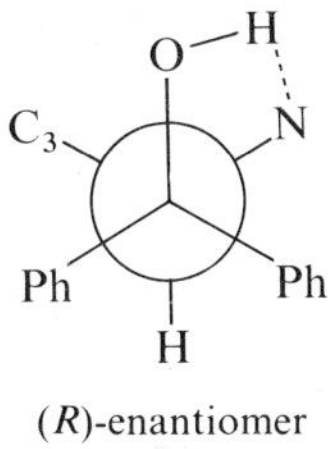 (*R*)-enantiomer 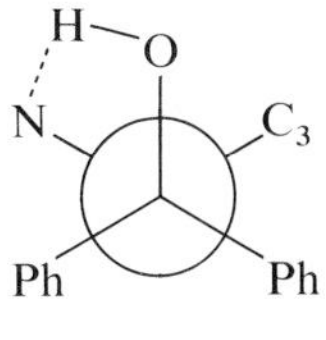 (*S*)-enantiomer

In 1,1-disubstituted compounds, such as *Dicyclomine Hydrochloride*, where the two substituents are attached to the same carbon atom, the preferred conformation will be that in which the larger of the two substituents is equatorial.

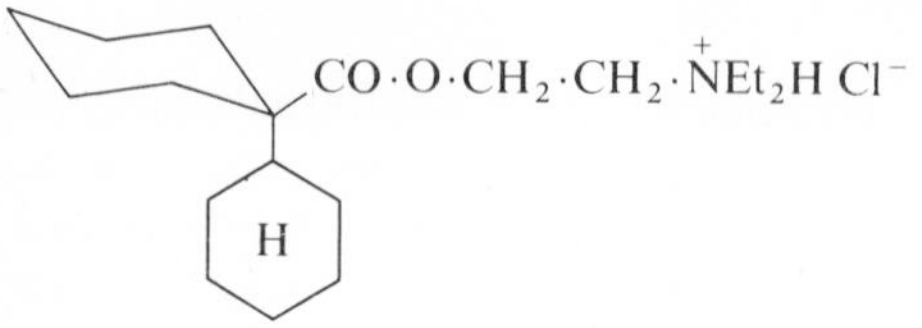

$CO \cdot O \cdot CH_2 \cdot CH_2 \cdot \overset{+}{N}Et_2H\ Cl^-$

H

Preferred conformation

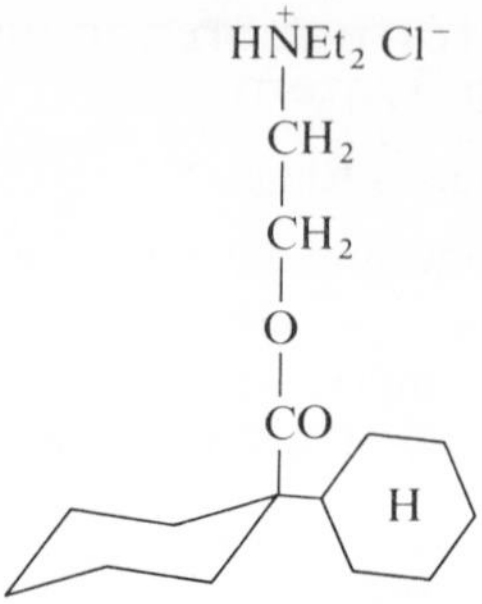

$H\overset{+}{N}Et_2\ Cl^-$ — CH_2 — CH_2 — O — CO

H

Dicyclomine Hydrochloride

Orientation and Reactivity of Substituents

The orientation (i.e. axial or equatorial) of a substituent in the more stable conformer is a reflection of the extent to which non-bonded interactions are relieved compared with those in less stable conformers. This finds expression in both physical and chemical properties. Thus, the infrared stretching frequency of axial hydroxyls (3637–3639 cm^{-1}) is generally marginally higher than that for the corresponding equatorial hydroxyl (3629–3630 cm^{-1}; Arthur, Cole, Thieberg and White, 1956; Allsop, Cole, White and Willix, 1956). Similarly, axially-orientated hydroxyl groups are often more difficult to esterify than the corresponding equatorial hydroxyls because of steric hindrance, and for similar reasons the corresponding axial esters, once formed, are more difficult to hydrolyse. For example, the axially-orientated 11β-hydroxyl of *Hydro-*

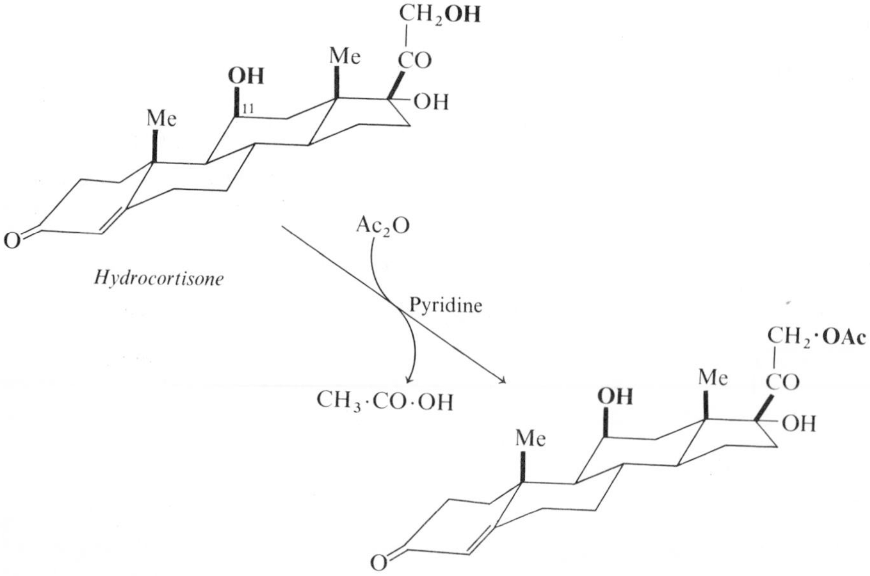

CH_2**OH**, CO, OH, Me, **OH**, 11, Me, O

Hydrocortisone

Ac_2O

Pyridine

$CH_3 \cdot CO \cdot OH$

$CH_2 \cdot$**OAc**, CO, OH, Me, **OH**, Me, O

Hydrocortisone Acetate

cortisone is unaffected by acetic anhydride (in pyridine) during conversion to *Hydrocortisone Acetate*, whereas the corresponding equatorial 11α-hydroxyl of *epi*-hydrocortisone is readily converted by the same reagent giving 11-*epi*-hydrocortisone-11,21-diacetate (Antonucci, Bernstein, Heller, Lenhard, Littell and Williams, 1953; Long and Gallagher, 1946).

Similar conformational criteria have been established for activity in the ergot alkaloid series, which exist as a series of isomeric pairs, due to stereoisomerism at C_8 of the ergoline nucleus. The parent lysergic and isolysergic acids show significant differences in pK_a due to the orientation of the carboxylic acid group in relation to the piperidine ring nitrogen, lysergic acid, with the equatorial carboxyl group being the stronger acid (Stenlake, 1953). The active oxytocic, *Ergometrine* (Ergonovine), retains the equatorial C_8-substituent of the parent acid, and differs from its pharmacologically inactive epimer (ergometrinine) in being soluble in alkali, the amide nitrogen of the latter being, presumably, inactivated by intramolecular N—H—N hydrogen bonding.

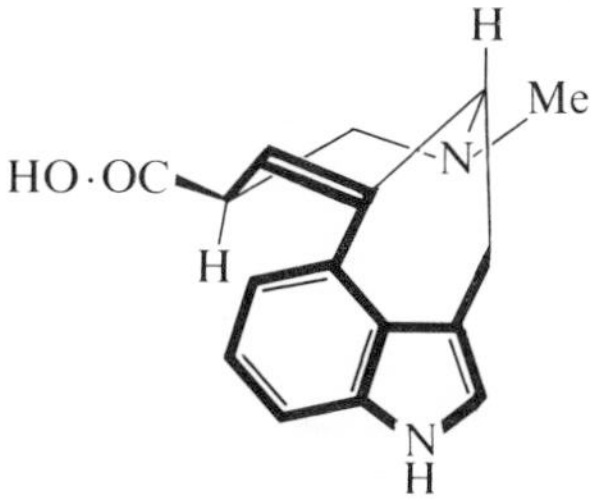

Lysergic acid
pK_a 7.86; 7.96

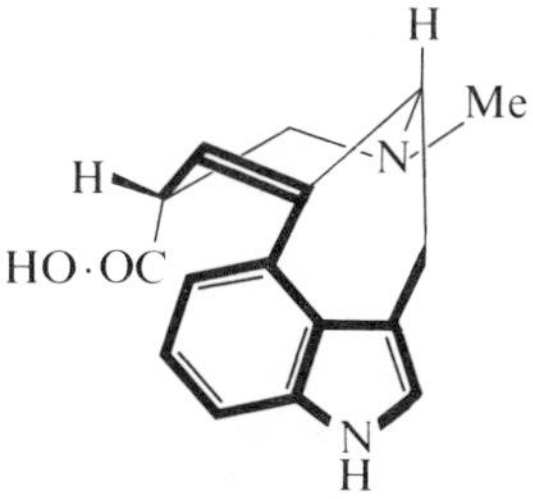

Isolysergic acid
8.31; 8.60

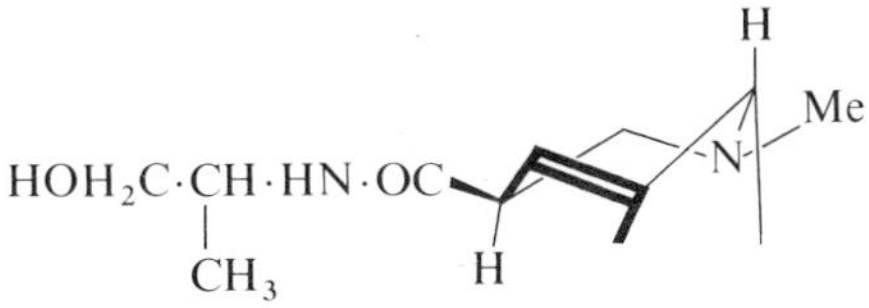

Ergometrine
pK_a 6.70
(Active oxytocic)

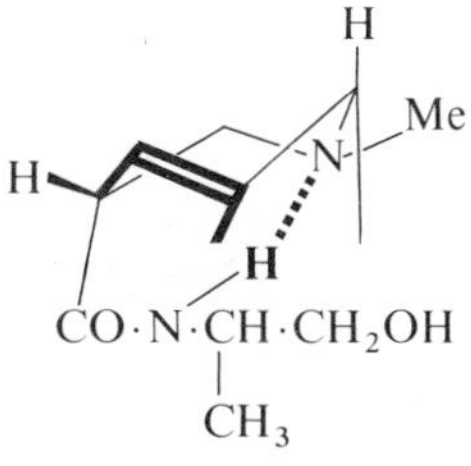

Ergometrinine
7.40
(Inactive)

The γ-isomer of the various benzene hexachlorides (*Gamma Benzene Hexachloride*) is the most effective as an insecticide. Not only do the isomers vary in insecticidal activity, but they also vary in chronic toxicity in a way which correlates significantly with the rate of mammalian metabolism, stereochemistry, and chemical reactivity. The first stage in the metabolism of *Gamma Benzene Hexachloride* in both mammals (Grover and Sims, 1965) and insects (Clark, Hitchcock and Smith, 1966) is dehydrohalogenation to γ-pentachlorocyclohexene (**2**, 5). It is an interesting observation, therefore, that *in vitro* rates

β-Benzenehexachloride

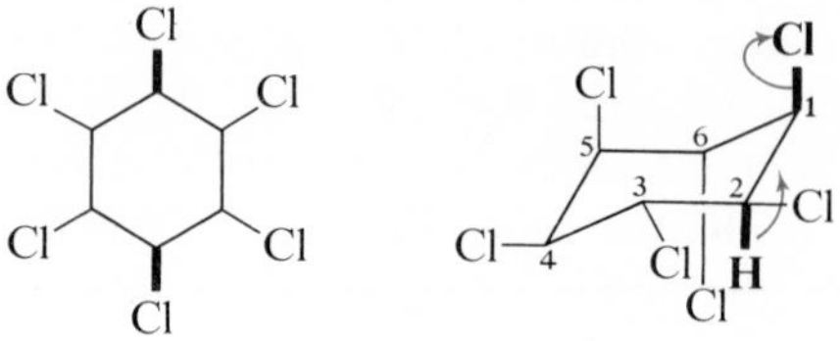

γ-Benzenehexachloride

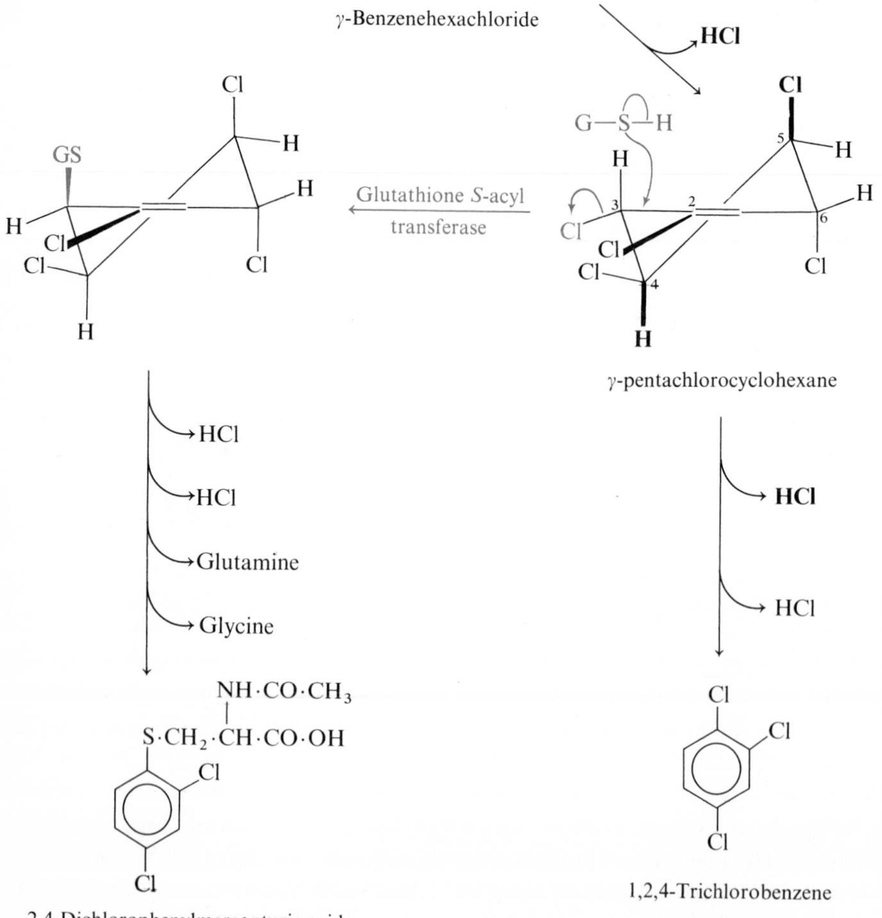

γ-pentachlorocyclohexane

2,4-Dichlorophenylmercapturic acid

1,2,4-Trichlorobenzene

for dehydrochlorination of benzenehexachloride isomers (Cristol, Hause and Meek, 1951) exactly parallel chronic toxicity (Parke, 1968; Table 24).

Table 24. Rates of Dehydrochlorination and Chronic Toxicities of Benzenehexachloride Isomers

Isomer	$k^{30°}$ (l.s^{-1}mol^{-1})	LD_{50}(g/kg)
α	0.5	1.5
β	2.11×10^{-5}	>2.0
γ	0.151	0.23
δ	0.182	0.75

The dehydrochlorination is an E2 elimination for which the stereochemical requirements are that the two leaving groups should be **trans** and **antiparallel**, i.e. with all four centres in the same plane. The *all-trans* β-isomer is unable to fulfil this requirement since all six of its C—Cl bonds are equatorial. In contrast, three of the C—Cl bonds in the γ-isomer are axial, and in consequence dehydrohalogenation is favoured. The E2 elimination forms the basis of the official method of assaying *Gamma Benzene Hexachloride* (**1**, 9). (See opposite.)

Conjugation with glutathione, prior to the formation of mercapturic acid conjugates, which is a feature of insect metabolism, occurs by nucleophilic attack at the relatively unhindered 3-position of γ-pentachlorocyclohexene.

1,4-Disubstituted Six-membered Saturated Ring Compounds

1,4-Disubstituted cyclohexanes are known in two geometrically isomeric forms. In the *cis*-isomer, in which the substituents are on the same side of the molecule, one is axial and the other equatorial. Two conformations A and B are possible, and the preferred conformation will be that in which 1,3-non-bonded interactions are least (**1**, 3), i.e. the one in which the larger of the two substituents is equatorial.

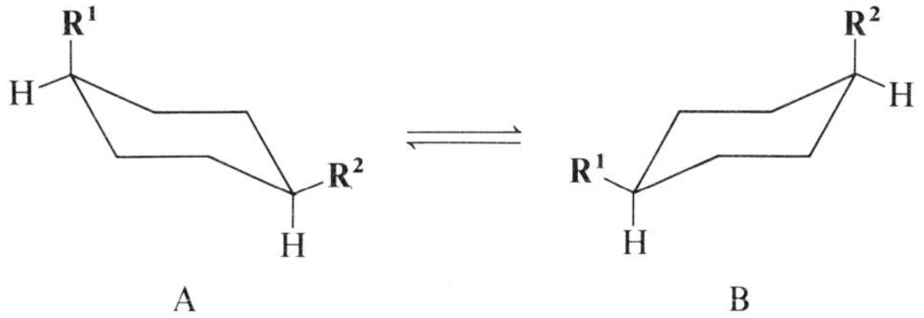

R¹ H R² H ⇌ R² H R¹ H

A B

cis

Of the two possible conformations for the *trans*-isomer, in which the substituents are on opposite sides of the molecule, conformation A with both substituents equatorial will be preferred, since 1,3-non-bonded interactions will again be appreciable in conformation B.

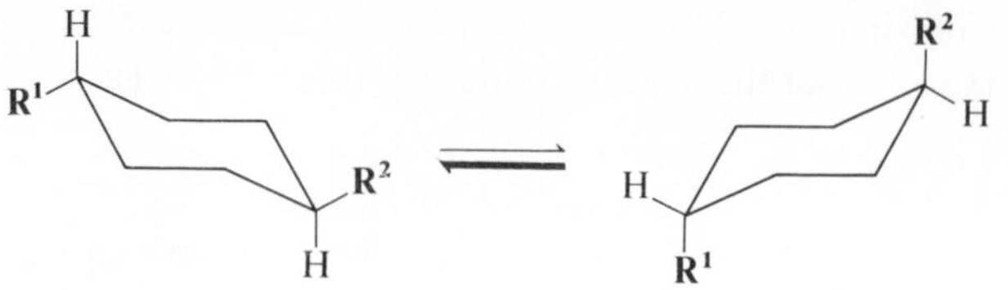

The *cis*-isomer, with one substituent axial irrespective of the conformation adopted, will experience a greater degree of non-bonded interaction than the *trans*-isomer with the substituents equatorial. As a result, the latter will be the thermodynamically more stable isomer.

Both *cis*- and *trans*-isomers are optically inactive, since both molecules have a plane of symmetry passing vertically through the 1 and 4-positions (in the plane of the paper as depicted). A similar situation arises in six-membered 1,4-disubstituted heterocyclics, such as *Chlorcyclizine*, *Haloperidol* and *Pethidine* (Meperidine), all of which have a plane of symmetry and are optically inactive. In general, conformation is dictated by substituent size, but where other factors such as 1,3-ring bridging intervene, as, for example, in *Atropine* and pseudoatropine, stable conformations with large axial substituents are possible. Pseudoatropine, in contrast to *Atropine*, has no mydriatic properties, but has about half the potency of *Atropine* in blocking the actions of acetylcholine on blood pressure (Liebermann and Limpach, 1892).

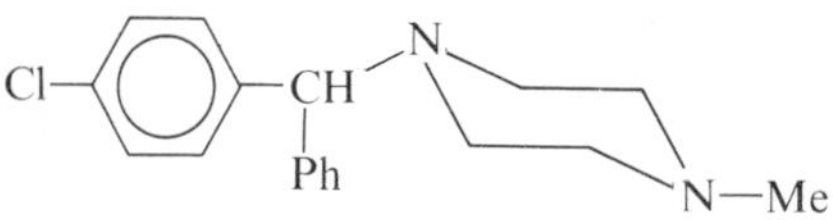

Chlorcyclizine

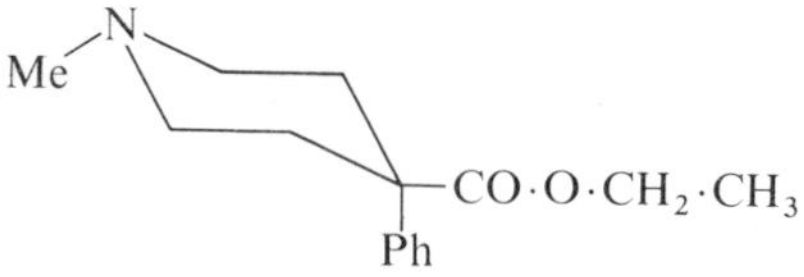

Pethidine

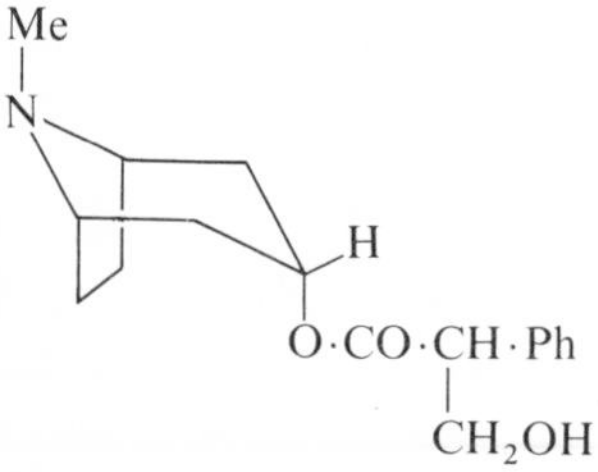

Atropine

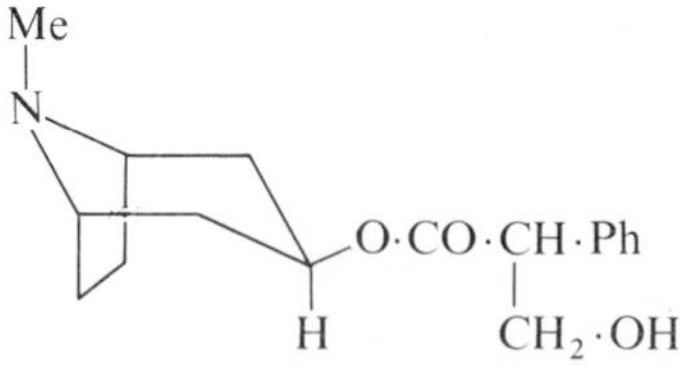

Pseudoatropine

Considerable controversy has raged over the relative space-filling capacity of the lone pair electrons and the N—H proton in piperidine. In the absence of hydration, however, the order of substituent size appears to be lone pair $< H < CH_3$. Nitrogen inversion also occurs in piperidine, so that reactions involving the lone pair, e.g. quaternisation, may give rise to products with both axially- and equatorially-orientated entering substituents.

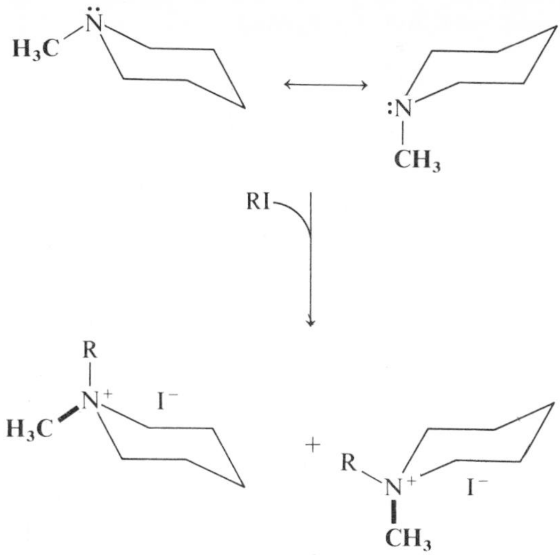

The equatorial orientation of the *N*-substituent is favoured in the tertiary bases, so that other things being equal, the entering group usually attacks axially. Whether or not the axial isomer predominates in the reaction product depends, however, on the relative stability of the isomers under the conditions of the reaction, i.e. whether or not equilibration occurs.

Considerations such as these determine the conformation and relative orientation of the two piperidinium substituents relative to the steroid nucleus in the bis-quaternary steroid neuromuscular blocking agent, *Pancuronium Bromide*.

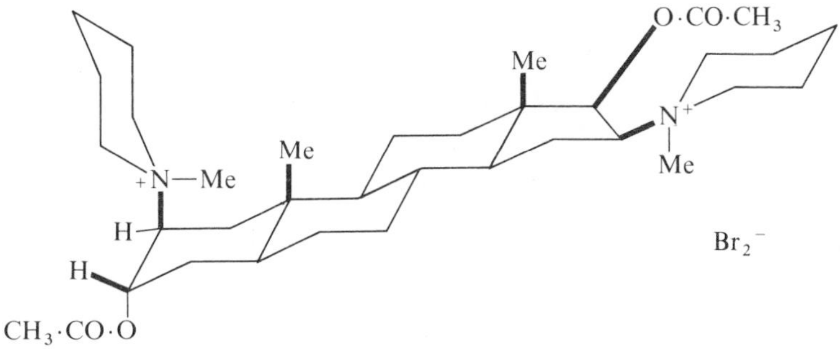

Pancuronium Bromide

Contrary to expectation, however, the *N*-methyl group in the tropane alkaloids has been established by quaternisation and other studies as axial to the six-membered ring. This accords with the fact that in this conformation, the N—CH_3 group is less hindered by 1,3-non-bonded interactions from the piperidine than from the pyrrolidine ring.

1,4-Bridged Six-membered Ring Compounds

1,4-Bridged six-membered ring compounds, such as bicyclo[2,2,2]octanes, bicyclo[2,2,1]pentanes and quinuclidines, have the six-membered ring constrained in a rigid boat conformation. This results in eclipsing of the C_2 and C_3, and of the C_5 and C_6 substituents with accompanying steric interactions.

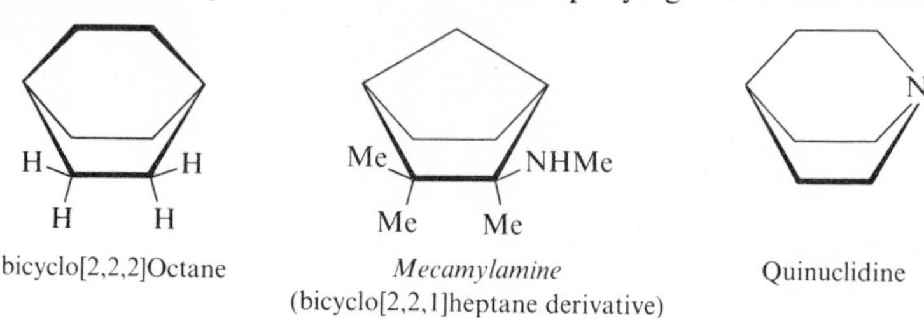

bicyclo[2,2,2]Octane

Mecamylamine (bicyclo[2,2,1]heptane derivative)

Quinuclidine

Structures of this type have been used extensively to prepare rigid conformational analogues of acetylcholine (Solter, 1965; Khokhar, 1971; Reid, 1972; Belleau and Pauling, 1970), such as (*R*)-(−)-3-acetoxyquinuclidine methiodide (p. 141).

1,2-Disubstituted Six-membered Saturated Ring Compounds

The conformation and relative stability of 1,2-disubstituted cyclohexanes is very similar to that which obtains for the 1,4-disubstituted compounds. Exceptionally, compounds are known in which the *trans*-isomer is stabilised in the diaxial conformation B as a result of electrostatic repulsion between substituents. Thus, dipole moment measurements show that *trans*-1,2-dibromocyclohexane exists at least in part in the diaxial conformation (Bender, Flowers and Goering, 1955). The bromo substituent in 2-bromocyclohexanone is also axial. Similarly, dissociation constant measurements show that cyclohexane-*trans*-1,2-dicarboxylic acid also partly adopts the diaxial conformation (Eliel, 1962). In general, however, the *trans*-isomers are stabilised in the di-equatorial conformation A, and the *trans*-isomers are more stable than the *cis*.

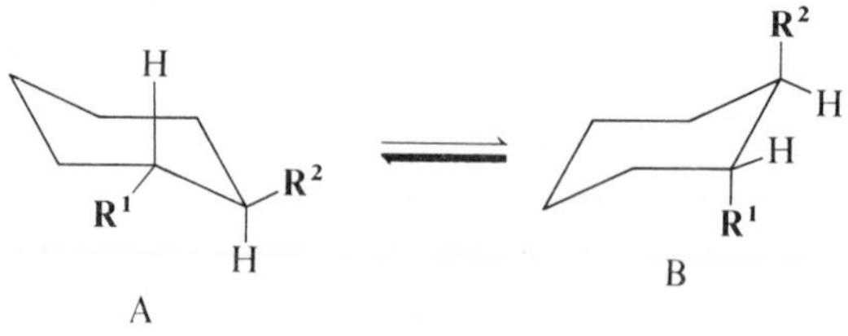

A ⇌ B

trans

The *trans*-isomers have neither plane nor centre of symmetry, but have reflection symmetry in that they are capable of existence in two non-superposable mirror image forms. *trans*-1,2-Disubstituted cyclohexanes are, therefore, capable of being resolved into optical isomers irrespective of whether the two substituents are the same or different.

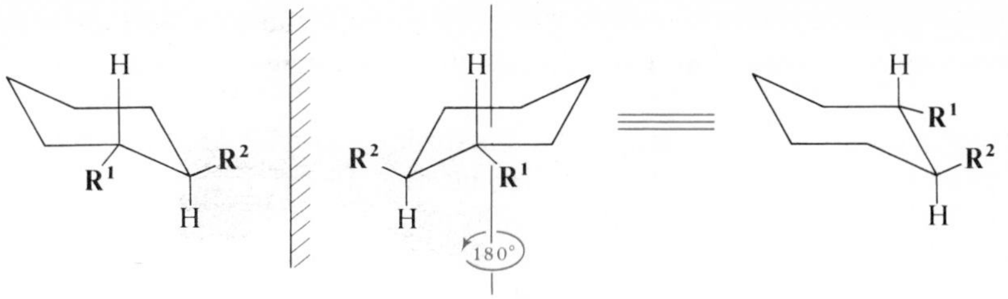

cis

Mirror image forms of *trans*-1,2-disubstituted cyclohexanes
(non-superposable)

In *cis*-1,2-disubstituted cyclohexanes in which both substituents are the same, the two possible interconvertible conformations A and B are in fact non-superposable, mirror image forms. The two forms, however, are energetically equivalent, and the rate of interconversion between the two is, therefore, such that they are incapable of separate existence, and in practice *cis*-compounds of this type are not resolvable. If, however, the two substituents are not identical, the preferred conformation is capable of resolution into two non-superposable mirror image forms.

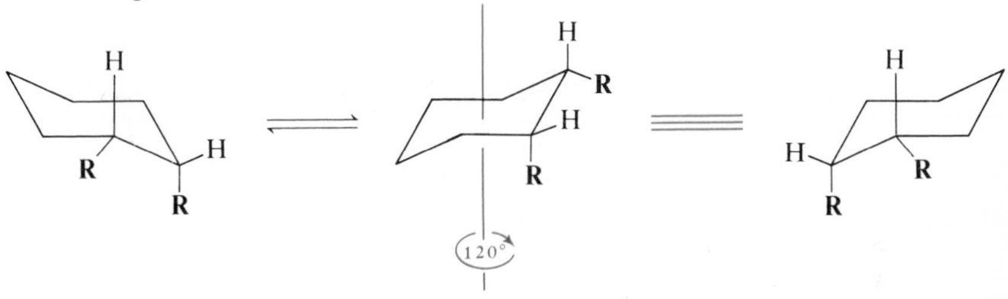

A *cis* B Mirror image of A

Configurations of the 3-methyl reversed esters of *Pethidine* (Meperidine), alphaprodine and betaprodine have been established unequivocally by X-ray crystallography as *trans* (3Me/4Ph) and *cis* (3Me/4Ph) respectively (Ahmed, Barnes and Masironi, 1963) with the two isomers in the following conformations in the solid state.

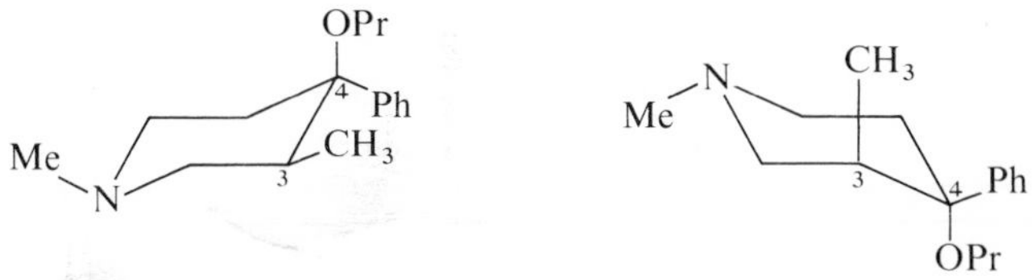

Analgesic potency (Morphine = 100)	Alphaprodine	Betaprodine
	(±) 97	(±) 550
		(−) 790
		(+) 350

(Randall and Lehmann, 1948, 1950)

These conformations do not, however, explain the differences in the rates of hydrolysis of alphaprodine and betaprodine (Beckett and Walker, 1955). The latter is hydrolysed more rapidly than alphaprodine, an observation which is more in keeping with the alternative betaprodine conformation in which the propionoxy substituent is equatorial and the phenyl group axial. Such a conformation, if present in the analgesically more potent betaprodine in solution, might more reasonably account for its activity because of its resemblance to the rigid conformation with the axial phenyl group which is present not only in *Morphine* itself, but also in the morphanan, *Levorphanol*, and the benzomorphan, *Phenazocine*.

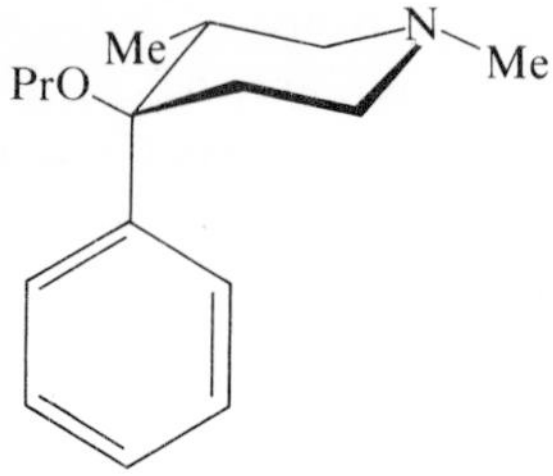

Betaprodine (alternative conformation)

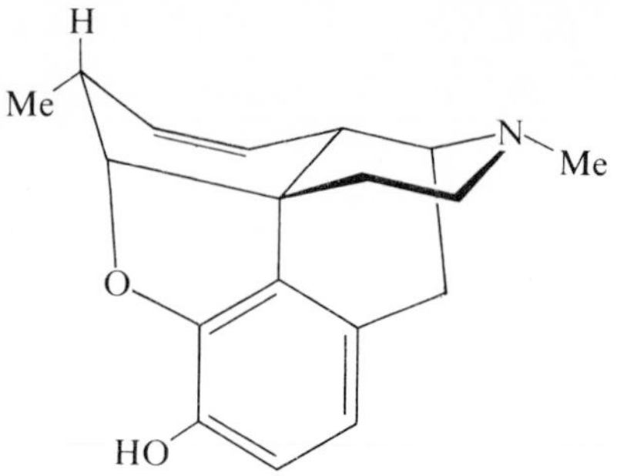

D-(−)-*Morphine*

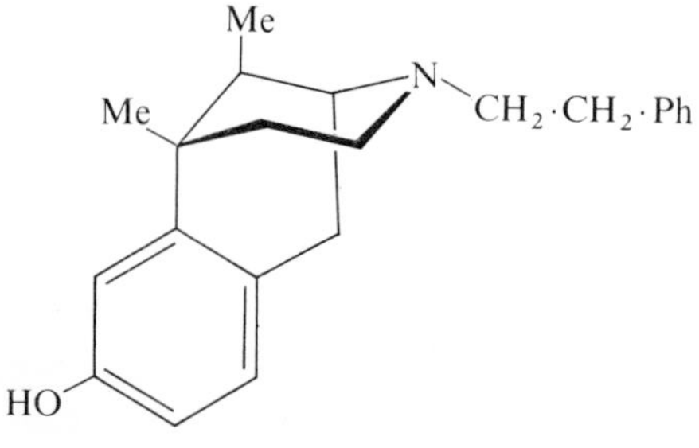

Phenazocine

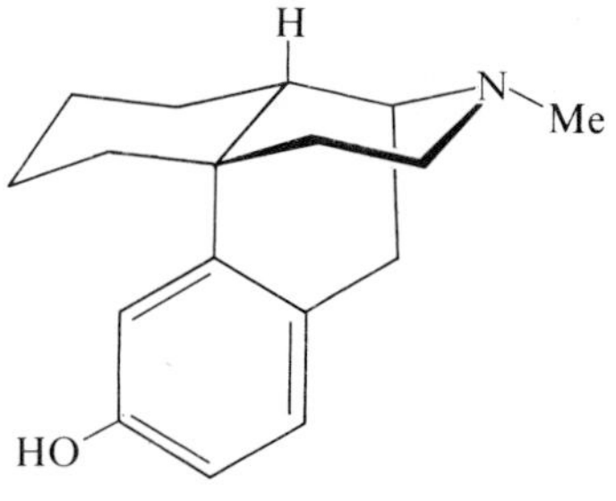

D-(−)-*Levorphanol*

According to Casy (1970), however, there is some evidence to suggest that *Pethidine* and its derivatives adopt differing modes of association with the analgesic receptor from the morphine, morphinan and benzomorphan group of analgesics. Thus, whilst some substituents, such as phenethyl, enhance potency in both groups, others have opposing effects. For example, *N*-phenacylnormorphinan is 6.5 times more potent than *Levorphanol*, whilst the corresponding *N*-phenacylnorpethidine has only one-tenth the potency of *Pethidine* (Janssen and Eddy, 1969). Similarly, *N*-allyl nor morphine and *N*-allyl-3-hydroxymorphinan are analgesic antagonists, whilst *N*-allylnorpethidine is an active analgesic and lacks antagonistic activity (Casy, Simmonds and Staniforth, 1968).

Some of these anomalies may well be resolved by further studies on the **enkephalins**, endogenous peptides which are able to function as pain suppressors. Two pentapeptides, differing only in the *C*-terminal amino acid, have been identified, H-Tyr-Gly-Gly-Phe-Met-OH and H-Tyr-Gly-Gly-Phe-Leu-OH, and shown to be present in brain tissue in a ratio of approximately 3:1 (Hughes, Smith, Kosterlitz, Fothergill, Morgan and Morris, 1975). Interestingly, the amino acid sequence present in Met-enkephalin, has also been identified as a fragment (amino acid residues 61–65) of β-lipoprotein, which is also apparently the parent in residues 41 to 58 of the natural melanocyte stimulating hormone (MSH).

Both enkephalins have been shown to exhibit analgesic activity when injected into mouse brain (Büscher, Hill, Römer, Cardinaux, Closse, Hauser and Pless, 1976), but morphine was some 75 and 240 times more potent than Met-enkephalin and Leu-enkephalin respectively. The analgesic effect of the enkephalins was much more rapid in onset (2 min) than that of morphine (peak activity in 15 min), but of much shorter duration (5 min), and less firmly bound to brain tissue than morphine, due to chain cleavage by enzymic hydrolysis. The synthetic tripeptide H-Tyr-Gly-Gly-OH and the analogous desamido pentapeptide (desamido-Tyr)-Gly-Gly-Phe-Met-OH were both devoid of analgesic activity. This confirms the importance of the intact peptide in orientating the terminal tyrosyl group towards the analgesic receptor in a similar steric configuration to that which is extant in morphine and morphine-like analgesics (Bradbury, Smyth and Snell, 1976). The enkephalins resemble morphine and other opiates in that they suppress the rate of electrical discharge by the cells (Bradley, Briggs, Gayton and Lambert, 1976; Gent and Wolstencroft, 1976).

The carrier function of larger peptides is demonstrated in fragment C from β-lipoprotein (Bradbury *et al.*, 1976) consisting of the terminal 30 amino acids which incorporate met-enkephalin. This is not only some fifty times more potent as an analgesic than morphine in the rat (Graf, Szekely, Ronai, Dunai-Kovaks and Bajusz, 1976) but also, in contrast to the enkephalins, is active when injected intravenously.

Fused Saturated Ring Systems

The 1,2-fusion of two cyclohexane rings in decalin gives rise to the possibility of two all-chair conformations in which the fusion of the ring is either *cis*- or *trans*.

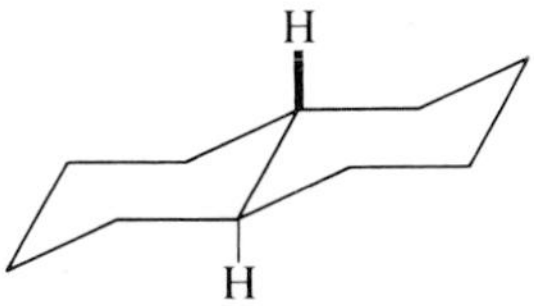

H

H

trans-Decalin

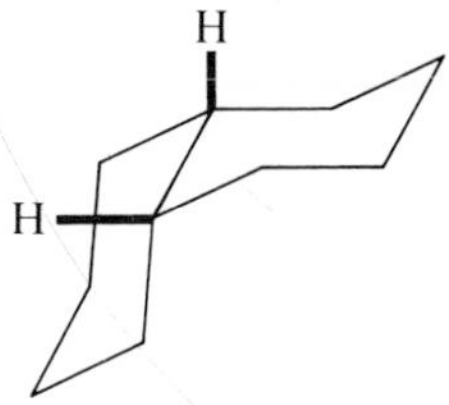

H

H

cis-Decalin

This type of ring fusion is found in the steroids (**1**, 22), and accounts for the general rigidity of this type of hydrocarbon skeleton. Only the bi-equatorial conformation is possible at the ring junction in *trans*-decalin, but the *cis*-compound with one substituent axial and one equatorial at the ring junction gives rise to the possibility of two alternative and equivalent conformations. Once, however, either ring is further substituted as in the steroid nucleus, the alternative *cis*-conformations are no longer equivalent and one will be preferred on conformational grounds to the other.

1,3-Disubstituted Cyclohexanes

1,3-Disubstitution of cyclohexanes is interesting in that it presents a situation, which in a number of respects is the opposite of that encountered in 1,2-disubstitution. Two conformations are theoretically possible for the *cis*-compound, of which the di-axial form B is unstable due to 1,3-non-bonded interactions. The bi-equatorial form A is, therefore, the preferred conformation, for most such compounds. However, form B is favoured in 1,3-dihydroxycycyclohexanes due to hydrogen bonding.

A B

cis-1,3-disubstituted cyclohexanes

With the substituents, R, identical, the molecule has a plane of symmetry, and hence is not resolvable into optical isomers, but with non-identical substituents, optical isomerism is possible. The diaxial *cis*-arrangement is, of course, stabilised in 1,3-bridged compounds, such as *Atropine* (p. 148), *Phenazocine* (p. 152), and also in the antihypertensive, *Pempidine* (Leonard and Hauck, 1957; Spinks and Young, 1958; Lee, Wragg, Corne, Edge and Reading, 1958). Examination of the three-dimensional structure of the latter shows it to be not nearly so sterically-hindered as might appear from its structural formula, due to the chair ring and axial disposition of two of the methyl substituents. The combined electron donor effects of the four α-methyl groups ensure that the molecule is a very strong base (pK_a 11.25), an important criterion for ganglion blocking activity. The corresponding 2,2,6-trimethylpiperidine has only about one-eighth the potency of *Pempidine*, although the corresponding 2,2,6,6-tetramethylpiperidine has about 75% of its activity (Spinks, Young, Farrington and Dunlop, 1958).

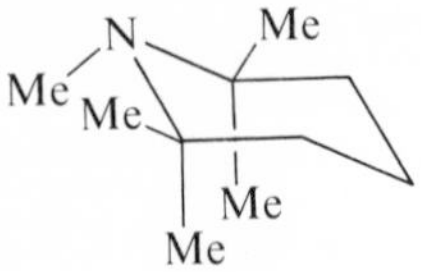

Pempidine 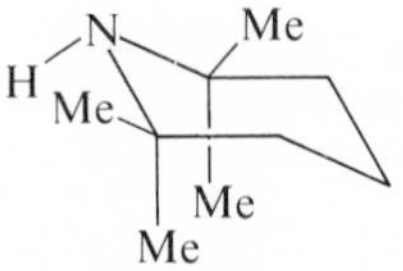 2,2,6,6-Tetramethylpiperidine 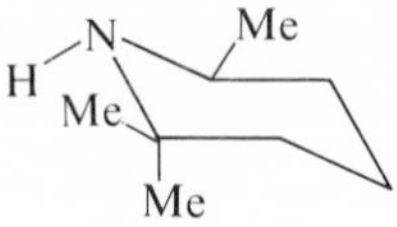 2,2,6-Trimethylpiperidine

The *trans*-1,3-conformation has a two-fold axis of symmetry and the two apparently alternative conformations are in fact seen to be identical by rotation through 180° about this axis. The compounds, provided the substituents are identical, are non-resolvable into optical isomers; if the substituents are different, then optical isomerism is possible. In contrast to the 1,2 and 1,4-di-substituted compounds, the 1,3-*cis*-compound with both substituents equatorial is more stable than the 1,3-*trans*-compound with one equatorial and one axial substituent.

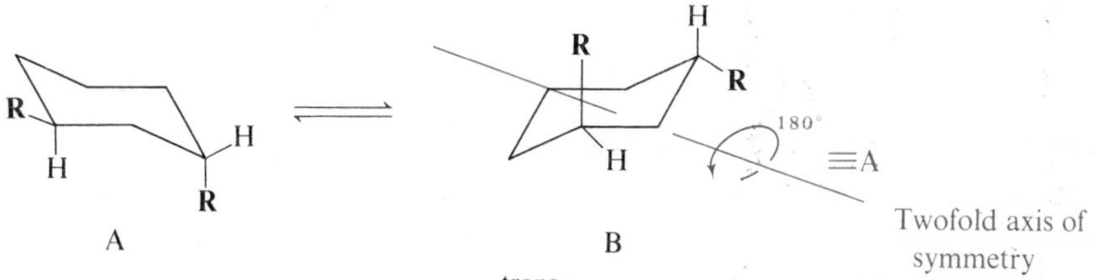

A B ≡A

trans

Twofold axis of symmetry

Conformation and Isomerism in Unsaturated Six-membered Rings

Cyclohexene

Insertion of a double bond, as in cyclohexene, causes deformation of the chair-ring, since carbon atoms 1, 2, 3 and 6 are necessarily planar owing to the geometry of the ethylenic bond (Raphael and Stenlake, 1953). As a result, a half-chair or skew conformation is formed in which C_5 and C_6 are out of plane, one above and one below the plane of the ring.

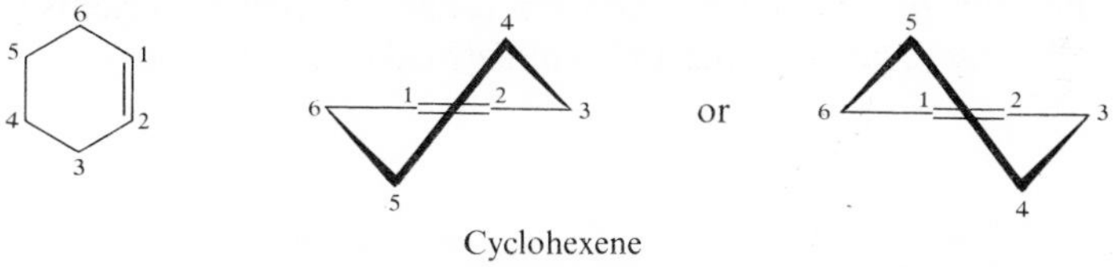

or

Cyclohexene

Substituents at C_3, C_4, C_5 and C_6 are neither truly axial nor truly equatorial, and the terms **pseudo-axial** and **pseudo-equatorial** are usually used.

Tetralin and Tetrahydropapavarine

Similar half-chair conformations exist in tetralin and tetrahydroisoquinoline, as a result of fusion of the saturated heterocyclic and benzene rings, and are specially relevant to consideration of the full three-dimensional structure of the neuromuscular blocking agent, *Tubocurarine Chloride* (Marshall, Murray, Smail and Stenlake, 1967; Stenlake, 1968a) and the ergoline nucleus of the ergot alkaloids (p. 145).

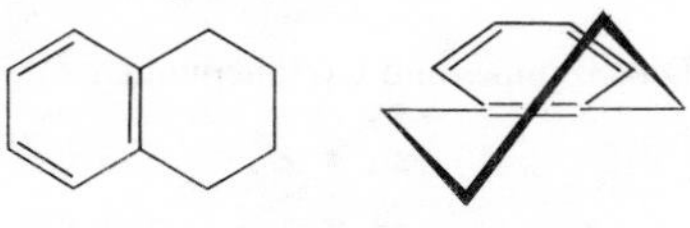

Tetralin

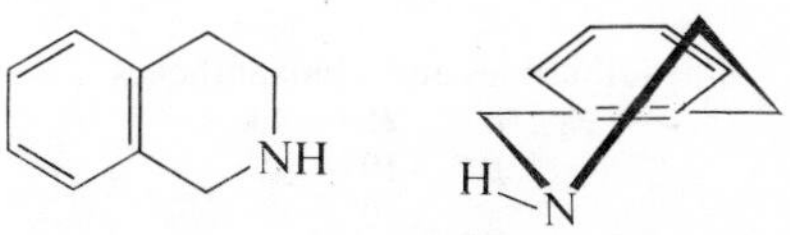

Tetrahydropapaverine

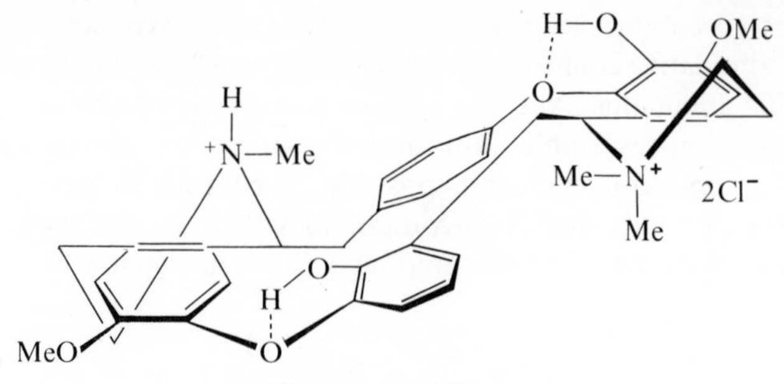

Tubocurarine Chloride

Tricyclic Tranquillisers and Antidepressants

The pharmacological action shown by members of this group of compounds, whether tranquilliser (neuroleptic) or antidepressant (thymoleptic) is largely determined by three factors (Table 25)

(a) the chemical structure of ring B,
(b) the overall conformation of the ring system,
(c) the side-chain conformation.

According to Wilhelm and Kuhn (1970), molecular shape in these molecules can be defined by three angles:

(a) α, the angle of flexure between the planes of ring A and C,
(b) β, the angle of annellation, which is the angle formed by the two lines of attachment (annellation) between rings A and B, and rings B and C respectively,
(c) γ, the angle of torsion between rings A and C, which represents the extent to which the ring system adopts an asymmetrical conformation.

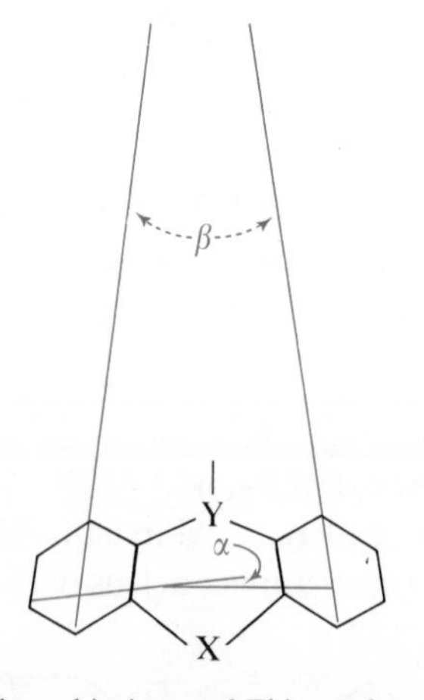

Phenothiazines and Thioxanthenes
α 25°
β 10°
γ 0°

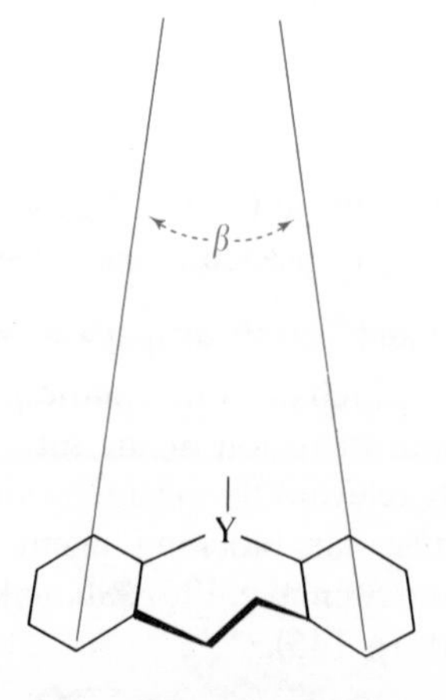

Dibenzapines and Cycloheptadienes
55°
40°
20°

Table 25. Classification of Tricyclic Tranquillisers and Antidepressants

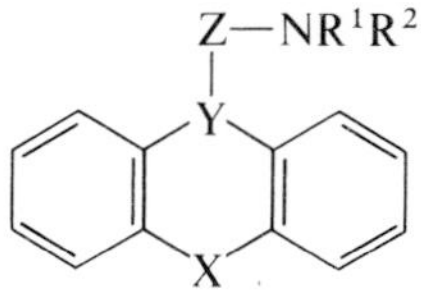

Z—NR¹R²

Class of Compound	X	Y	Z (optimum)	Action
Phenothiazine	S	N	$—CH_2·CHMe·CH_2—$	Tranquilliser
Thiaxanthene	S	C	$—(CH_2)_3—$	Tranquilliser
Dibenzapine	$—CH_2·CH_2$	N	$—(CH_2)_3—$	Antidepressant
Dibenzocycloheptadiene	$—CH_2·CH_2$	C	$—(CH_2)_3—$	Antidepressant

Essentially compounds, such as the phenothiazines and thiaxanthenes with relatively flat molecules ($\alpha = 0.25°$), small angles of annellation ($\beta = 0–10°$) and zero torsion ($\gamma = 0°$) between rings A and C are tranquillisers. The more angled structures of the dibenzapines, *Imipramine* and *Desipramine*, and the cycloheptadienes, *Amitriptyline* and *Nortriptyline*, are antidepressants. The molecules are essentially butterfly-shaped with a large angle of flexure ($\alpha = 55°$) with large angles of annellation ($\beta = 40°$), and with appreciable torsion angles ($\gamma = 20°$) between rings A and C (Bente, Hippius, Poeldinger and Stach, 1964). The overall molecular shape, and both orientation and conformation of the side-chain are largely determined by the constitution of ring B. As a result, a simple relationship exists between chemical structure and pharmacological action (Table 25). X-ray crystallographic studies of *Chlorprothixene* (Dunitz, Eser and Strickler, 1964) and electron-spin resonance studies of phenothiazines (Fenner, 1970) show that the following side-chain conformation is favoured in tranquillisers.

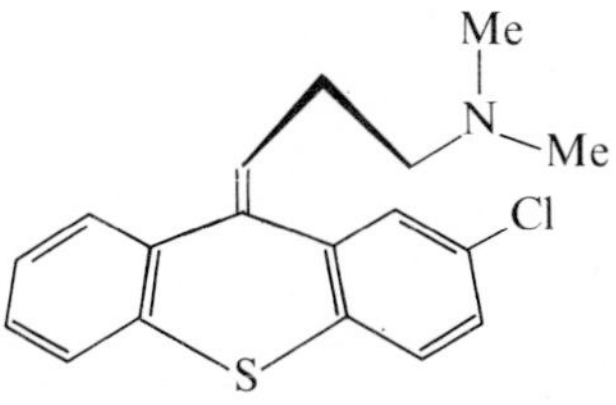

Chlorprothixene

This conclusion is reinforced in studies on a series of dibenzo[*be*]bicyclo [2,2,2]octadienes, which show that compounds such as *Benzoctamine* with side-chains forced to adopt similar conformations, act as tranquillisers (Wilhelm

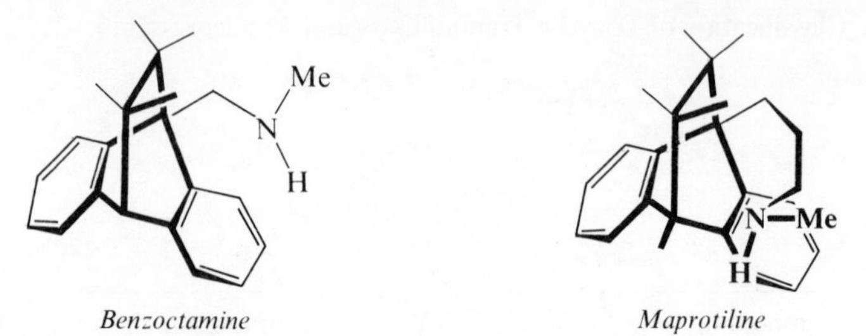

Benzoctamine *Maprotiline*

and Schmidt, 1969). Analogues with longer side-chains capable of adopting an alternative conformation in which the nitrogen lies more nearly over the benzenoid ring, such as *Maprotiline*, on the other hand, are antidepressants.

4 Drug Ingestion, Transport and Excretion

THE BASIS OF DRUG TRANSPORT

Introduction

Drug action depends not merely on the ability to involve a specific pharmacological response, but also on the possession of properties which assist the drug to reach the required site of action. A number of factors are involved, the most fundamental of which relate to the physico-chemical properties of the drug. These determine the ability or otherwise of the drug to diffuse across biological membranes, and whether or not it is intensively bound at depôt sites in fat, on the serum proteins or in other tissues for which it has special affinity. Other important factors are the route and form in which the drug is administered. These may be interdependent, in that drugs which are not absorbed, or which are metabolised to inactive forms in the gastro-intestinal tract, must either be suitably modified chemically, or pharmaceutically by use of an appropriate formulation, or failing this be administered by an alternative route. Thus, *Benzylpenicillin* (Penicillin G), which is unstable under the acidic conditions of the stomach, is generally administered by injection, but the acid-stable *Phenoxymethylpenicillin* (Penicillin V) is effective when administered orally. *Streptomycin* is not appreciably absorbed from the gastro-intestinal tract because of its strongly basic characteristics, and must also be administered by injection. Other compounds, which would not withstand the acidic and enzymic conditions of the stomach, may be formulated in enteric-coated (acid-resistant) capsules which only break down and release the medicament under the more alkaline conditions which prevail in the intestines.

Where the speed of response to medication is important as, for example, in the administration of heart stimulants, antispasmodics, or anaesthetics, either injection of liquid preparations or inhalation of gases and aerosols offers the most appropriate route. But, by whatever route the drug enters, it is entrained in a dynamic system, which in addition to transporting it to the required site, is equally if not more highly geared to its ultimate removal by one or more of the available excretory pathways, a process which is frequently though not always assisted by metabolic transformation (**2**, 5) to more readily excretable compounds. The more important pathways for drug absorption, distribution and excretion are shown in Fig. 25.

The Nature of Cell Membranes

Animal cell membranes are specialised tissues forming a boundary layer through which metabolites and other substances pass in and out of the cell. They vary in

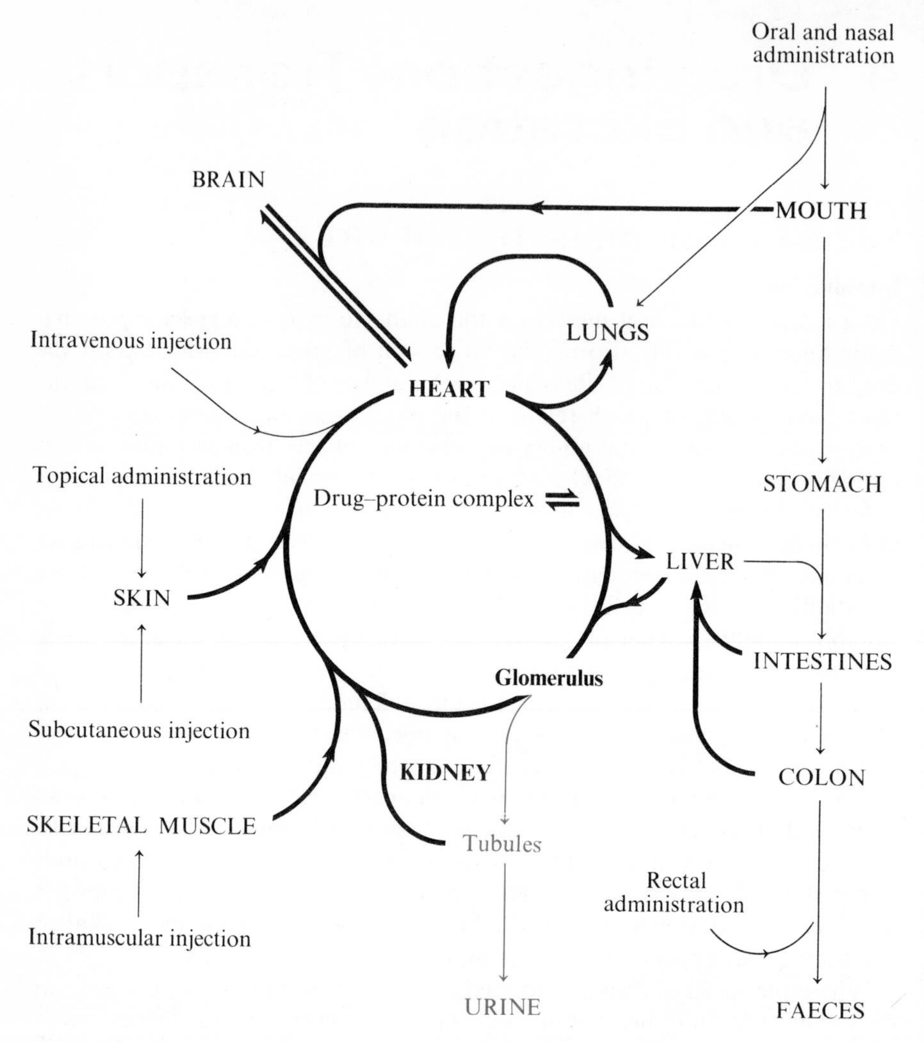

Fig. 25 Diagrammatic representation of drug absorption, distribution and excretion pathways

thickness, but are usually of the order of 100 Å. They appear to exist as layered colloidal lipoprotein gels enclosing a bimolecular sandwich of phospholipid molecules, principally lecithins and cephalines. These are arranged in such a way that their long paraffinic chains face inwards from either side to form a central core, which is interspersed with molecules of cholesterol (**1**, 22). The outer surfaces of the lipid layer are composed of the polar water-soluble heads of the phospholipids linked, possibly by polar attraction, to the outer protein films (Fig. 26).

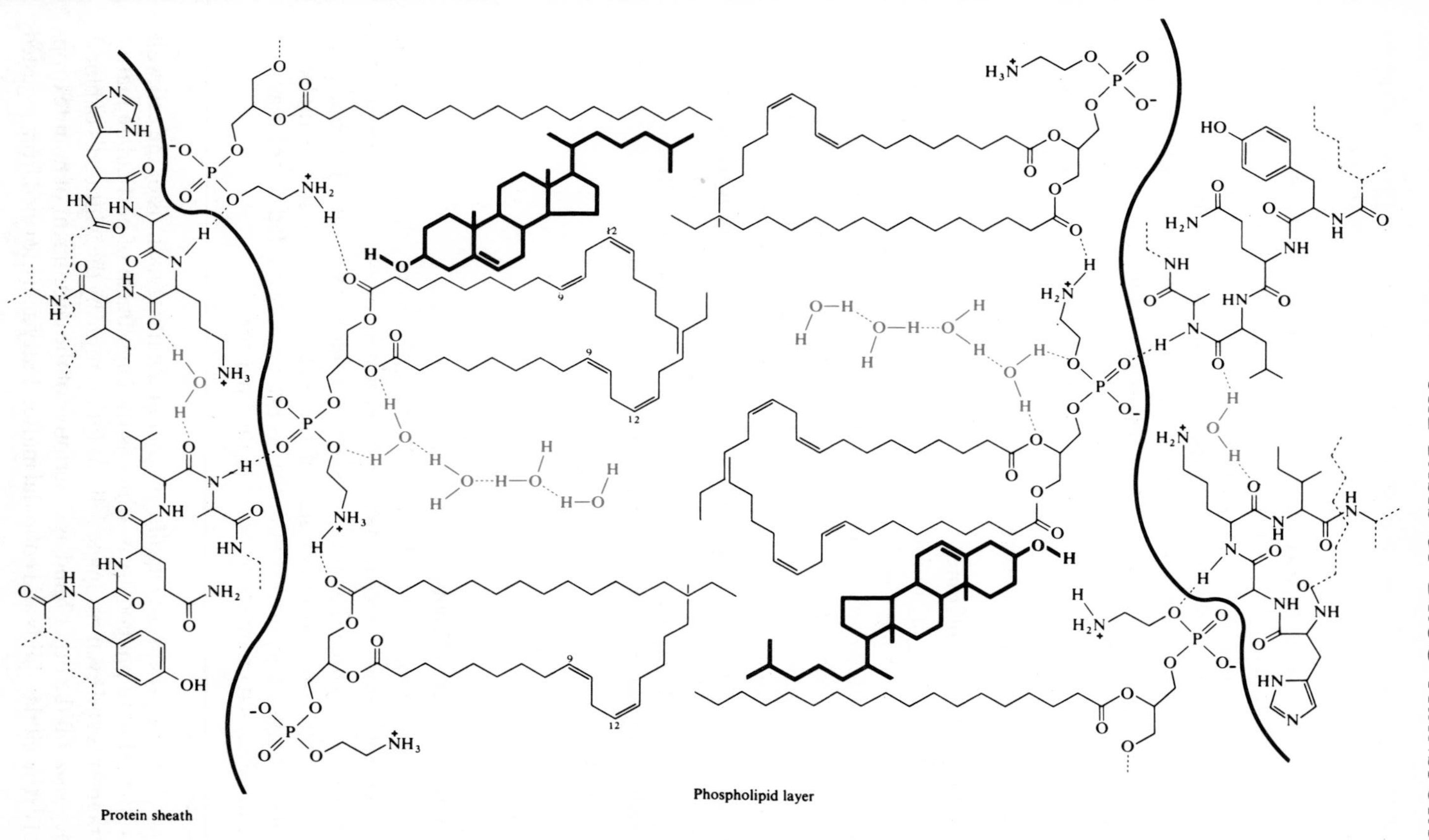

Fig. 26 Representation of cellular membrane structure

The hydrated globular protein structures provide an affinity for aqueous solutes, which in the case of small molecules and ions are then able to perfuse into and through the lipid core. The mechanism of ion transport is not clear, but artificial bimolecular films between salt solutions have been shown to become permeable to cations in the presence of certain antibiotics, such as valinomycin and gramicidin A (Katz, 1973). These cyclic polypeptides are able to complex with selected cations (**1**, 19) and hold them in the central cavity. It is possible that extensions of the surrounding protein film may actually penetrate the bimolecular phospholipid core forming conducting channels through which the ions may pass through the membrane.

The hydrophobic regions of the globular protein, similarly, provide an entré for lipid molecules into the membrane for subsequent transmission through the essentially phospholipid core.

Penetration of Cell Membranes

Penetration of a drug in suitable concentration to its locus of action depends upon its ability, or otherwise, to traverse a number of tissue-cell barriers. For example, it is essential that a drug administered orally, but required for its effect outside the alimentary tract, should be capable of passing through the stomach or gut wall into the plasma. Conversely, substances required to act within the alimentary canal should possess physico-chemical properties, which provided the compounds are administered orally, will limit diffusion.

This ability, or otherwise, to traverse cell membranes is closely linked with the chemical and physical properties of the compound. Penetration of the membranes can occur in one of the following ways:

(a) filtration,
(b) pinocytosis,
(c) passive lipid diffusion,
(d) specialised transport mechanisms.

Filtration (aqueous diffusion)

Filtration is the term used to describe the passage across cell barriers of small molecules, such as water, hydrated ions and water-soluble molecules of small dimension, such as urea. The ready passage of water is responsible for equalising the osmotic pressure on either side of the membrane. The size of molecule capable of being transported in this way is presumably limited by the diameter of the undeformed channels or pores in the lipid core.

Pinocytosis

Pinocytosis is a process of engulfment by minute invaginations of the cell wall. It is rather akin to phagocytosis, which is the engulfment of particulate matter, cell fragments and even complete cells, as for example by the white cells of blood. Pinocytosis, on the other hand, is descriptive of the engulfment of minute droplets of extracellular fluids and colloidal matter. Fatty acids, formed from ingested fats by hydrolysis, are absorbed by pinocytosis in the small intestine. Bile salts are essential for this process on account of their surface activity, which reduces the size of the fat droplets, and stabilizes the emulsified fat droplets. The importance of pinocytosis in drug absorption, however, is not established.

Passive Diffusion

Most drugs cross cell membranes and permeate tissues by a process of simple diffusion. This is essentially the movement of a substance from an area where it is present at high concentration to one of low concentration. Provided the system on either side of the membrane is isolated, an ideal state which does not normally obtain in complex biological systems, diffusion will continue until the concentration of drug on either side of the membrane is equalised. The rate at which this equilibrium is attained depends on a number of factors, which may be expressed in terms of Fick's Law:

$$\text{Rate of diffusion} = \frac{k \cdot A(c_1 - c_2)}{d}$$

where A = area of diffusing surface
d = thickness of diffusing membrane
c_1 = concentration of diffusing compound outside the membrane
c_2 = concentration of diffusing compound inside the membrane
k = diffusion constant

The diffusion constant (k) depends on a number of factors, including the membrane characteristics, and the molecular weight, steric configuration, lipid solubility, and, if ionisable, the dissociation constant of the drug. For any given drug, the rate of diffusion across a particular membrane, therefore, depends solely on the concentration gradient across the membrane. This varies exponentially with time in accordance with the standard **first order rate equation**,

$$\ln(a) - \ln(a - x) = kt$$

where a is the initial concentration, and $(a - x)$ the concentration remaining (undiffused) after time (t). Hence,

$$2.303 \log(c_t - c_\infty) = 2.303 \log(c_0 - c_\infty) - kt$$

According to this equation, a semilog plot of $2.303 \log(c_t - c_\infty)$ against time (t) will be linear, with a slope of $-k$ and an intercept equal to $2.303 \log(c_0 - c_\infty)$, as in Fig. 27.

Solubility in water is the first requirement for transport of a drug to the membrane surface, but diffusion across the membrane with its lipid core is dependent on its lipid solubility. The rate of diffusion of a neutral molecule will, therefore, be dependent not only on the concentration gradient $(c_1 - c_2)$ on either side of the membrane, but will also be related to the lipid/water partition coefficient

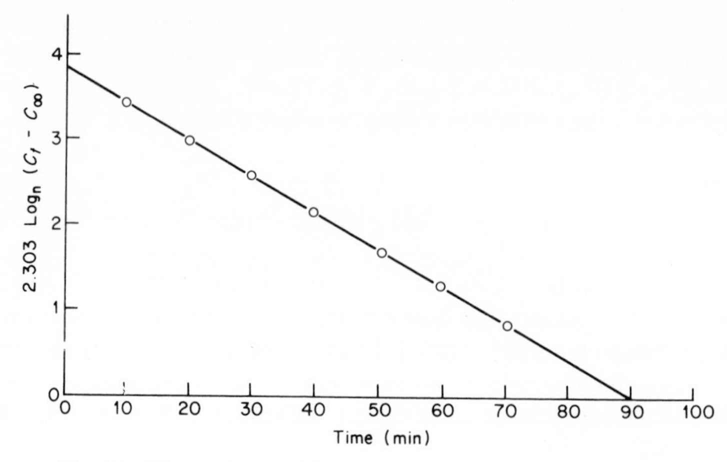

Fig. 27 First order semilog plot of concentration against time

of the drug (Fig. 28). In general, the higher the partition coefficient, the greater the affinity for the lipid membrane, and the higher the rate of diffusion into the membrane.

It is not possible to determine lipid/water partition coefficients which are wholly appropriate to natural biological membranes. Although partition coefficients of drugs between octanol and water are becoming increasingly available (Leo, Hansch and Elkins, 1971, and other studies reported in **2**, 1),

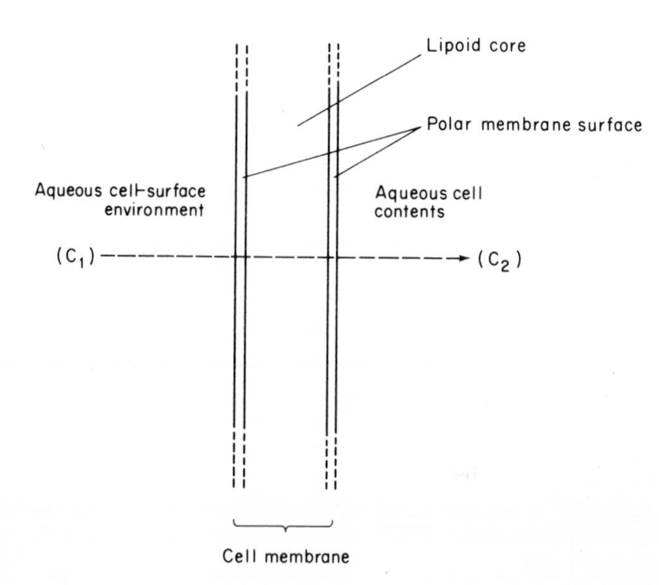

Fig. 28 Diagrammatic representation of passive diffusion across a cell membrane

data are still not available for many modern drugs. In the absence of such information, some idea of the likely distribution and hence rate of absorption can be gained from consideration of the solubilities of the drug in water (or aqueous buffer) and an immiscible organic solvent, such as chloroform. Such deductions will, however, be only approximate in anything other than dilute solutions and would be invalidated if either molecular association or dissociation occurs in one of the solvents and not the other. Solubilities of some non-ionic drugs are given in Table 26 (below). In general, high lipid solubility is associated with structures embodying strongly electronegative substituents, which generate large bond dipoles (C—O; C—S; C—Halogen), though this effect is largely

Table 26. Approximate Solubilities of some Non-ionic Drug Substances (at 20°)

Compound	Solubility[1]		Compound	Solubility[1]	
	Water	*Chloroform*		*Water*	*Chloroform*
Acetomenaphthone	In[2]	S[3]	Fludrocortisone Acetate	In	50
Betamethasone	In	1100	Fluocinolone Acetonide	In	15
Betamethasone Valerate	In	2	Fluoxymesterone	In	200
Caffeine	60	5.5	Hydrocortisone Acetate	In	150
Calciferol	In	FS[4]	Mephenesin	100	12
Carbimazole	500	3	Mephenesin Carbamate	In	50
Carbromal	In	2	Methandienone	In	S
Chloral Hydrate	FS	3	Methylprednisolone	In	530
Chlorbutol	130	FS	Methylprylone	14	FS
Chloramphenicol	400	In	Methyltestosterone	In	S
Chlorpropamide	In	9	Oestradiol Benzoate	In	S
Chlorphenesin	200	S	Oestrone	In	110
Chlortrianisene	In	1.5	Phenacetin	1310	14
Colchicine	S	S	Prednisolone Pivalate	In	16
Cortisone Acetate	5000	4	Prednisone	In	200
Dexamethasone	In	165	Prednisone Acetate	In	6
Dienoestrol	In	S	Progesterone	In	FS
Digitoxin	In	40	Tolbutamide	In	FS
Digoxin	In	In	Triamicinolone	500	In
Dihydrotacchysterol	In	FS	Triamcinolone Acetonide	In	40
Dimethisterone	In	FS	Urea	1	In
Erythromycin	1000	6			
Ethchlorvinyl	In	FS			
Ethinyloestradiol	In	20			
Ethionamide	In	20			
Ethisterone	In	110			

[1] Expressed as 1 part of solute in x parts of solvent
[2] In = Insoluble or almost insoluble
[3] S = Soluble
[4] FS = Freely Soluble

annulled and water solubility greatly increased in compounds where the electronegative substituent is also attached to hydrogen (C—O—H).

Active Transport Mechanisms

Specialised or active transport (Wilbrandt and Rosenberg, 1961) of nutrients and possibly medicaments is said to occur when compounds are transported across cell membranes against a concentration gradient, or in the case of charged fragments (ions) against a potential gradient. Active transport is, therefore, an energy-demanding process, and energy must be put into the system. Active transport is thought to involve carrier substances, which combine with the substance to be transported, move across the membrane in combination, and then dissociate. It has been suggested that it is facilitated by specific enzymes known as permeases.

In contrast to passive diffusion, the rate of active transport processes is dependent on drug concentration, and hence is first order only at concentrations which are less than those required to saturate the transport mechanism. When functioning at maximum capacity, the rate of active transport is constant, and independent of concentration, following **zero order kinetics** in accordance with the general rate equation in which:

$$x = kt$$

Hence,

$$\mathrm{d}t(c_t - c_\infty) = (c_0 - c_\infty) - kt$$

From this, it is apparent that a plot of the concentration $(c_t - c_\infty)$ against time (t) will be linear, with a slope of $-k$ and an intercept $(c_0 - c_\infty)$.

Active transport mechanisms show a high degree of chemical specificity. Well-established examples include the reabsorption of Na^+, K^+ and glucose from the tubular epithelium of the kidney. Here the cations are reabsorbed in the proximal tubules from the glomerular filtrate, against a potential gradient of about 20 mV. The same potential gradient assists the accompanying reabsorption of Cl^- and PO_4^{3-} ions. Active transport is also involved in the absorption of sugars and amino acids in the gut. Drugs which resemble natural metabolites, such as *Methyldopa*, are absorbed by this mechanism.

Active transport also plays a particularly important part in the uptake of amino acids into brain tissue. Separate transport systems appear to exist for the uptake of neutral and basic amino acids (Richter and Wainer, 1972). Rate studies *in vitro* on brain slices also show distinct differences in the transport of large and small neutral amino acids, basic amino acids and acidic amino acids (Table 27), which suggests that acidic amino acids may also have a separate transport mechanism (Blasberg, 1968). There is also some evidence that separate transport mechanisms exist for ω-amino acids, such as gamma-aminobutyric

Table 27. Transport Rates for Amino Acid Classes in Brain Tissue (Blasberg, 1968)

Class of Amino Acid	V_{max}(μ mol/ml intracellular water/min)
Acidic	4.0–4.5
Small neutral	3.0–4.0
Large neutral	1.0–1.5
Large basic	0.5–1.0

acid (GABA), for imino acids (proline and hydroxyproline) and for amido derivatives (glutamine). All show a high degree of stereoselectivity for the natural L-series amino acids (Guroff and Udenfriend, 1962), and each shows considerable evidence of competition between similar amino acids for the system (Lajtha and Cohen, 1972).

Within each transport class, considerable modification of structure can be made without affecting the transport properties, though in general the rate of transport decreases with increasing molecular complexity. Thus, *Malphalan*, L-4-di(2′-chloroethyl)-aminophenylalanine is thought to be transported into tumour cells by the L-phenylalanine carrier mechanism, though its D-isomer is not so transported and is reputed to be less active.

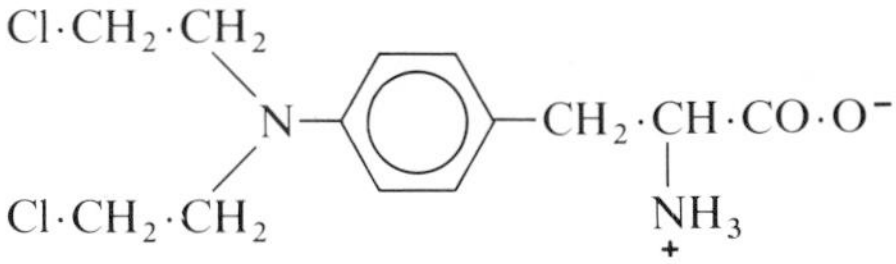

$$(Cl{\cdot}CH_2{\cdot}CH_2)_2N{-}C_6H_4{-}CH_2{\cdot}CH(NH_3^+){\cdot}CO{\cdot}O^-$$

Melphalan

$$C_6H_5{-}CH_2{\cdot}CH(NH_3^+){\cdot}CO{\cdot}O^-$$

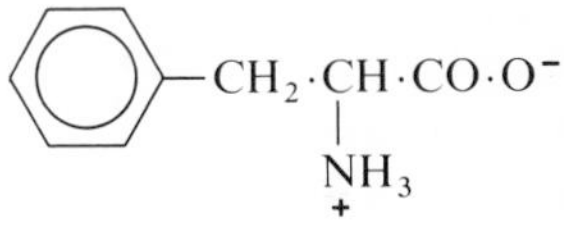

Phenylalanine

Similarly, the transport of *Levodopa* across the blood-brain barrier is inhibited competitively by phenylalanine and tyrosine (Yoshida, Kaniike and Namba, 1963). Its stereoisomer, D-dopa, is not transported to any appreciable extent. The active transport of sugars into red blood cells is also highly stereospecific (**1**, 21).

Other active transport mechanisms also show competitive inhibition between carrier substrates, as for example, in the inhibition of *Benzylpenicillin* excretion in the proximal tubule of the kidney by *Probenecid*. All specialised transport processes can also be saturated and brought to a halt if the solute concentration becomes excessive. They are also subject to non-competitive inhibition by substances which inhibit cell metabolism.

Facilitated diffusion is a form of carrier transport mechanism in which the substrate does not move against a concentration gradient (Danielli, 1954; Le Fevre, 1961).

DRUG ABSORPTION AND BIOAVAILABILITY

Gastro-intestinal Absorption

Absorption of Neutral Compounds

Absorption of neutral compounds from the gastro-intestinal tract into the blood stream depends essentially on their lipid affinity (p. 199). Thus, the ready chloroform and fat solubility of compounds such as *Clofibrate* and *Tolbutamide* ensure their systemic action when administered orally.

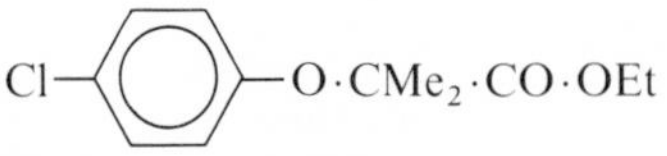

$Cl-C_6H_4-O\cdot CMe_2\cdot CO\cdot OEt$

Clofibrate

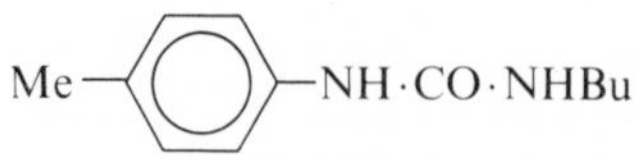

$Me-C_6H_4-NH\cdot CO\cdot NHBu$

Tolbutamide

Similarly, neutral salts, such as quaternary ammonium compounds, which are fully ionised and in general highly water-soluble rather than lipid-soluble, show absorption characteristics which closely parallel their lipid solubilities (Table 53, **1**, 16).

Absorption of Ionisable Salts of Acids and Bases

The majority of organic compounds used in medicine are either weak acids or weak bases, capable of dissociation into ions. Only a few such compounds form stable ion pairs in solution and, with one or two exceptions, transport across tissue boundaries occurs largely in their neutral, non-ionic forms. Absorption, as is now well-established, therefore depends not only on the lipid solubility of the non-ionic form, but also on the dissociation constant of the acid or base, and the pH of the media on either side of the membrane (Brodie and Hogben, 1957), as dictated by the Henderson equation.

In the case of weak acids such as *Aspirin* and many of the non-steroidal anti-rheumatic drugs, this relationship may be expressed as follows:

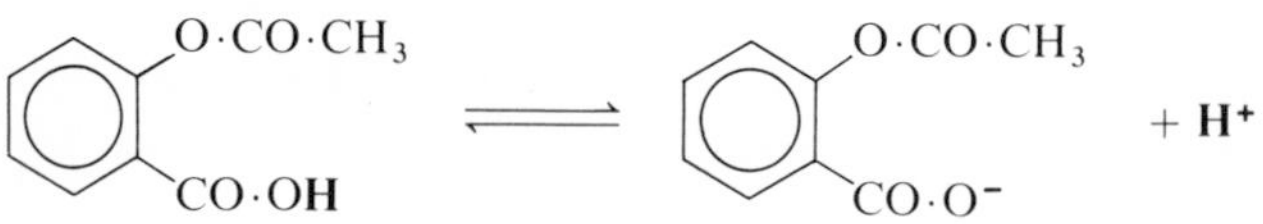

$$C_6H_4(O\cdot CO\cdot CH_3)(CO\cdot OH) \rightleftharpoons C_6H_4(O\cdot CO\cdot CH_3)(CO\cdot O^-) + H^+$$

Aspirin

The relationship between pH and pK_a and the ionisation of *Aspirin* may be represented diagrammatically, as in Fig. 29.

Figure 29 shows that weak acids, such as *Aspirin* (pK_a 3.6), are present almost completely in the non-ionic form in the acid conditions of the stomach, and for this reason are readily absorbed. Once present in the plasma, however, *Aspirin* is almost completely in the ionised form, so that irrespective of the dynamics of the circulation, there is virtually no tendency for diffusion back into the stomach.

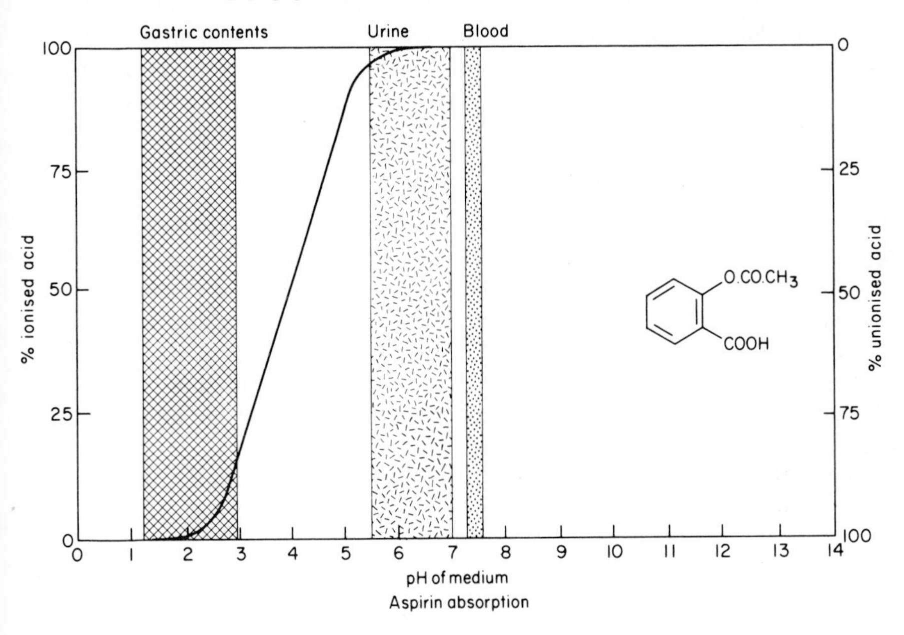

Fig. 29 Aspirin absorption

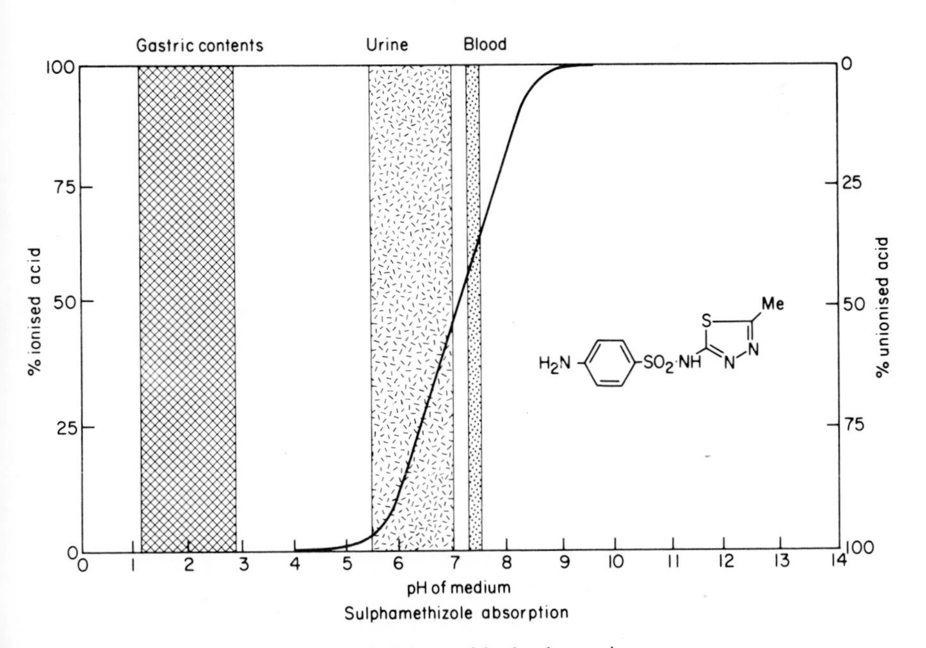

Fig. 30 Sulphamethizole absorption

Similar diagrams may be used to predict the absorption and transport characteristics of any other acid (or base) of known pK and water–lipid partition characteristics across natural tissue boundaries. Thus, weaker acids, e.g. barbiturates and sulphonamides (pK_a 7–8), are in contrast completely unionised in gastric contents (Fig. 30), but only about 50 % ionised in plasma. In consequence, any such acids should be readily absorbed from the stomach into the plasma, but since a significant proportion of the drug is unionised in plasma, it will also be capable of diffusion back from the plasma into the stomach. In the absence of other factors, an equilibrium situation would result. Stomach contents, however, are not static, being subjected to increasing amounts of acidic gastric juice, whilst a dynamic situation is also present in the plasma, by virtue of the circulation of blood, by binding of drugs to plasma proteins, by diffusion of drugs outwards into other tissues, by glomerular filtration, and even by actual metabolism of the drug.

For basic drugs, such as *Codeine Phosphate* (pK_a 7.95), ionisation may be expressed as follows:

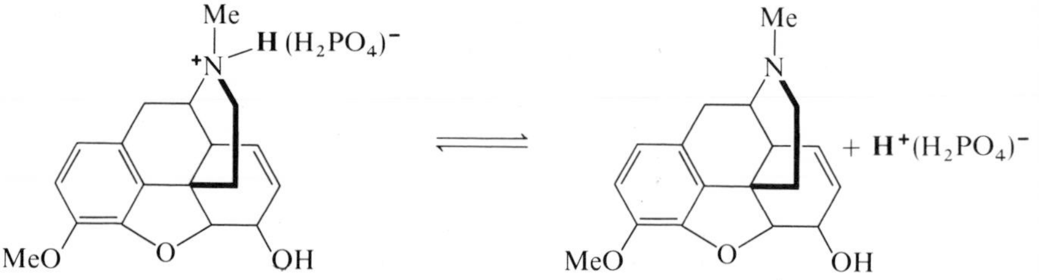

$$BH^+ \rightleftharpoons B + H^+$$

$$pK_a = pH + \log \frac{[BH^+]}{[B]}$$

The relationship between pH, pK_a and the ionisation of *Codeine Phosphate* may be represented diagrammatically, as in Fig. 31.

Figure 31 shows that basic drugs of pK 6.0 and above are almost completely ionised in gastric contents, and hence are not readily absorbed from the stomach. Dissociation occurs, however, in the more alkaline conditions of the intestines with absorption across the intestinal mucosa.

The influence of partition coefficient is superimposed on that of ionisation. Thus, *Barbitone* (Barbital) with a $CHCl_3/H_2O$ partition coefficient of 1.0 will be much less readily absorbed than *Thiopentone* (Thiopental) which has a $CHCl_3/H_2O$ partition coefficient of 102.

Supporting evidence for the pH-partition hypothesis of drug absorption has been obtained (Schanker, Shore, Brodie and Hogben, 1957) by measurement of the gastric absorption of drugs from rats' ligated stomachs one hour after the introduction of the drug in *0.1N* hydrochloric acid (Table 28). Similar results

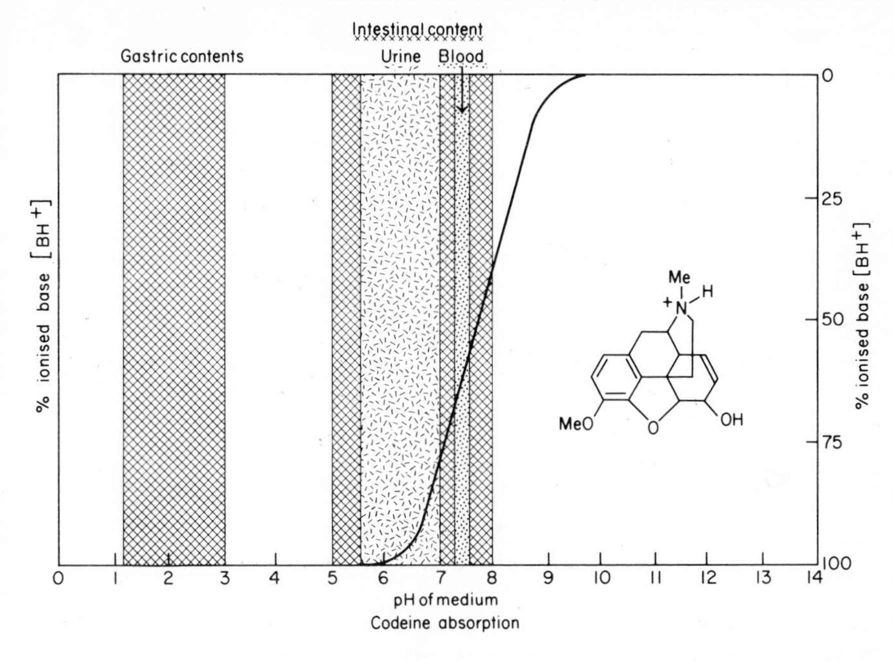

Fig. 31 Codeine absorption

have also been demonstrated for the absorption of drugs from the human stomach (Fig. 32).

The results of these experiments are supported by another series (Shore, Brodie and Hogben, 1957) which showed the distribution of drugs between plasma and gastric juice after administration by intravenous infusion. The appearance of basic drugs in gastric juice in amounts which increase in

Table 28. Gastric Absorption of Drugs from Rat Ligated Stomach[1]

Drug	pK_a	Solubility in water at 20° (%)	Solubility in $CHCl_3$ at 20° (%)	Drug distribution in	
				Gastric contents (%)	Plasma (%)
Salicylic Acid	3.0	0.20	2.2	39	61
Acetylsalicylic Acid	3.5	0.30	5.9	65	35
Thiopentone	7.6	Slight	Freely soluble	54	46
Barbitone	7.8	0.62	1.3	96	4
Dextrorphan	9.2			100	0
Mecamylamine Hydrochloride	11.2	Freely soluble	Freely soluble	100	0

[1] One hour after introduction in *0.1N* HCl (From Schanker, Shore, Brodie and Hogben, 1957)

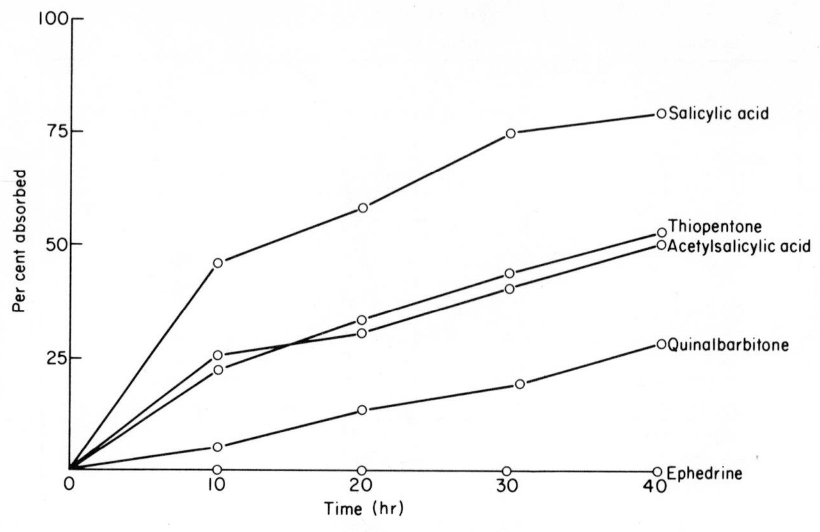

Fig. 32 Drug absorption from the human stomach

proportion to their pK_a (Table 29) demonstrates the inherent reversibility of the diffusion process under appropriate pH conditions.

Whilst the pH-partition hypothesis provides a useful basis for postulating the absorption pattern of most compounds, some caution is necessary. Thus, it is known that some base hydrochlorides, and even quaternary salts, such as laudanosine methiodide (Stenlake, 1968b), have quite high solubility in solvents such as chloroform. One such compound, *Dextromethorphan Hydrochloride* (pK_a 7.97) has been shown to be absorbed quite readily from rat stomach at pH 2.0 (Fiese and Perrin, 1968). *Tetracycline* is similarly transported across the gut wall partly in the positively-charged state as an ion pair (salt) at a rate which is dependent on the surface activity of the salt (Perrin and Vallner, 1970).

Table 29. Distribution of Acidic and Basic Drugs between Plasma and Gastric Juice when Administered to Dogs by Intravenous Infusion

Drug	pK_a	Solubility in water at 20° (%)	Solubility in $CHCl_3$ at 20° (%)	Drug Distribution in	
				Gastric contents (%)	Plasma (%)
Salicylic Acid	3.0	0.2	2.2	0	100
Probenecid	3.4	Slight		0	100
Phenylbutazone	4.4	Slight	80	0	100
Thiopentone	7.6	Slight		10	90
Barbitone	7.8	0.62	1.3	37.5	62.5
Dextrorphan	9.2			97	3

(From Shore, Brodie and Hogben, 1957)

Evidence is also available to show that the ionised form of *Tetracycline* is absorbed from the gut.

It seems likely that ion-pairing is important in the absorption of low molecular quaternary salts, such as *Neostigmine Bromide* and *Pyridostigmine Bromide*, and that it accounts for their somewhat erratic uptake. Ion-pairing with trichloracetic acid has also been shown to increase the rate of absorption of quaternary salts after oral administration (Irwin, Kostenbauder, Dittert, Staples, Misher and Swintosky, 1969), the lipid solubility of the ion-pair being no doubt enhanced by the halogen substituents.

Miscellaneous Factors Affecting Gastro-intestinal Absorption

A number of other factors, less easily quantified, also materially affect the pattern of gastro-intestinal absorption. Thus, the vastly greater surface area of the intestinal mucosa compared with that of the stomach ensures extensive intestinal absorption of acidic compounds, such as *Aspirin*, despite the fact that they are almost completely ionised under the more alkaline conditions which prevail in the intestines. Similarly, the rate of absorption of *Theophylline* is increased in the presence of ethanol and other alcohols, including propyleneglycol and, to a lesser extent, mannitol and sorbitol, due to the increase in water-flux which they effect from the intestines (Houston and Levy, 1975). Again, passage from the stomach may be quite rapid, perhaps no more than 30 min if medicines are administered under fasting conditions. Conversely, medicines taken after food may remain in the stomach for some hours. Food, however, stimulates gastric secretion and hence exposes the drug to the predictable acidic conditions of the stomach, unless of course the patient suffers from achlorhydria (lack of hydrochloric acid secretion).

Rapid passage from the stomach into the intestines can adversely affect the absorption of some compounds with poor aqueous solubility in acid media. Thus, acidic drugs, such as *Phenoxymethylpenicillin* which are primarily absorbed from the stomach, may be swept into the more alkaline medium of the intestines and absorption delayed. Equally, the solution of amphoteric salts (e.g. tetracycline antibiotics) in the stomach, which is an essential prerequisite for subsequent intestinal absorption of the undissociated base itself, may be incomplete with a consequent failure to reach the normal clinically effective blood level. The mucopolysaccharide, mucin, which is also normally secreted in the stomach to protect the gastric mucosa from the action of hydrochloric acid, is itself strongly acidic (**1**, 21), and hence capable of ion-pairing with strongly basic quaternary ammonium and guanidino compounds, thus delaying their onward passage. Bioavailability of the diuretic, *Acetazolamide*, is some 50% less from sustained release capsules compared with the equivalent dose in tablets (Bayne, Rodgers and Crisologo, 1975). It is suggested that this is due to the formation of a non-absorbable 1:1 complex with carbonic anhydrase (Schoenwald, and Ward, 1976) which has maximum stability at pH 7, and minimum stability in the acid conditions of the stomach (Coleman, 1967) where the enzyme is normally concentrated. All these factors, however, are only likely to be of major significance for single dose administration of medicines, and are unlikely to seriously modify the adsorption of drugs administered in any prolonged course of treatment over a period of days or of longer duration, when plasma and tissue levels are usually built up to a steady state.

Bioavailability from Oral Dosage Forms

Dissolution

Although adequate lipid solubility characteristics are essential for diffusion from the gastro-intestinal tract into the bloodstream, there is abundant evidence that the absorption process consists essentially of two steps, aqueous solution in the gastric medium followed by aqueous–lipid partition.

$$\text{Drug} \; \text{Aqueous GI media} \xrightarrow{k_1} \text{Drug solution (aqueous)} \xrightarrow[\text{lipid membrane}]{k_2} \text{Drug solution (plasma)}$$

Either aqueous solubility or water–lipid partition if sufficiently low can be rate-limiting for the absorption process. In accord with this concept, there is substantial evidence of poor or slow absorption characteristics in a variety of medicaments, all of which have low water solubility in common. These include, among others, *Iopanoic Acid* (Holmdahl and Lodin, 1959), *Spironolactone* (Gantt, Gochman and Dyniewicz, 1961), *Griseofulvin* (Atkinson, Bedford, Child and Tomich, 1962), *Phenylbutazone* (Phenbutazone; Searl and Pernarowski, 1967), *Tolbutamide* (Varley, 1968), *Phenytoin* (Martin, Rubin, O'Malley, Garagusi and McCauley, 1968), *Nitrofurantoin* (McGilveray, Matlock and Hossie, 1971) and *Digoxin* (Lindenbaum, Mellow, Blackstone and Butler, 1971). Rapid dissolution of otherwise insoluble weak acids can, however, usually be achieved by use of their sodium and potassium salts.

Aqueous dissolution rates are dependent upon the solid state characteristics of the compound, particularly crystal packing, crystal habit and the specific surface area of the solid. The wetability of the surface and solvation of the dissolving molecules or ions are also important factors in determining the rate of solution of the compound.

Crystal Packing

The lattice energy of the crystal, which is a function of crystal packing is an important determinant of solubility; the tighter the crystal packing, the lower the solubility of the compound. Melting points are a reflection of crystal packing, and for this reason melting points of a group of chemically related substances often parallel their solubilities (Table 30).

Table 30. Melting Points and Aqueous Solubilities of some Sulphonamides

Sulphonamide	m.p.	Solubility in water (approx.)
Sulphanilamide	165–166.5	1 in 170
Sulphamethoxypyridazine	182–183	1 in 700
Sulphapyridine	191–193	1 in 3000
Succinylsulphathiazole	192–195	1 in 5000
Sulphafurazole	195–198	1 in 7000
Sulphadiazine	252–256	1 in 13 000

Compounds in the amorphous state may be expected to dissolve more readily than crystalline materials, since amorphous solids are not subject to the strong cohesive forces which exist between molecules in a closely packed crystal lattice.

However, even where active ingredients are readily obtainable in an amorphous form, this is not always the more stable form, and reversion to a less soluble crystalline form could well occur *in vivo*, either prior to or in parallel with absorption. Thus, *Novobiocin*, which is available in an amorphous form, is metastable in aqueous suspension and reverts to a crystalline form from which it is less readily available. The usual solid state form of *Novobiocin Calcium*, on the other hand, is stable and therefore preferable in solid dosage products (Mullins and Macek, 1960).

Crystal Form and Habit

The surface characteristics of crystalline materials, and hence their solubility, are dependent on the number, size and shape of the crystal faces. The external shape or crystal **habit** depends on the relative development of the different faces, which can vary according to the crystallisation conditions employed. Thus, one and the same compound may crystallise either as cubes or octahedra, which despite their markedly different shape have the same interfacial angles, and the same elements of symmetry. Crystalline **forms** are distinguished by symmetry differences, which are in turn reflected in different shapes of crystal face and interfacial angles.

A number of compounds exhibit **polymorphism**, the ability to crystallise in different forms, and numerous examples of solubility differences due to polymorphism have been recorded. These include, *Cortisone Acetate* (Carless, Moustafa and Rapson, 1968; Mesley, 1968), *Sulphathiazole* (Sulfathiazole; Mesley, 1971), *Sulphamethoxydiazine* (Moustafa, Ebian, Khalil and Motawi, 1971), *Chloramphenicol* (Martin, Rubin, O'Malley, Garagusi and McCauley, 1968) and *Cephaloridine* (Chapman, Page, Parker, Rogers, Sharp and Staniforth, 1968).

In general, the lower the thermodynamic activity of the polymorph, the lower is its apparent solubility and the slower its absorption (Higuchi, Lau, Higuchi and Shell, 1963). Thus, the two *Chloramphenicol Palmitate* polymorphs, which differ in solution free energy (ΔG_{30°) by some 3.2 kJ mol^{-1}, show a tenfold difference in absorption rate (Aguiar, Krc, Kinkel and Samyn, 1967). On the other hand, the *Sulphamethoxydiazine* (Sulfameter, polymorph II, which has a solution free energy ($\Delta G_{30°}$) greater than its less soluble polymorph III by 1.22 kJ mol^{-1}, has an absorption rate only 1.4 times that of the latter (Moustafa, Ebian, Khalil and Motawi, 1971; Khalil, Moustafa, Ebian and Motawi, 1972). Similarly, *Mefenamic Acid* polymorphs, which differ in solution free energy by 1.05 kJ mol^{-1}, show only small differences in their absorption rates (Aguiar and Zelmer, 1969). *Cephalexin* (Pfeiffer, Yang and Tucker, 1970), *Cephaloridine*, and certain of the *Cortisone Acetate* polymorphs are really **pseudopolymorphs**, in which lattice strain arising from the inclusion of crystallisation solvents (water, methanol and ethanol) leads to enhanced solubility. This effect illustrates the need to standardise crystallisation solvents in the preparation of medicaments with poor water solubility, which are used in low dosage.

Specific Surface Area

Specific surface area, which is often expressed in terms of particle size, is another important determinant of dissolution rate. Reduction in particle size, which increases the surface area exposed to the solvent, has been shown to increase aqueous solubility and hence the absorption of a number of important medicaments. Thus, Duncan, MacDonald and Thornton (1962) have shown (Table 31) that doses of 0.5 g *Griseofulvin* of surface area 0.35 m^2/g give similar blood concentrations in human subjects as doses of 0.25 g *Griseofulvin* of surface area 1.5 m^2/g.

Similarly, studies by Shaw, Carless, Howard and Raymond (1973) have confirmed that particle size is an important factor in controlling plasma levels of *Digoxin* administered in cachets. The rate of aqueous dissolution over a 2 hour period was some three times faster with powder of mean diameter 3.7 μm than with that of 22 μm mean diameter, and mean plasma *Digoxin* levels in eight patients treated with a single daily dose for one week were 31% higher for 3.7 μm than with 22 μm powder. It is perhaps significant that a ninth patient on the 3.7 μm powder developed *Digoxin* toxicity. Mean plasma–*Digoxin* levels following single doses of 0.5 mg *Digoxin* of differing particle size are shown in Fig. 33.

Table 31. Effect of Surface Area of Griseofulvin Particles on Absorption in Human Volunteers

Formulation	Dose	Surface Area (m^2/g)	Blood Levels (0–49 hr) (μg/ml/hr)
Suspension	0.5	0.35	17
	0.25	1.5	13.4
Tablets	0.5	0.35	20.8
	0.25	1.5	17.2

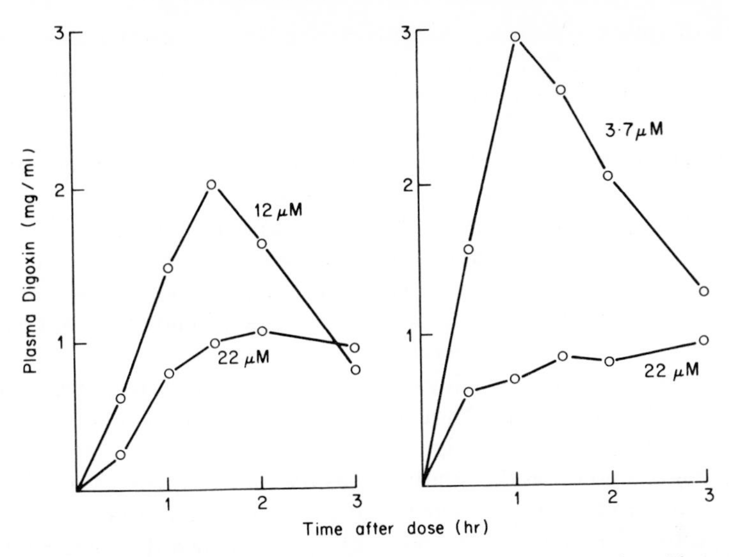

Fig. 33 Mean plasma levels following single doses of 0.5 mg *Digoxin* (from Shaw, Carless, Howard and Raymond, 1973)

There is now some evidence to suggest that the method by which size reduction is achieved is also important. The usual method is milling, either alone or in admixture with a solid diluent. X-Ray and other studies on *Digoxin* before and after hand grinding (Florence, Salole and Stenlake, 1974) indicate that control of particle size alone is not enough. Dissolution profiles (Fig. 34) show the appearance of a freely soluble amorphous layer on the crystals, which is rapidly removed after about 30 min in the dissolution medium. Thereafter,

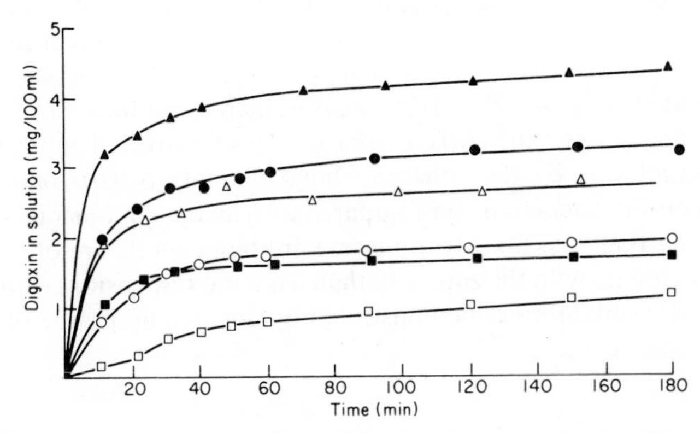

Fig. 34 Accelerated dissolution profiles of three samples of *Digoxin* in *Polysorbate 80* solution (0.005%) at 37°

△ ○ □ original samples
▲ ● ■ mortar-ground samples

the rate of solution decreases sharply as the remaining crystalline material dissolves much more slowly. This view is substantiated by the formation of an essentially amorphous product on ball-milling for 5 h a micronised sample of *Digoxin*.

X-ray diffraction studies showed differences between milled and unmilled samples, which could be accounted for by differences in grain size or in crystallinity. Solid state infrared spectra, however, showed the growth of a band of 1780 cm^{-1} on grinding. Differential thermal analysis traces of ground samples also showed endothermic phase transitions around 170°, well below the melting points. These were present in unground material, and could be accounted for by the presence of amorphous material.

Precipitation *in vivo* of insoluble acids and bases from their water-soluble salts in a finely divided amorphous form offers an alternative method for the preparation of products with improved dissolution and bioavailability characteristics. Thus, studies on the relative bioavailability of *Iopanoic Acid*, precipitated *in vivo* from *Sodium Iopanate* (Iopanate Sodium) and from the crystalline acid respectively (Holmdal and Lodin, 1959) have shown that precipitation of the insoluble acid in a finely divided (amorphous) form from its more soluble salt *in situ* (by the action of gastric secretion in the gastro-intestinal tract itself) is a viable method of enhancing bioavailability. It is evident, therefore, that pH control of dissolution test media to something approaching gastric and/or intestinal conditions can also be an important factor in the assessment of bioavailability from water-soluble salts if they are liable to precipitate an insoluble acid or base *in vivo*.

Liophilisation of solids from solution, which is used extensively in the preparation of powders intended for re-solution and injection, offers a further alternative method for the preparation of more readily soluble forms of drugs with enhanced surface area for incorporation of solid dosage forms. Particle size measurements of powders prepared in this way do not offer a realistic means of assessing the increased solubility which is attained, because of their porosity. Measurement of their specific surface areas is, therefore, much more significant. The importance of specific surface area is also effectively demonstrated in a totally different way by the enhanced bioavailability of *Griseofulvin* from a micronised oil-in-water emulsion compared with an oily or aqueous suspension (Carrigan and Bates, 1973). Mean plasma antibiotic levels are obtained some 1.5–2.3 times higher with the emulsion than from the suspension, with a 2.5 fold increase in bioavailability as demonstrated by the area under the plasma concentration–time curve.

Solvation and Wetability

Solvation and hence solution of solids in water is dependent on their ability to form hydrates, a property which derives from the hydrogen bonding capability of the molecule. Solubility in water is, therefore, favoured in structures containing

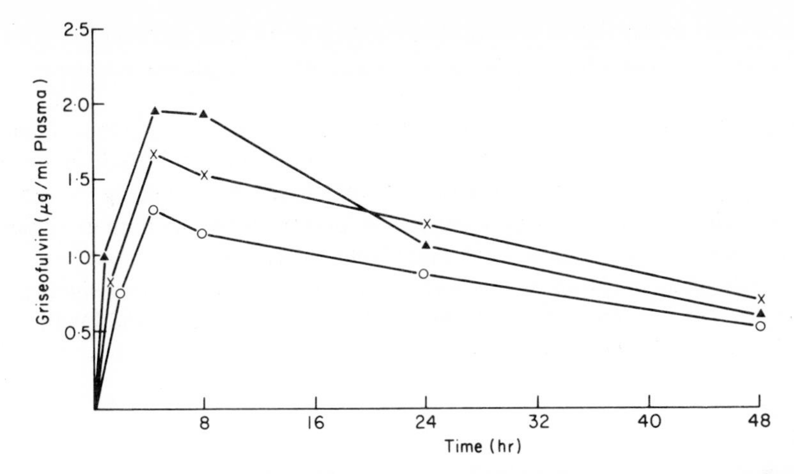

Fig. 35 Plasma levels of *Griseofulvin* after administration of a single oral dose of one gram with and without *Sodium Lauryl Sulphate* (200 mg) (from Marvel, Schlichting, Denton, Levy and Cahn, 1964)

○——○ Griseofulvin 0.41 m^2/g)
×——× Griseofulvin (0.41 m^2/g) + Sodium Lauryl Sulphate
▲——▲ Griseofulvin (1.0 m^2/g)

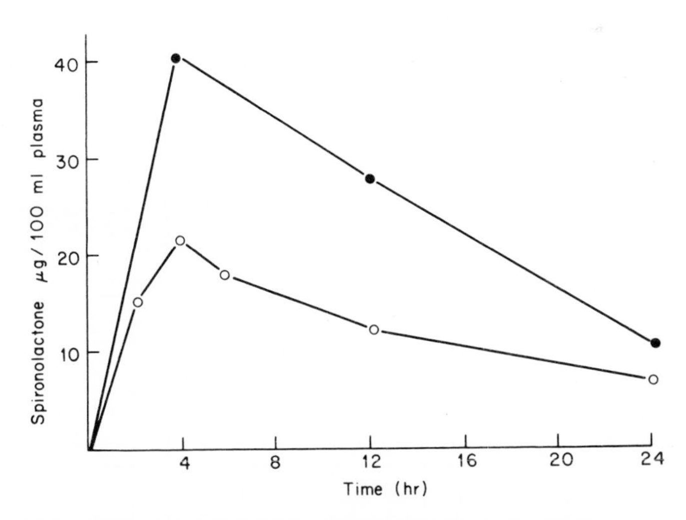

Fig. 36 Plasma levels of *Spironolactone* after administration of single doses with and without *Polysorbate 80* (from Gannt, Gochman and Dyniewicz, 1961)

○——○ Spironolactone (400 mg)
●——● Spironolactone (200 mg) + Polysorbate 80 (200 mg)

OH and NH groups, and also in other nitrogenous and oxygenated groups capable of forming hydrogen bonds involving their lone pairs of electrons.

The dissolution of relatively insoluble substances can also be enhanced by reducing the solid–liquid interfacial tension by means of surfactants, such as *Sodium Lauryl Sulphate* and *Polysorbates*. Formulation with *Sodium Lauryl Sulphate* is just as effective as reduction of particle size in enhancing the release of *Griseofulvin* from tablets and increasing plasma levels in human subjects (Marvel, Schlichting, Denton, Levy and Cahn, 1964; Fig. 35).

Similarly, incorporation of *Polysorbate 80* into *Spironolactone Tablets* gave a fourfold enhancement of peak plasma levels (Gantt, Gochman and Dyniewicz, 1961; Fig. 36) with a concomitant increase in urinary sodium excretion (Fig. 37).

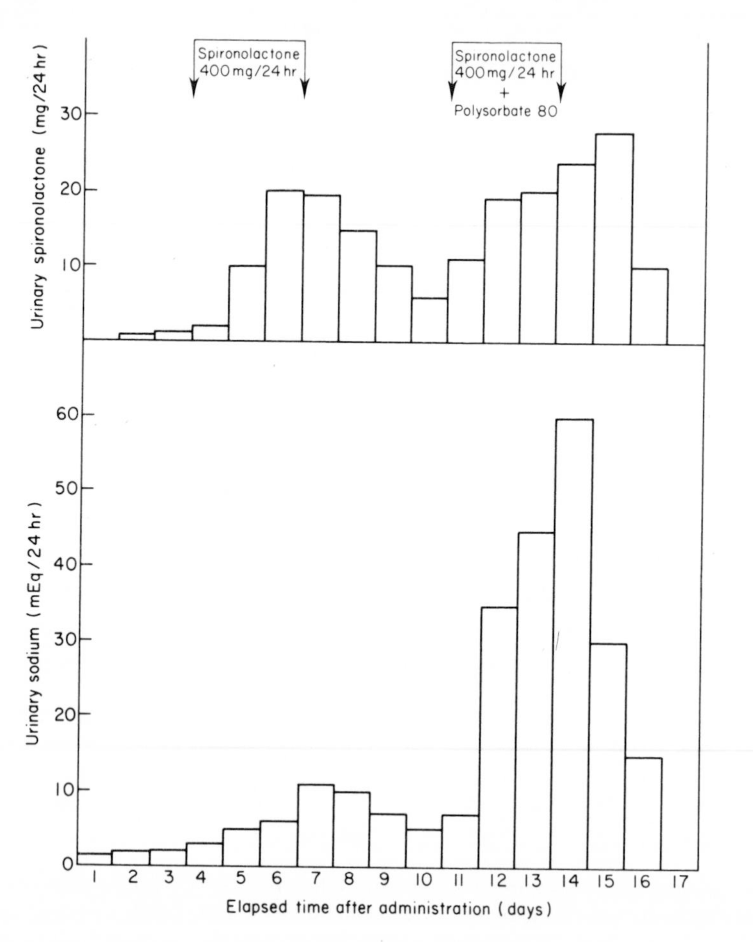

Fig. 37 Urinary sodium excretion after administration of *Spironolactone* with and without *Polysorbate 80* (from Gannt, Gochman and Dyniewicz, 1961)

Drug Interactions

Co-administration of drugs, formulation with excipients capable of interaction to form non-absorbable complexes, or concurrent medication with antacids or absorbents, can also significantly modify the pattern of absorption. Thus, spray-dried *Lactose*, which is frequently used as a filler in tablet manufacture, often contains appreciable amounts of 2-hydroxymethylfurfural, which is capable of interaction with primary and secondary amines and active methylene compounds, such as *Haloperidol*. Manufacturing losses have been reported in direct compression tabletting of *Haloperidol* with spray-dried *Lactose* amounting to some 5–8% of the active ingredient (Janicki and Almond, 1974) to form a condensation product from which the parent drug is unlikely to be released by metabolism (Stenlake, 1975).

F—C₆H₄—CO·**CH₂**·CH₂·CH₂·N(piperidine)(OH)(C₆H₄Cl)

Haloperidol

CHO / furan / CH_2OH

$\mathbf{H_2O}$

F—C₆H₄—CO·**C**(=**CH**—furan—CH_2OH)·CH₂·CH₂·N(piperidine)(OH)(C₆H₄Cl)

5-Hydroxymethylfurfural

The absorption of tetracycline antibiotics, similarly, is hindered in the presence of calcium or magnesium salts, which form non-absorbable metal ion antibiotic complexes (**1**, 22). Adhesives, such as *Sodium Alginate* and *Sodium Carboxymethylcellulose*, are also capable of forming insoluble salts or complexes which can delay the absorption and local action of organic bases from otherwise water-soluble salts.

The alkalising effect on gastro-intestinal media of antacids, such as *Sodium Bicarbonate* and *Magnesium Carbonate*, which are often self-prescribed without thought for their effect on the action of prescribed medicines, are known to delay the absorption of acidic medicaments, including penicillins, *Phenylbutazone* and *Nitrofurantoin*. Gas- and toxin-adsorbents, such as *Charcoal*,

Kaolin, Attapulgite and *Bentonite*, can also act as potent binders, delaying the uptake of otherwise readily absorbable compounds.

Buccal and Sublingual Absorption

Many drugs are capable of being absorbed via the buccal glands which are present in the mucous membrane lining the mouth. Absorption also occurs simultaneously and usually much more rapidly, through the sublingua, the area beneath the tongue. As indicated by Fig. 25, sublingual absorption provides a route for the administration of drugs which leads to direct absorption into the venous circulation, without prior passage through the liver, as in the case of gastro-intestinal absorption. Drugs, such as *Glyceryl Trinitrate* (Nitroglycerin), which are metabolised by the liver and hence inactivated when absorbed from the gastro-intestinal tract, are more active and more rapidly effective when administered by the buccal route (Bogaert and Rosseel, 1972). Drugs readily absorbed and effective by this route include not only cardiovascular drugs, but also steroids and barbiturates (Gibaldi and Kanig, 1965).

Beckett and his colleagues (1967; 1968a and b) have shown that both basic and acidic compounds are absorbed in the mouth from buffer solutions between

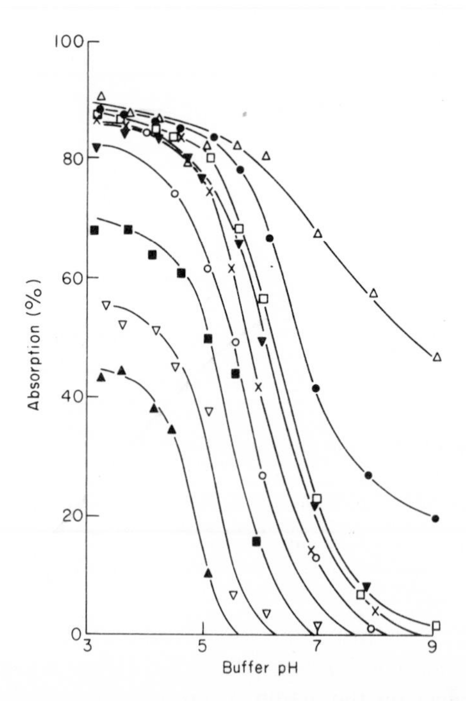

Fig. 38 The buccal absorption of straight chain fatty acids (Subject ACM) △, dodecanoic; ●, undecanoic; □, decanoic; ▼, nonanoic, ×, octanolic; ○, heptanoic; ■, hexanoic; ▽, valeric; ▲, butyric (from Beckett and Moffat, 1968)

pH 8 and 9 in accord with the pH–partition hypothesis. Thus, absorption of bases increases with increasing buffer pH, whilst the converse is true of long-chain fatty acids. The absorption of fatty acids also increases with increasing chain length in parallel with their increasing fat solubility (Fig. 38). The experimental technique used in these studies was to circulate a solution of the drug in an appropriate buffer round the mouth by movement of the tongue and cheeks for five minutes. The amount of drug remaining in the solution was determined by gas-liquid chromatography and the amount absorbed then calculated by difference. The process is reversible, since subsequent washing of the mouth with a buffer of appropriately different pH releases drug to the buffer solutions. It was concluded that this reversal was not merely a de-adsorption, but release of drug actually absorbed through the membrane, since others have observed that sulphonamides and antibiotics may also be actively secreted in saliva (Borzelleca and Cherrick, 1965). In a later study, Beckett and Moffat (1970) demonstrated a positive correlation between the rate constants for buccal absorption of a series of ten acids in a single human subject and the logarithm of their n-heptane–*0.1N* hydrochloric acid partition coefficient.

Pulmonary Absorption

Drugs administered by aerosol and inhalent cartridges are primarily intended for direct and speedy local effect, particularly for the relief of bronchial muscle spasm. Systemic absorption also occurs, and is controlled by the same physicochemical parameters as apply in gastro-intestinal absorption. Because of the very large surface area of lung tissue and its excellent blood supply, inhalation also provides a route for the rapid absorption of volatile lipid–soluble general anaesthetics and for emergency treatment with volatile heart stimulants, such as *Ethyl Nitrite* and *Octyl Nitrite*. Even high molecular weight compounds, such as *Sodium Cromoglycate* (Cromolyn Sodium; MW 512) are absorbed quite rapidly (Cox *et al*., 1970).

Compounds for the treatment of bronchial spasm are more often than not administered either in the solid state directly in very finely divided (micronised) form from inhalent cartridges or as solid–liquid suspensions with a fluorocarbon aerosol propellant. A considerable proportion of the administered dose is, therefore, caught in the mouth and upper airway passages and absorbed directly therefrom. One method of measuring the proportion of the administered dose actually penetrating the lungs is based on comparison of urinary excretion data for the same dose administered by inhalation and intravenous injection (Kirk, 1972). Thus, Cox and his colleagues (1970) have calculated in this way that only 8.2% of *Sodium Cromoglycate* administered from a spinhaler actually reaches the lungs, and is absorbed systemically. Much of the product is, therefore, deposited in the adapter, the mouth, throat and trachea, from which it is not absorbed, owing to the highly polar lipid–insoluble characteristic of the material.

In practice, however, it is usual to assess airway penetration by use of a compartmentalised glass or plastic simulated lung apparatus, several of which have been described (Cox *et al.*, 1970; Karig, Peck and Sperandio, 1973), coated internally with a suitable gel to trap the drug particles.

The major physico-chemical parameter of products administered from cartridge powder is particle size. Theoretically, the smaller the particle size, the more effective is lung penetration, since larger particles are readily trapped in the mouth and upper airway passages. For aerosol suspensions, both droplet size and particle size of the suspended solid are similarly important. A lower limit to particle size is necessary for all solid active ingredients administered by inhalation, since particles of less than 3 μm in diameter are more prone to aggregation due to electrostatic attraction. Bioavailability is also very much affected by the characteristics of the administration device and the mouth adapter.

Topical Absorption

Penetration of human skin may occur by one of three routes, through the hair follicles, the sweat ducts, or the stratum corneum. The relative importance of these alternative pathways depends very much on the extent to which the medicament is able to penetrate the stratum corneum. This consists essentially of layers of keratinised cells, which are heavily hydrated (*ca* 75% water) with localised arrangements of phospholipid (*ca* 5%) between the hydrated keratin fibrils. The outer layers of the stratum corneum are progressively less hydrated and more horny. Diffusion into and through these outer layers is purely passive and relatively unselective. As the lower layers are reached, diffusion of lipid-soluble materials through the lipid elements of the tissue is favoured, whilst the progress of hydrophilic substances capable of hydrogen bonding to water, such as hydroxysteroids, is positively impeded. In contrast, the hair follicles and sweat ducts are much less selective, though, because of their comparatively small cross-sectional area compared with the stratum corneum, they are of minor importance for absorption, except for hydrophilic compounds which are normally impeded by the stratum corneum.

Removal of the horny layer of the stratum corneum artificially by stripping with adhesives allows ready penetration of the skin by all medicaments, including those to which it is normally relatively impermeable. Skin damaged by abrasion or disease, similarly, loses much of its barrier effect. Hydration also increases skin penetration generally, irrespective of whether it arises from perspiration under occluded dressings or by the use of humectants (e.g. glycols; **1**, 20) in skin medication. Changes in water permeability can also be effected with markedly increased penetration by the use of such water-miscible vehicles as dimethyl sulphoxide and dimethylformamide in high concentration (*ca* 80%), which lower the dielectric constant. It has been suggested that these compounds also remove skin acids, but the ready reversibility of their effects on removal suggests that this explanation is unlikely. Dimethyl sulphoxide has, nonetheless, been shown to increase the skin penetration of a wide range of medicinal compounds including *Griseofulvin* and various steroids.

As might well be predicted, the physico-chemical characteristics of the medicament, which affect both its penetrability and availability from the medium in which it is applied, are important criteria of topical activity and skin permeability. Thus, although the inherent corticosteroid activity of *Triamcinoline* is some five times that of *Hydrocortisone*, both are relatively ineffective when applied topically in the treatment of skin disorders, because both have negligible water and lipid solubilities (Table 26). In contrast, *Triamcinolone Acetonide*, which has appreciable lipid solubility (1 in 40 in chloroform), shows topical corticosteroid activity several times greater than that of the parent steroid. This is possibly attributable to a depôt or reservoir effect (Malkinson and Ferguson, 1955; Vickers, 1963) with slow release of the parent steroid by hydrolysis of the protective acetonide link in the normal acidic milieu of the skin surface, (*ca* pH 6.0).

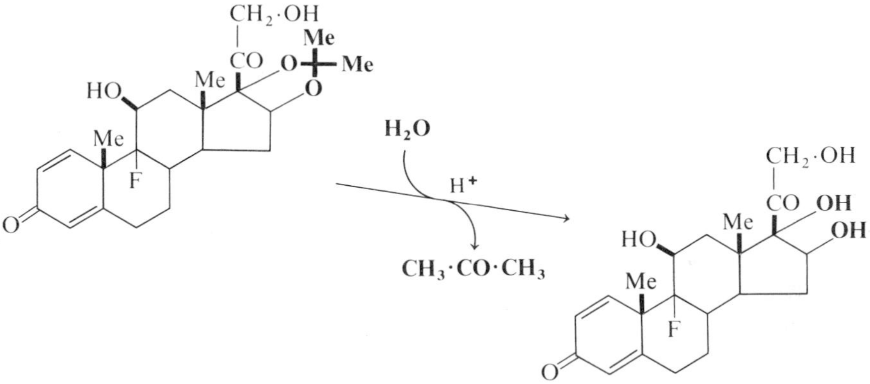

Whilst there is some evidence of systemic effects arising from topically administered corticosteroids, this is generally not widespread or serious (Champion and Goldin, 1975). Highly lipid-soluble materials are, however, readily absorbed and can easily reach systemically effective levels, as for example in the case of the topically administered antileprotic agent, *Ditophal* (**1**, 14). Various kinetic studies of controlled kinetic absorption have been reported (Idson, 1975; Shaw, Chandrasekaran and Taskovich, 1975).

The solid-state characteristics of the medicament if this is insoluble are also important, and extensive studies, particularly of corticosteroid preparations, have clearly established the importance of particle size. The rôle of the vehicle in determining the extent of skin penetration has also been clearly established (van Abbé, Spearman and Jarrett, 1969). Essentially, vehicles based on triglyceride fats and oils are more suited to ensure skin penetration of lipid-soluble medicaments than paraffin hydrocarbons, whilst hydrophilic bases and aqueous creams assist absorption of water-soluble compounds. For example, a base from which *Fluocinolone Acetonide* is readily released and clinically effective has been shown to markedly impair the release and effectiveness of the corresponding acetate, *Fluocinonide* (Burdick, 1972), which is intrinsically about five times as potent as the former. Skin permeability is also favoured by the use of vehicles incorporating non-ionic surfactants due to their disrupting effects on membrane structure (Mezei and Ryan, 1972).

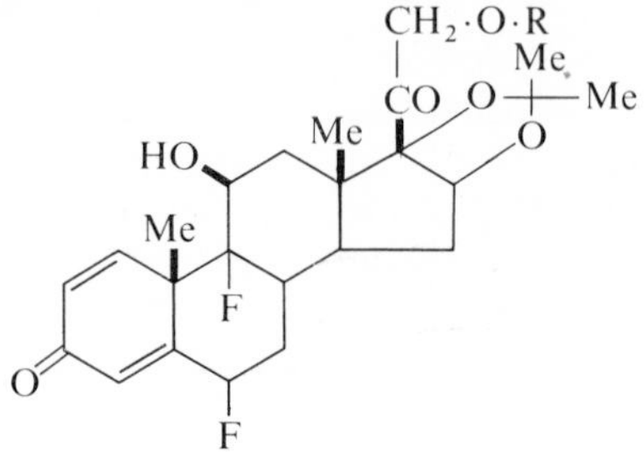

R=H *Fluocinolone Acetonide*
R=Ac *Fluocinonide*

DRUG DISTRIBUTION

Blood

Carrier and Distribution Capability

The blood circulation is not only central to life support as the carrier of oxygen and nutrients, but is also the key vehicle playing a central rôle in conveying drugs to their site of action. Fig. 39 illustrates the dynamic depôt, carrier and metabolic functions of blood which are central to the absorption, retention, distribution and excretion of drugs in man. It will be evident, therefore, that blood levels of drugs merely indicate the balance at any one moment in time of a multiplicity of factors, which in addition to those determining absorption and elimination, include

(a) distribution of the drug between plasma and blood cells,
(b) plasma protein binding,
(c) muscle protein binding,
(d) phospholipid affinity and fat deposition, and
(e) metabolism in the plasma, liver and kidney.

A qualitative impression of the individual and relative importance of these factors can be gleaned from consideration of their relative physical capacities for drugs and metabolites, and where applicable, fluid turnover rates. Thus, the total blood volume of about 5 litres, needs firstly to be seen against a normal daily fluid intake of about 2.5 litres, and comparable losses by excretion via the urine (1500 ml), sweat (500 ml), lungs (400 ml) and faeces (100 ml). Distribution and excretion of drugs, however, is materially influenced by biliary excretion (700–1200 ml/day), and plasma filtration by the glomeruli in the kidney.

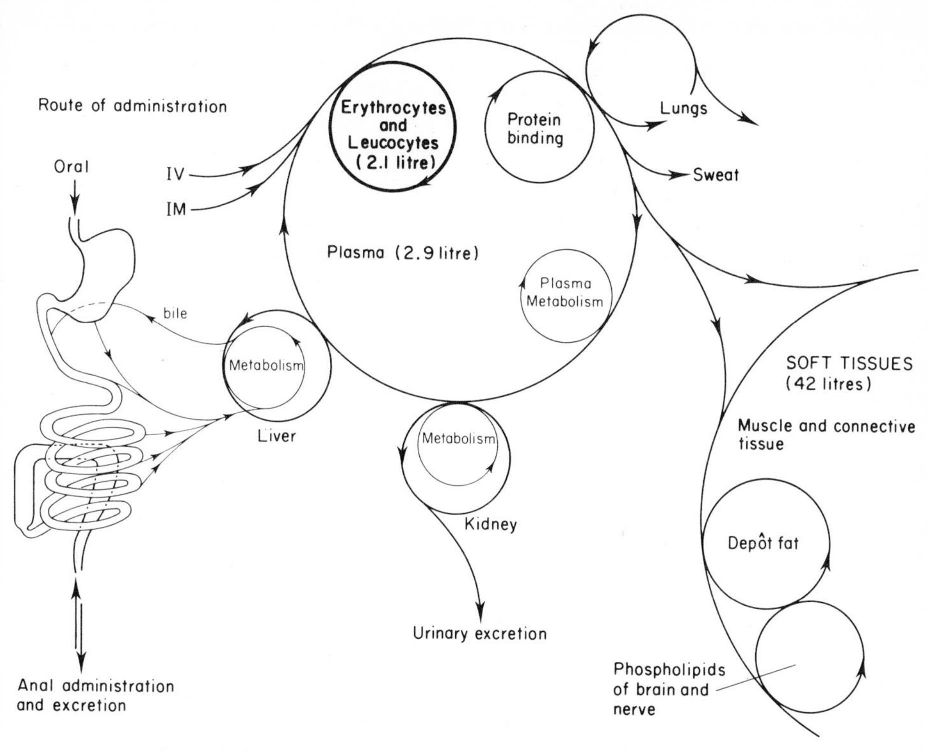

Fig. 39 Pathways in drug absorption distribution and excretion

Glomerular filtration is estimated to be of the order of 137 litres of plasma per day, so that on this basis, the plasma volume of *ca* 2.9 litres is filtered and recycled some 47 times every 24 h. The relative volumes of blood plasma (*ca* 2.9 litres) and the soft tissues of the body (*ca* 42 litres), of which some 4–6% is fat in the normal lean adult, also indicate their relative potential as drug storage depôts. The pattern of distribution between plasma, storage sites, and excretory pathways is largely dependent upon the physico-chemical characteristics of the drug, though active transport processes are often structurally specific.

Blood Levels

More accurate assessments of potential drug bioavailability can only be obtained by calculation of the area under the plasma concentration–time curve (AUC), plotted from a series of measurements standardised in terms of the patient's body weight, according to the formula:

$$\text{Standard area} = \frac{W}{75}\int_0^{72\text{h}} c \cdot \mathrm{d}t$$

where c = concentration/in μg ml^{-1} h^{-1}
W = patient's body weight in kg

Two-compartment Open Model Distribution

In practice, the shape of the plasma concentration curve depends on the route of administration, the extent and rate at which the drug is transferred to the tissues and storage depôts, and the rate at which the drug is excreted. In pharmacokinetic terms, such a system is described as a **two-compartment open system**, the two compartments being the blood and tissues respectively. The system is considered as an **open** one, since drug is removed continuously by excretion from the central (blood) compartment (Teorell, 1937; Riegelman, Loo and Rowland, 1968).

$$\begin{array}{ccc} \text{A} & & \text{B} \\ \boxed{\text{Blood}} & \underset{k_r}{\overset{k_f}{\rightleftharpoons}} & \boxed{\text{Tissues}} \\ \downarrow k_e & & \\ \text{Excretion} & & \end{array}$$

k_f, k_r and k_e are rate constants for storage, release and excretion processes.

The initial phase of the plasma concentration–time curve depends upon the method of administration. Intravenous injection (Fig. 40) gives an immediate initial high plasma drug concentration, which falls rapidly as the drug is distributed to the tissues. Excretion also commences immediately, so that in this initial phase, blood levels fall due to both distribution and excretion. Once blood and tissue concentrations are equalised, further excretion depletes both blood and tissues. The rates of depletion depend upon the transport processes

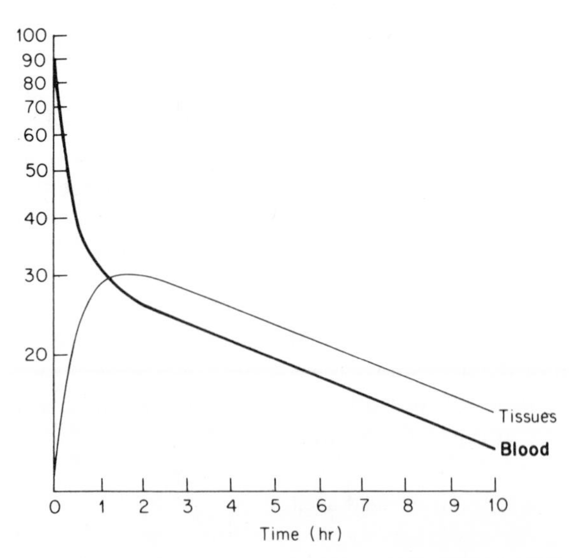

Fig. 40 Semilog plot of blood and tissue drug levels against time after intravenous injection

involved, but provided passive diffusion is the sole process, semilog plots of both the plasma concentration–time and tissue concentration–time curves are linear and parallel with negative slopes in the second phase. In such a situation, the **elimination phase** of the plasma concentration–time curve, therefore, gives an accurate indication of total bodily excretion.

Such **biphasic** plasma concentration–time curves fit the following equation (Riegelman, Loo and Rowland, 1968) which is characteristic of the proposed two-compartment open system:

$$C_p = Ae^{-\alpha t} + Be^{-\beta t}$$

In this, the exponents α and β, can be obtained from the slope of the semilog plot of the plasma concentration–time curve in the distribution and elimination phases respectively (slope distribution plot $= \alpha/2.303$; slope elimination plot $= \beta/2.303$). These exponents are complex functions derivable from the three rate constants k_f, k_r and k_e for the processes involved. The constants A and B are given by intercepts of the distribution and elimination plots with the ordinate.

One-compartment Open Model Distribution

It will also be evident that if the distribution of the drug between blood and tissues is sufficiently rapid for the exponent α to become very large compared to β, the first term in the equation reduces to a limiting value of zero and the equation then approximates to:

$$C_p = Be^{-\beta t}$$

Such a mono-exponential expression, which describes a one-compartment open system, is seen with the rapid distribution and onset of action of neuromuscular blocking agents.

$$\boxed{\text{Tissues} \underset{k_r}{\overset{k_f}{\rightleftharpoons}} \text{Blood}} \xrightarrow{\beta} \text{Excretion}$$

(α labels the Tissues–Blood exchange.)

Uptake of the drug is so rapid that the α-phase (distribution phase) is barely evident in the plasma concentration–time curve. Onset of paralysis is virtually immediate and dose related according to first order kinetics, so that the log dose–response curve is linear (Levy, 1967), and the plasma concentration–time curve shows only the characteristics of the β-phase (excretion).

Three-compartment Open Model Distribution

Similarly, drugs which are held in depôt sites by plasma or muscle protein binding, deposition in fatty tissue, or which undergo extensive enterohepatic circulation, usually show triphasic plasma concentration–time curves following intravenous injection. Examples include *Digoxin* (Doherty and Perkins, 1962; Doherty, Perkins and Flanigan, 1967), *Tubocurarine Chloride* (Gibaldi, Levy and Hayton, 1972) and hydroxycoumarin anticoagulants (Nagashima, Levy and O'Reilly, 1968). The plasma concentration–time curves show the characteristics of a three-compartment open system, which in the case of *Tubocurarine Chloride* may be represented as:

Tissues $\underset{k_r = 0.0458}{\overset{k_f = 0.0389}{\rightleftharpoons}}$ Blood $\underset{k_r = 0.0057}{\overset{k_f = 0.0104}{\rightleftharpoons}}$ Depôt site

$k_e = 0.0072$ ↓

Excretion

In accordance with this concept, the blood level data fit a tri-exponential equation, in which the exponents α, β and γ relate to the distribution, steady-state (fast) elimination and depôt site (slow) elimination phases, respectively.

$$C_p = Ae^{-\alpha t} + Be^{-\beta t} + Ge^{-\gamma t}$$

Absorption into Two-compartment Open Model Systems

When a drug is administered orally instead of intravenously, the plasma concentration–time curve reflects the effect of the additional absorption phase. The simple two-compartment open system must, therefore, be extended as follows.

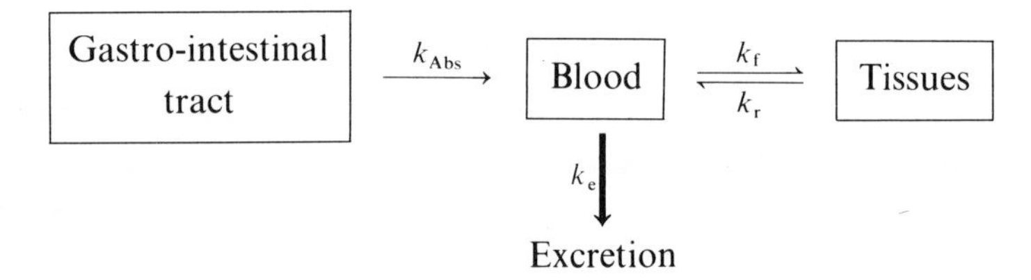

The absorption phase is normally a first order process. The effect on the plasma concentration–time curve of superimposing such an absorption phase onto a two-compartment distribution with elimination at fixed rates is shown in Fig. 41. This shows that the peak blood level rises and is achieved more rapidly as the rate of absorption increases. Also, the higher peak blood levels lead to more rapid elimination. The total area under the plasma concentration–time curve, however, remains the same, irrespective of the rates of absorption and elimination. Such differences may be seen with the same drug when administered in, for example, two different solid states and solution.

Apparent Volume of Distribution

The value of blood level studies and their pharmacokinetic compartmental analysis lies in the facility which they provide for the precise control of dosage regimens. This applies not only in simple blood level–response related situations, such as obtain with oral anti-diabetic drugs, but also where there is substantial

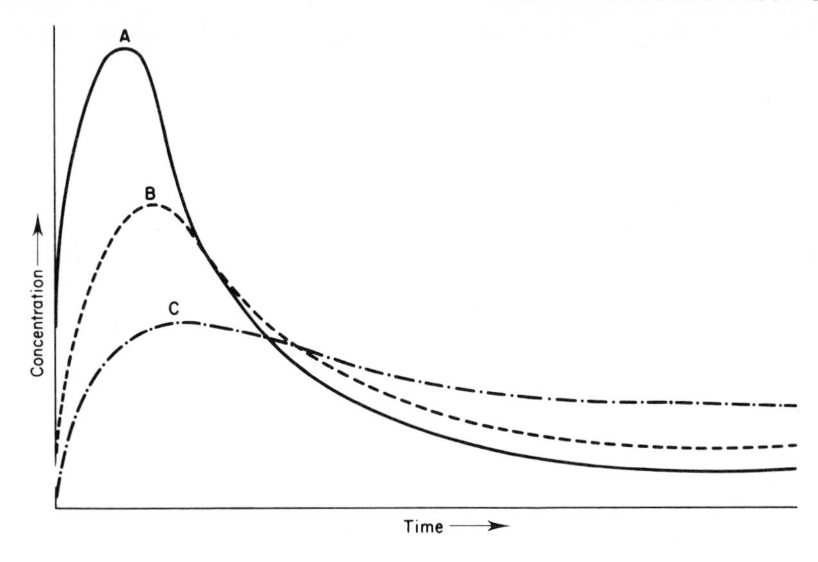

Fig. 41 Effect of absorption phase on blood levels of drug administered orally in varying doses (A > B > C)

deep compartmental storage, as in the case of *Digoxin*, and even where there is no direct correlation between blood levels and pharmacological effect, as for example in treatment of blood clotting disorders with anticoagulants (Levy, 1973).

The **apparent volume of distribution** (V_d) provides a means of calculating compartmental distribution of drugs. It is equal to the volume of body water which the compartment is calculated to contain if the drug was uniformly distributed between the plasma and the compartment in the same concentration. For a drug distributed according to a one-compartment model, in which the α-phase is very short, the apparent volume of distribution is given by the following expression in which AUC is effectively that of the β-phase.

$$V_d = \frac{D_0}{(AUC)}$$

where D_0 = the dose of the drug
AUC = the area under the plasma concentration–time curve following intravenous injection of the drug

The apparent volume of distribution can be calculated similarly for drugs distributed according to a two-compartment model by extrapolating the β-phase of the semilog plot of the plasma concentration–time curve back to zero time. Values calculated by this method are high because they underestimate the effect of excretion. Other methods of calculating V_d in two-compartment systems have been described, which give lower values.

The apparent total volume of distribution is a useful indicator of factors such as drug deposition, renal failure or competition for active secretion, all of which reduce excretion. Normal values of V_d for drugs distributed according to simple one- or two-compartment models equate with the normal volume of total body water (45 to 50 l). Antibiotics, such as *Lincomycin* and *Cephalexin*, show markedly reduced values in patients with renal failure (Gibaldi and Perrier, 1972). V_d is also lowered by some 40% when *Probenecid* is used to saturate tubular secretion and achieve retention of *Benzylpenicillin* (Gibaldi and Schwartz, 1968). In contrast, high values ($V_d = 500$) are characteristic of tissue binding in deep sites. Such V_d is calculated as the free (unbound) plasma drug concentration. V_d is also high for drugs which are extensively bound to plasma proteins, and appropriate corrections must be applied to give a more accurate assessment of the apparent volume of distribution.

Blood Cell Deposition

Some limited idea of the quantitative aspects of cell deposition can be derived from the knowledge that the total blood volume in a normal adult man (70 kg) is about 5 l, of which only 2.9 l consists of plasma, and the remaining 2.1 l blood cells. The cell membranes are lipid in character and semi-permeable, and most lipid-soluble neutral drugs, undissociated acids and unionised bases pass readily into the cells and reach concentrations in equilibrium with free (unbound) drug in the plasma. Thus, 40% of the blood radioactivity from the antithyroid drug [^{35}S]-methimazole (the plasma metabolite of *Carbimazole*) is associated with the blood cells (Marchant and Alexander, 1972), and a similar uptake of a number of other thioamide compounds has also been reported (Giri and Combs, 1972). The uptake of the anti-arrythmic drug, *Disopyramide*, in which blood cell partitioning is favoured has been shown to be concentration-dependent (Hinderling, Bres and Garrett, 1974) in accordance with the lipid-diffusion hypothesis. The influence of drug dissociation is evident in the equilibrium between blood plasma and cells *in vitro* which is reached in about 4 min with *Sulphathiazole* (Sulfathiazole; pK_{a_2} 7.0) compared with one hour for *Sulphacetamide* (pK_{a_2} 5.3) (Langecker and Schulz, 1954). Similarly, the antimalarial bases, *Chloroquine* and *Pyrimethamine*, which have pK_a's $\ngtr$ 7.0 and strongly lipophilic characteristics, are also readily taken up by erythrocytes, whether infected or not.

Concentration and actual retention of some drugs in blood cells due to localised binding or metabolism has also been observed. Thus, sulphonamides have been found in blood cells several days after administration, at a time when they are no longer present in plasma (Hansen, 1940; Maren, Mayer and Wadsworth, 1954), and the radioactive label from 2-[^{14}C]-methimazole, though not that from [^{35}S]-methimazole, shows a small, but distinct concentration gradient in favour of the cells some 4 to 5 days after administration, indicative of drug metabolism (Stenlake, Williams and Skellern, 1973). Concentration of *Disopyramide* and its des-alkyl-metabolite in red cells by factors of 1.2 and 1.4 over plasma concentration (1.0) have also been reported (Hinderlang, Bres and Garrett, 1974) indicating some form of localised cell-binding.

Red blood cells, in particular, have a high capacity for the absorption of small water-soluble molecules, such as urea (Jacobs, Glassman and Parpart, 1935). Hydroxyurea ($H_2N \cdot CO \cdot NHOH$), which has anti-tumour activity (Stearns, Losee and Bernstein, 1963), and is used in the treatment of myeloblastic leukaemia (Rosner, Rubin and Parise, 1971) is taken up preferentially by leucocytes (white cells).

Plasma and Muscle Protein Binding

Plasma Protein Binding

The binding of drugs to plasma, tissue and enzymic proteins is an important factor affecting their distribution, metabolism and excretion. In particular, binding to plasma proteins, if appreciable, can markedly affect the amount of free drug available for equilibration with the rest of the system. Human blood plasma normally contains on average about 6.72 g protein/100 ml, consisting of albumin 4.04 g/100 ml, globulins (2.34 g/100 ml) and fibrinogen (0.34 g/100 ml; Table 32). Drugs are, therefore, normally bound mainly to the major component, albumin, and mainly by processes which are relatively non-specific and readily reversible. It is important to realise that the levels of plasma protein may be significantly altered in abnormal physiological states or in disease. Thus, plasma albumin levels are often significantly reduced in advanced rheumatic disease (Ballantyne, Fleck and Carson Dick, 1971) and levels of corticosteroid binding globulin are enormously raised in pregnancy.

Table 32. Human Plasma Proteins

Protein	Normal concentration (g/100 ml plasma)		Isoelectric point (approx.)
Albumin		4.04	4.9
Globulins			
α_1	0.31		5.2
α_2	0.48		4.9
β	0.81		5.4–5.8
γ	0.74		5.8–8.2
		2.34	5.2
Fibrinogen		0.34	
	Total	6.72	

Other proteins, besides albumin, are also involved in protein binding, as for example, with *Dicoumarol* (Dicumarol; Weiner, Shapiro, Axelrod, Cooper and Brodie, 1950; Table 33). A number of steroid hormones, notably *Oestradiol* (Estradiol; van Baelen, Heyns, Schonne and de Moor, 1968), *Testosterone*

(Steeno, Heyns, van Baelen and de Moor, 1968) and *Hydrocortisone* (Daughaday, 1956) are also specifically and preferentially bound to particular globulin fractions.

Table 33. Binding of *Dicoumarol* to Plasma Proteins

Dicoumarol

Protein fraction (0.5% solution)	% *Dicoumarol* bound to Protein
Albumin	99
β-Globulin	20
γ-Globulin	20

It is interesting to note, too, that *Dicoumarol* is almost insoluble in water at physiological pH, but extensive binding to plasma protein makes it approximately one hundred times more soluble in plasma than in saline.

In general, a dynamic equilibrium exists between bound and unbound drug. Hence, binding is a function of free drug concentration, taking the form of the Langmuir absorption isotherm (Fig. 42), so that with rising drug concentration, the protein approaches saturation.

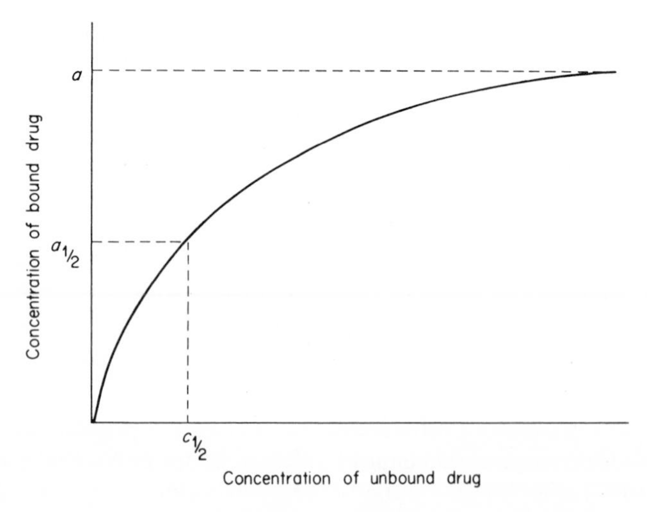

Fig. 42 Protein binding as a function of concentration

The interaction of drug (D) with an unoccupied binding site of protein (S) may be considered as a reversible reaction obeying the Law of Mass Action:

$$D + S \underset{k_2}{\overset{k_1}{\rightleftharpoons}} DS$$

This equilibrium can, therefore, be characterised in terms of an apparent association constant, $K(k_1/k_2)$, the concentration of unbound drug (D), the concentration of the protein (P), and the number of binding sites per protein molecule (n). The relationship between these values, and the fraction of total drug which is bound to protein (β) is expressed according to Goldstein (1949) as:

$$\beta = \frac{1}{1 + 1/KnP + D/nP}$$

Hence, the fraction bound increases with increase of K, n and P, and decreases with increase in D.

For example, *Phenylbutazone* (Phenbutazone), which has pK_a 4.4 and hence is 99.9% ionised at pH 7.4, is bound to the extent of 98% to plasma proteins at therapeutic levels; at supra-therapeutic levels, however, the fraction of total drug bound decreases, because the protein has passed its saturation point for the drug (Table 34). Thus, a 2.5 fold increase in the total loading produces a fifteen-fold increase in the free drug concentration. In contrast, changes of total drug concentration of a relatively weakly bound drug, such as *Digoxin* which is only about 35% bound, produces a change in free drug concentration which much more nearly parallels the change in total concentration.

Table 34. Binding of Phenylbutazone to Human Plasma Proteins at Therapeutic and Supra-therapeutic Levels

Phenylbutazone Concentrations (mg/100 ml)			Fraction Bound (β) (%)
Total	Bound	Unbound	
10[1]	9.8	0.2	98
15	14.5	0.5	96.6
22.5	20.5	2.0	91.8
25	22	3.0	99

[1] Therapeutic level (From Brodie and Hogben, 1957)

A number of more refined treatments of association constants have been carried out by Scatchard and his collaborators (1949, 1950, 1957).

Binding Parameters

Binding forces between drugs and tissue proteins are, in general, primarily electrostatic. Binding characteristics are, therefore, to some extent determined by the isoelectric point (iep) of the protein involved. For most plasma proteins, this is below physiological pH (7.4). They, therefore, carry a negative charge at that pH. This does not mean that only cations can be bound, but merely that a preponderance of such ions can be bound. Thus, albumin (iep 4.9) has a net negative charge at pH 7.4 of 16 electron units, composed of 100 negative and 84 positively-charged groups, so that it is capable of binding both anions and cations (Tanford, Swanson and Shore, 1955). Both acidic and basic drugs as, for example, *Probenecid*, *Sulphinpyrazone* (Sulfinpyrazone), *Phenylbutazone* and *Chlorpromazine*, are strongly protein-bound.

Normally, the high buffering capacity of the blood, which is exercised by bicarbonate, haemoglobin and plasma proteins, retains blood pH within narrow limits. In exceptional circumstances, acidaemia (blood pH <7.36) may result from, for example, excessive protein metabolism giving rise to excessive amounts of sulphuric acid, or from the ingestion of ammonium chloride, which is metabolised in the liver with the conversion of ammonia to urea, leaving an excess of hydrochloric acid. Similarly, alkalaemia (blood pH greater than 7.4) may arise from ingestion of sodium bicarbonate, magnesium carbonate, or alkali metal salts of hydroxyacids which are themselves metabolised to CO_2 and hence bicarbonate. In such exceptional conditions, therefore, drug protein binding may be modified by the effect of pH on the dissociation of either drug or binding protein, or both. Thus, protein binding of *Disopyramide* is unchanged with increasing plasma pH, but is decreased significantly at pH 6.7 due to the increased net positive charge on the protein, which inhibits binding of the increasingly protonated *Disopyramide* (Hinderling, Bres and Garrett, 1974).

Neutral molecules, especially large, stretched or flattish structures, such as paraffin hydrocarbons and steroids, which are able to fit the non-polar groups of the protein, are bound by van der Waals forces, hydrophobic interactions and/or hydrogen bonds (**2**, 2). Thus, chloroform, ether and a number of other volatile compounds were shown as far back as 1904 to be more soluble in serum than in saline solution (Moore and Roaf, 1904); similar solubility differences have been observed with *Cyclopropane* (Orcutt and Seevers, 1937). Skipski and his collaborators (1967) have also shown by experiments using gas-liquid chromatography that human plasma lipids contain some 4.4% of paraffin hydrocarbons of which between 18 and 50% are of the n-alkane series. The high plasma solubilities of other non-ionic compounds, such as naphthoquinones, cholesterol and steroid hormones (Heymann and Fieser, 1948; Eik-nes, Schellman, Lumry and Samuels, 1954) are similarly explicable on the basis of protein binding due to appropriate combinations of hydrophobic interactions and hydrogen bond formation (Table 35).

Cortisone and *Hydrocortisone* are good examples of non-ionisable substances which are both extensively bound (*ca* 80%) to plasma proteins (Peterson, Wyngaarden, Guerra, Brodie and Bunin, 1955). Binding occurs to albumin and to one of the α-globulin fractions (Daughaday, 1956) which has been named transcortin (Slaunwhite and Sandberg, 1959). Attempts to isolate this corticosteroid binding fraction by anion-exchange chromatography and gel-filtration (Muldoon and Westphal, 1967; Chader and Westphal, 1968) have proved only

Table 35. Binding of Steroids to Bovine Serum Albumin at 37°

Steroid	Solubility in Buffer (pH 7.3) (μ mol l^{-1})	Solubility in 5% Bovine Serum Albumin (μ mol l^{-1})
Cortisone	600	1400
Deoxycortone	180	1030
Testosterone	125	2150
Methyltestosterone	120	1750
Oestradiol	0.6[1]	310[1]

[1] measured at 25° (From Eik-nes, Schellman, Lumry and Samuels, 1954)

partially successful. The purest corticosteroid binding globulin isolated, using added 4-[^{14}C]-cortisol as a marker, contained 1 mole of corticosteroid per 40 700 g of protein. It had one high affinity binding site with association constants at 4° and 37° of $k = 9 \times 10^8 M^{-1}$ and $4.7 \times 10^7 M^{-1}$ respectively, in accord with the marked temperature dependence observed by DeMoor and his colleagues (1962). Removal of bound corticosteroid from the isolated corticosteroid binding globulin by gel filtration at 45°, however, resulted in loss of binding activity.

The influence of hydrophobic interactions on the plasma protein binding of ionisable compounds, such as barbiturates and fatty acids, is discussed elsewhere (**2**, 2). Penicillins, similarly, are bound to human plasma proteins to an extent which is largely dependent on the hydrophobic character of the side-chain (Bird and Marshall, 1967). The binding of *Chlorpromazine*, also, which is some 90% bound to human plasma protein, is only marginally affected by pH changes indicating that factors other than ionic bonding are significantly involved (Curry, 1970).

Displacement from Protein Binding

It has long been known that oleate is able to displace protein-bound phenol red (Rosenthal, 1926) and L-tryptophan (McMenamy and Oncley, 1958). The latter may well be expected from a comparison of the relative association constants of oleic acid (1.1×10^8) and ^{a}-tryptophan (1.25×10^5) for human serum albumin. Thorpe (1964) has suggested that release of fatty acids into the blood by physiological stimuli may displace protein-bound drugs in the same way. Similarly, changes of drug concentration can also occur as a result of the displacement of other strongly bound drugs. Thus, *Aspirin* enhances the action of some drugs by displacing them from their plasma–protein binding sites. Important groups of drugs affected include the anticoagulants, *Ethyl Biscoumacetate*, *Nicoumalone* (Acenocoumarol) and *Warfarin Sodium*, and the oral antidiabetics, *Acetohexamide*, *Chlorpropamide* and *Tolbutamide*, free plasma levels of which can undergo sudden and dangerous enhancement following treatment with *Aspirin*, *Phenylbutazone*, *Flufenamic Acid* and other strongly acidic drugs.

Thyroxine and endogenous steroids may also be displaced from protein binding when they are present at abnormally high plasma levels. Particularly high levels of endogenous corticosteroids are reached after injections of Depot *Tetracosactrin* (Cosyntropin) and also normally in women in the third trimester of pregnancy. Displacement of 11-hydroxysteroids is observable in the former on treatment with *Aspirin*, but is not observed in the pregnancy cases because the level of the specific corticosteroid binding protein is also raised, so that the critical binding capacity (CBC) of plasma for 11-hydroxysteroids, which is normally about 26 μg/100 ml, is exceeded (Stenlake, Williams, Davidson, Downie and Whaley, 1969). In contrast, drug displacement of more weakly bound drugs, such as *Digoxin*, which is only about 35% bound to plasma proteins is unlikely to precipitate a sudden change in bioavailability.

Few substances actually bind to plasma proteins by covalent bond formation. There is some evidence, however, that *Aspirin* acts in this way. Experiments with *Aspirin* labelled with [^{14}C] on both the acetyl and salicylic carboxyl groups have shown that it acetylates human serum albumin *in vitro* (Hawkins, Pinckard and Farr, 1968) with an uptake of *ca* one acetyl group/mole HSA at pH 7.3. Human serum albumin is permanently altered *in vivo* as a result of *Aspirin* therapy, and this alteration is accompanied by a permanent increase in capability for binding *Sodium Acetrizoate*. *Aspirin* has also been shown to acetylate fibrinogen, hyaluronidase, RNA and DNA (Pinckard, Hawkins and Farr, 1968), and it is suggested that this ability of *Aspirin* to acetylate natural biopolymers may explain some of the differences between the biological effects of *Aspirin* and salicylates. The effect of modification of plasma proteins by *Aspirin*, or any other drug binding through covalent bonds, is dependent on the extent and continuity of medication. Unless medication with the binding agent is continued, the modified protein will only persist, and then to a dwindling extent through the normal plasma albumin turnover period, which is normally of the order of 21–28 days.

Muscle Protein Binding

Muscle protein, which accounts for some 30 kg of the total body weight in a normal adult, also forms an important natural depôt site for some drugs. Thus, the long period (7–14 days) required to stabilise patients on *Digoxin* therapy is due to extensive deposition of the drug in skeletal muscle protein in addition to heart muscle. Thus, Coltart, Howard and Chamberlain (1972) reported mean concentrations and standard deviations of 1.2 ± 0.8 ng/ml, 11.3 ± 4.9 ng/g and 77.7 ± 43.3 ng/g for plasma, skeletal muscle, and cardiac muscle respectively. General diffusion into muscle and other proteinous organs together with fat deposition clearly plays a very significant rôle in drug bioavailability, but there are as yet few other examples where this factor has been assessed other than by general drug distribution studies.

Location and Deposition in Lipid Tissues and Body fat

Disposition and Nature of Body Lipids

Fatty tissues account for some 15–20 % of body weight in the normal adult, but in obese individuals, this can rise to as much as 50 %. The brain, spinal cord and nervous tissue together account for only about 1.6 kg of phospholipid tissue, whereas even in the average lean 70 kg adult, something like 10 kg of other fatty tissue is present. Nervous tissue consists mainly of phospholipids. Other fatty tissues are mainly composed of triglycerides, i.e. glycerol esterified by natural long-chain fatty acids, but probably also include some cholesterol fatty acid esters as well. The fatty hydrocarbon chains of these esters promote the solubility of hydrocarbons and other molecules incorporating paraffinic (alkyl) and aromatic (aryl) substituents. The higher fat solubility of lengthening hydrocarbon chains is reflected in the greater solubility of *Butobarbitone* (Butethal; R = Bu) in chloroform (1 in 3) and ether (1 in 10) compared with that of *Barbitone* (Barbital; R = Et) which has solubilities of 1 in 75 and 1 in 40 in the two solvents respectively.

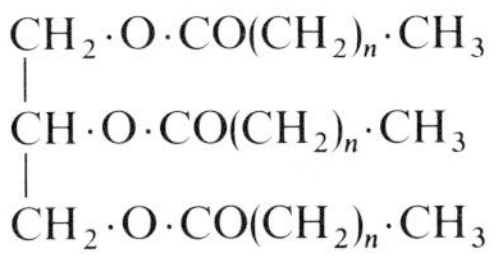

$$\begin{array}{l} CH_2{\cdot}O{\cdot}CO(CH_2)_n{\cdot}CH_3 \\ | \\ CH{\cdot}O{\cdot}CO(CH_2)_n{\cdot}CH_3 \\ | \\ CH_2{\cdot}O{\cdot}CO(CH_2)_n{\cdot}CH_3 \end{array}$$

Triglyceride

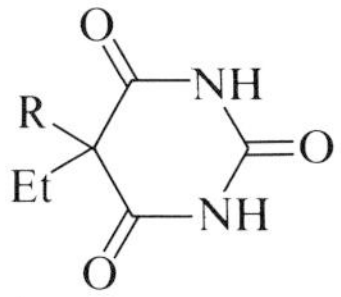

Barbitone R = Et
Butobarbitone R = Bu
Phenobarbitone R = Ph

Both chloroform and ether are commonly used as solvents and have polar characteristics which reflect to a greater or lesser extent the polar and solubility characteristics of natural fatty tissue (and lipid membranes) and in this respect provide a rough, but useful guide to the level of fat solubility *in vivo.*

Chemical Structure and Lipid Solubility

In addition to their hydrocarbon-like characteristics, fatty acid esters, although neutral and non-ionic, have polar characteristics which arise from the electronegativity of the C—O and C=O bonds. This gives rise to bond dipoles which in turn cause the whole molecule to exhibit a turning moment (dipole moment) when placed in an electric field. The dipole moment is the physical expression and measure of the unequal distribution of electrons over the molecule as a whole. It is this polarity of natural fatty acid esters which facilitates the fat solubility of covalent compounds that are themselves polar, but still non-ionic in character. Thus, as shown in Table 26 (p. 165), high lipid solubility of neutral compounds is associated with chemical structures embodying strongly electronegative substituents (O, N, S, Halogens), and hence bond dipoles (Table 13, **2**, 2) in such a way that they generate large dipole moments.

The influence on fat solubility of incorporating hydrocarbon and polar characteristics into a neutral lipid-insoluble molecule is well illustrated by comparing the solubilities of *Glycerol* (Glycerin) and the ether derivatives, *Mephenesin* and *Chlorphenesin*. Thus, whereas *Glycerol* is completely miscible with water and almost completely insoluble in ether and chloroform, both *Mephenesin* and *Chlorphenesin* have low water solubilities and appreciable solubility in both ether and chloroform. The greater polarity contributed by the C—Cl bond in *Chlorphenesin* is reflected in an increase in its solubility in organic solvents. The effect of opposing bond dipoles is seen in the lowered solubility of *Mephenesin Carbamate* in chloroform (1 in 50) compared with that of *Mephenesin* (1 in 12). These factors influence the ability of such compounds to concentrate in the central nervous system where they function as central muscle relaxants.

$CH_2(OH)\cdot CH(OH)\cdot CH_2\cdot OH$

Glycerol

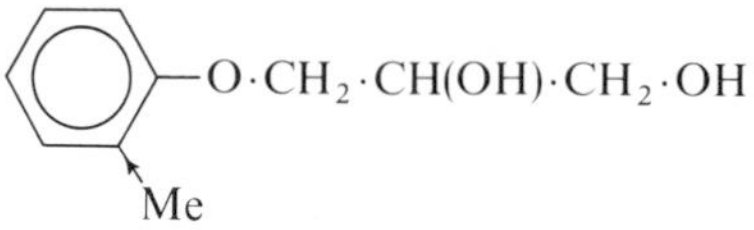

$C_6H_4(Me)$—$O\cdot CH_2\cdot CH(OH)\cdot CH_2\cdot OH$

Mephenesin

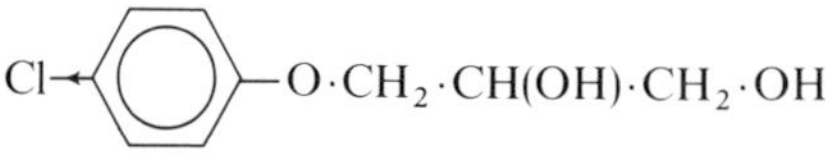

Cl—C_6H_4—$O\cdot CH_2\cdot CH(OH)\cdot CH_2\cdot OH$

Chlorphenesin

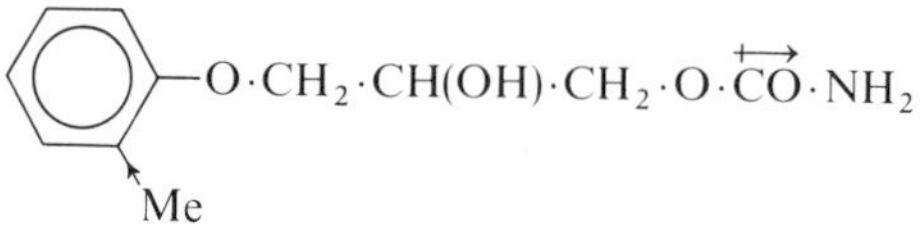

$C_6H_4(Me)$—$O\cdot CH_2\cdot CH(OH)\cdot CH_2\cdot O\cdot \overleftrightarrow{CO}\cdot NH_2$

Mephenesin Carbonate

The effectiveness of phenothiazine tranquillisers is also markedly enhanced in those compounds, such as *Chlorpromazine* and *Trifluoperazine*, which incorporate halogen dipoles capable of enhancing concentration in the lipid tissues of the central nervous system. Halogen substituents in the antimalarial compounds, *Chloroquine*, *Pyrimethamine* and *Proguanil* (Chloroguanide), similarly assist their passage into the red blood corpuscles where the malarial parasites are concentrated (Hitchings, 1952).

The action of iodine-containing X-ray contrast media depends both on the opacity of iodine to X-rays and on the lipid solubility which the halogen confers on the molecule. The latter facilitates diffusion and hence transport and excretion. Thus, lower molecular weight compounds, such as *Diodone* (Iodopyracet), *Iothalamic Acid* and *Sodium Diatrizoate*, are readily excreted by the kidney, and in consequence are most useful in pyelography and urography.

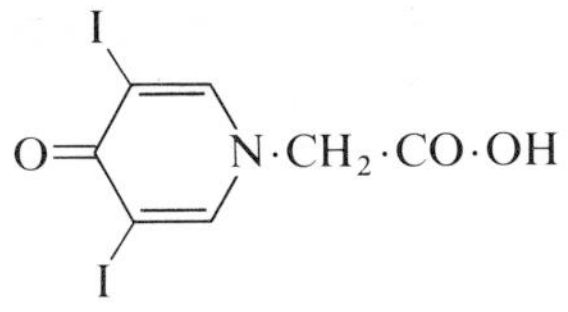

I

O= N·CH_2·CO·OH

I

Diodone

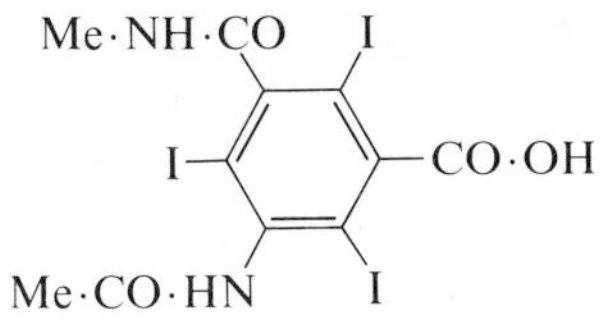

Me·NH·CO I

I— —CO·OH

Me·CO·HN I

Iothalamic Acid

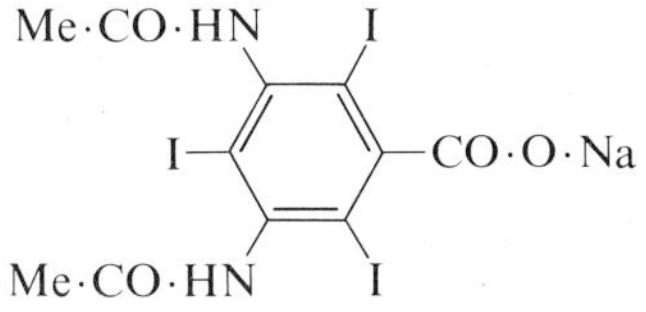

Me·CO·HN I

I— —CO·O·Na

Me·CO·HN I

Sodium Diatrizoate

Long-chain fatty acid ester derivatives, such as *Iophendylate*, have special affinity for nervous tissues and are used in myelography, whilst higher molecular weight compounds, i.e. *Iodipamide* (MW 1140) are excreted in the bile and are useful for examination of the gall bladder (cholexystography) and bile duct (cholangeography).

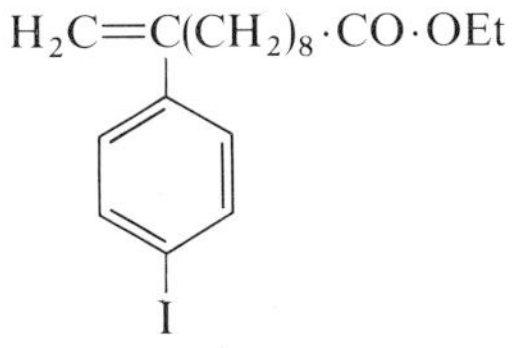

$H_2C{=}C(CH_2)_8$·CO·OEt

I

Iophendylate

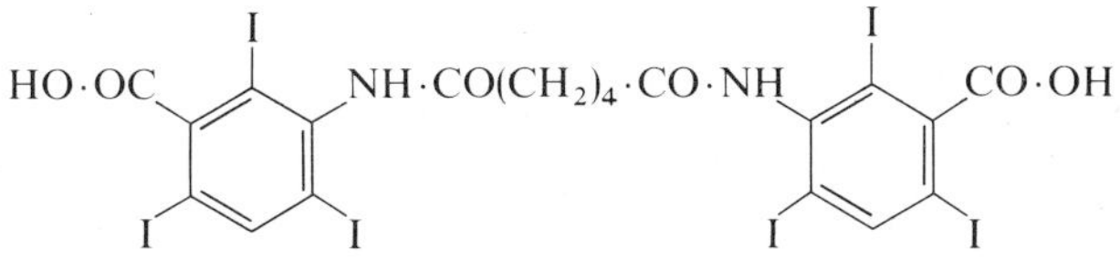

I I

HO·OC NH·CO$(CH_2)_4$·CO·NH CO·OH

I I I I

Iodipamide

Modification of Drug Action by Body Fat Deposition

Neutral drugs with high lipid solubility, e.g. general anaesthetics, for which the tissues of the central nervous system have a high affinity, are also extensively deposited in body fat. The intravenous anaesthetic, *Thiopentone*, which is almost instantaneous in its effect, is also slowly and steadily taken up by body fats, reaching a maximum concentration of some 10 times the plasma concentration about three hours after administration when levels in other tissues are falling rapidly. This factor accounts for the short duration of sleep, and other effects described below.

The very high lipid affinity of *Thiopentone* results from the incorporation of a number of favourable structural parameters in the molecules of certain barbiturates. Thus, the phenyl substituent in *Phenobarbitone* increases the solubility in chloroform only slightly over that of *Barbitone*, since the enhancing effect of hydrocarbon substituent size, which is clearly evident in the higher chloroform solubility of *Butobarbitone*, is almost completely annulled in *Phenobarbitone*, by the dipole effect of the aromatic π-electron system in opposition to that of the molecule in general. Similarly, since the C=S bond has greater double-bond character and hence lower bond polarity than the C=O bond, the opposing effect of the thiocarbamyl group on the overall molecular dipole of thiobarbiturates is less than that of the carbamyl group in their corresponding oxygen analogues. Hence, *Thiopentone* has a much higher solubility in chloroform and lipid tissues than its oxygen analogue, *Pentobarbitone*.

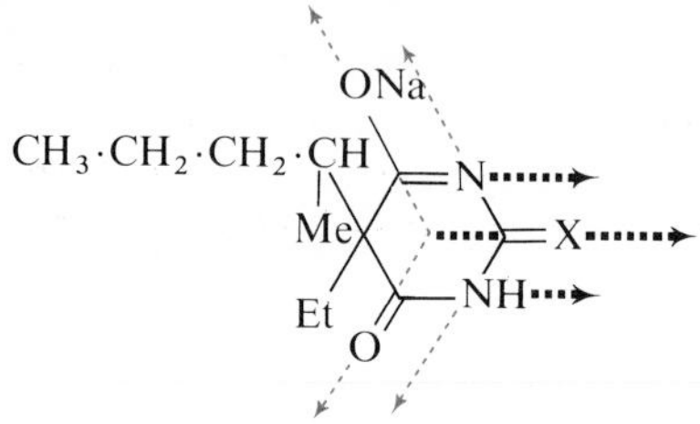

ONa
$CH_3 \cdot CH_2 \cdot CH_2 \cdot CH$
Me
Et
N
X
NH
O

Thiopentone, X = S
Pentobarbitone, X = O

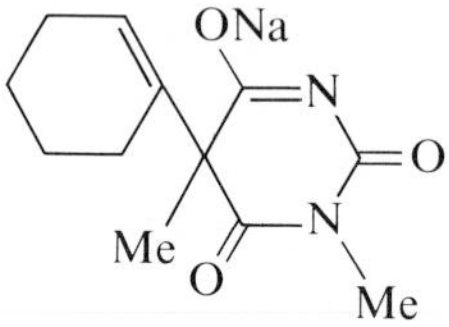

ONa
N
O
N
Me
Me
O

Hexobarbitone

Figure 43 shows the distribution of *Thiopentone* into various tissues of the dog after intravenous injection (Brodie, 1952). In contrast to other tissues, fat takes up *Thiopentone* relatively slowly, and the maximum concentration is reached only after about 3 hr, when the concentration in most other tissues is well past the maximum and falling steadily. The maximum fat concentration is then some ten times that of plasma. A similar distribution of *Thiopentone* occurs in man (Brodie, Mark, Papper, Lief, Bernstein and Rovenstine, 1950), and this would appear to account adequately for the observed effects of the drug, namely the short duration of sleep produced; the cumulative effect seen with repeated dosage at times when the concentration in fat is already high and the drug is being released from depôt sites; and also for the persistent somnolence seen in some persons after large doses.

Plasma pH is also necessarily an important factor in fat storage of ionisable fat-soluble drugs. This has been demonstrated with *Thiopentone* in dogs under inhalation with carbon dioxide. The latter lowers plasma pH to about 6.8, increasing the proportion of unionised *Thiopentone*, thereby causing a corresponding increase in fat deposition.

Pentobarbitone, the oxygen analogue of *Thiopentone*, has a much lower fat solubility than the latter, and deposition in body fat is no higher than in other tissues. Intravenous injections, therefore, produce plasma levels in man about three times those with equivalent doses of *Thiopentone*. In contrast, *Hexo-*

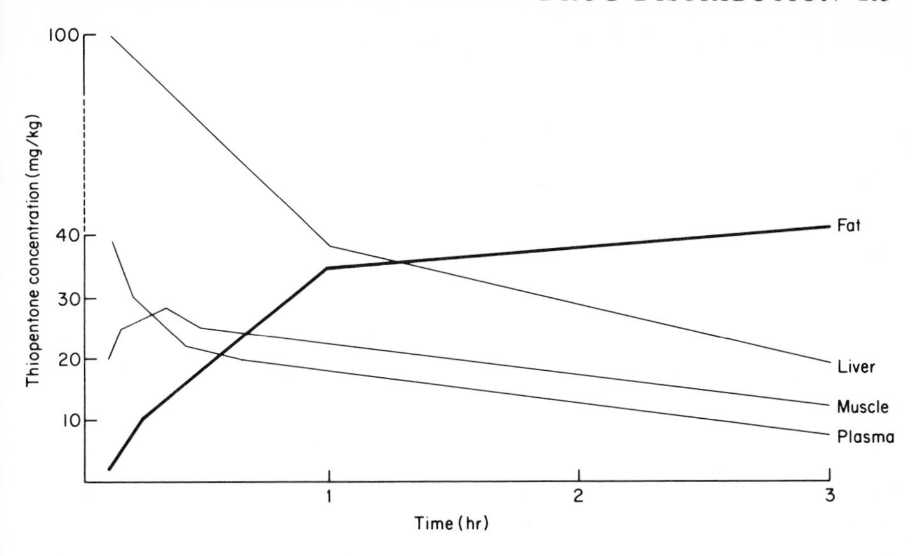

Fig. 43 Tissue distribution of *Thiopentone* in the dog (from Brodie, 1952)

barbitone, like *Thiopentone*, is also extensively localised in fatty tissue. It is, however, an *N*-methyl substituted barbiturate and also a much weaker acid (pK_a 8.4) than most other barbiturates. In consequence, some 91 % is undissociated at physiological pH compared with only about 60 % for other barbiturates. Accordingly, fat solubility is promoted. Water solubility is also decreased, since the N—H ···· O hydrogen bonding characteristics present in other barbiturates are lost in the *N*-methyl compound. The very short duration of action of *Hexobarbitone* (Hexobarbital) is in part a consequence of its initial deposition in body fat.

Penetration into the Cerebrospinal Fluid and Brain Tissue

The inability of certain substances to penetrate from the blood into the cerebrospinal fluid (CSF) and into brain tissue has led to the concept of the blood–CSF and blood–brain barriers. Whilst, however, there is undoubted evidence of the existence of these barriers, their precise location is still a matter of some uncertainty. The blood–CSF barrier appear to be primarily sited in the epithelium of the choroid plexuses of the brain, whilst the blood–brain barrier seems to lie either in the brain capillary wall or the surrounding layer of glial cells. The arteries leading to the brain and spinal cord actually blend with the glial membranes, so that the basement membranes of cerebral capillaries are almost completely covered by neuroglial cells. It is significant not only that this neuroglial cell layer is still incomplete at birth, but also that the blood–brain barrier is less effective in the newborn child than in adults. Also, full development of the blood–brain barrier coincides with completion of the neuroglial membrane.

Table 36. Penetration of Drugs into the Cerebrospinal Fluid

Drug	pK_a	% Undissociated at pH 7.4	Distribution coefficient (Chloroform/buffer pH 7.4)	Penetration time[1]
Thiopentone	7.6	61.3	102	2
Barbitone	7.8	71.5	4.8	40
Acetanilide	13.0	99.9	3.7	120
Salicylic acid	3.0	0.01	0.02	360

[1] Penetration time = time in minutes required to attain a cerebrospinal fluid/plasma ratio 1

Since plasma and cerebrospinal fluid are at approximately the same pH, lipid-soluble substances rapidly reach approximately the same free concentration on both sides of the barrier. The rate at which drugs penetrate into the CSF and the brain is, therefore, dependent on their lipid solubility and lipid/water partition coefficient (Brodie and Hogben, 1957; Table 32). Mayer, Maickel and Brodie, 1957). Substances of high lipid solubility like the anaesthetic gases, *Chloroform*, *Ether* and *Halothane*, will therefore penetrate rapidly. Since also only the undissociated forms of drugs are lipid-soluble, the penetration of weak acids and bases is a function of their dissociation constant (Mayer, Maickel and Brodie 1957; Table 36).

The Placental Barrier

The placenta naturally not only regulates the transfer of essential nutrients to the developing foetus, but also acts as a screen for its protection from harmful materials which might otherwise be transmitted directly from the mother. In this, it is only partially successful, since whilst it acts as a barrier for high molecular weight compounds (MW > 1000), it is clearly established that many drugs and toxic substances are transported across the placenta. Thus, whilst most essential nutrients enter the foetus by active transport processes (see below) against a concentration gradient, simple diffusion of other substances also occurs (Schanker, 1962), and the same physico-chemical factors apply for foetal absorption as for gastro-intestinal absorption. Fully ionised drugs, such as *Suxamethonium* (Moya and Kvisselgaard, 1961) and *Tubocurarine* (Crawford and Gardiner, 1956) are usually poorly absorbed, unless present in exceptionally high concentration in the maternal circulation. Dissociable acids and bases, however, are translated across the placenta at rates depending upon the lipid solubility of the undissociated acid or base and its dissociation constant.

In accordance with these factors, general and local anaesthetics, analgesics, barbiturates, sulphonamides, phenothiazines and tetracycline antibiotics all cross the placenta. Thus, the effects of *Morphine* and related narcotic analgesics, and also *Nalorphine*, are apparent in the evidence of respiratory depression and withdrawal symptoms in newborn babies of addicts, and in the levels of addictive drugs which have been identified in their blood and excreta (Joelsson and

Adamsons, 1966). Similarly, absorbed barbiturates have been shown to stimulate foetal hepatic glucuronyltransferase activity. Normal delay in onset of this enzymic activity in the foetus and the newborn child can, however, cause problems not only with narcotics, but also when the mother is dependent on life-preserving drugs, such as oral anti-diabetics, because of the enormous difference in the biological half-life of the drug in the mother (1–2 hr) and newborn child (40 hr).

Mammary Secretion

Few systematic studies of the concentration and elimination of drugs in milk have been made. The dangers of contaminating cow's milk by *Penicillin* following treatment for mastitis and with pesticide residues from feedstuffs (Sisodia and Stowe, 1964) are well known. Such evidence as is available suggests, however, that drugs become concentrated in milk following systemic administration according to the pH–partition hypothesis. Concentration of acidic drugs, such as *Penicillin*, and the more acidic sulphonamides in milk is, therefore, not strongly favoured (Table 37), whilst concentration of basic drugs is much more likely.

Table 37. Distribution of Acidic and Basic Drugs between Plasma and Milk

Compound	pK_a	Milk/Plasma Ratio
Penicillin	2.7	0.25
Sulphadimethoxine	6.0	0.20
Sulphanilamide	10.4	1.00
Sulphathiazole	7.14	0.43
Tetracycline	9.7	1.60
Quinine	8.4	4.8
Erythromycin	8.8	8.7

More recent human studies (Knowles, 1965) have shown that stimulants (alcohol and *Caffeine*), analgesics (*Aspirin* and *Morphine*), hypnotics (barbiturates) and purgatives (*Phenolphthalein*), and even nicotine from tobacco smoking can reach the child indirectly by this route. The transfer of *Diazepam* and *Oxazepam* to breast milk has also been reported (Patrick, Tilstone and Reavey, 1972).

Biliary Excretion and Entero-hepatic Circulation

The secretion of bile in man provides the normal route for the release of the bile acids which are essential for the absorption of fats from the intestines. The bile acids, which are biosynthesised in the liver from cholesterol, are secreted in the bile as their conjugates with glycine and taurine (**1**, 22) which increase the molecular weight above the critical level for biliary excretion. The conjugates are cleaved by microbial metabolism in the gut, releasing the unconjugated bile acids to complex with and assist the absorption of fats into the circulation. Bile acids so absorbed are returned for re-secretion once again in the bile to complete the process of entero-hepatic cycling, which it is estimated occurs at least twice with each meal (Bergström, Dahlqvist, Lundh and Sjóvall, 1957).

Bile also acts as a vehicle for the excretion of drugs from the liver. For each species, however, there appears to be a minimum molecular weight, below which little biliary secretion occurs. Thus, although the cut-off molecular weight is about 300 in the rat, this is significantly higher in other species, and about 500 in man. This is illustrated in Table 38, which shows the percentage of the administered dose of *Morphine* (MW 285, secreted as its glucuronide MW 461; March and Elliott, 1954) and *Chloramphenicol* (MW 323, secreted as its 3-glucuronide MW 499; Glazko, 1966) secreted in bile in various species.

Table 38. Percentage of Administered Dose Secreted in Bile

Species	Morphine	Chloramphenicol
Rat	63	80
Guinea Pig	50	—
Dog	35–50	60
Rabbit	8	—
Rhesus Monkey	5–20	15
Man	1–7	3

Biliary excretion of drugs is modelled on the excretion of bile acid conjugates, and therefore is geared primarily to the excretion of water-soluble anions. Carboxylic acids, which are not of sufficient molecular weight for secretion as such, are conjugated as higher molecular weight ester- or *O*-glucuronides. Thus, *Indomethacin* (MW 358) undergoes biliary excretion as the ester glucuronide (MW 534), but only to the extent of 15% of the administered dose in man (Hucker, Zacchei, Cox, Brodie and Cantwell, 1966). On the other hand, the sulphonic acid, *Sulphobromophthalein* (Sulphobromphthalein; MW 794), which is administered intravenously as the sodium salt as a test of hepatic function, is of sufficiently high molecular weight to be excreted to the extent of 60% in bile in man (Oliverio and Davidson, 1962).

In contrast, cationic substances are not appreciably excreted in bile. Thus, biliary secretion of *Tubocurarine Chloride* (MW 696) is low in dogs, some 36% of an intravenous dose appearing in the urine within 3 hr and 75% in 24 hr. This pattern of excretion is confirmed in bile cannulation studies which showed bile secretion of the drug to be 6% in 3 hr and 11% in 24 hr (Cohen, Brewer and Smith, 1967). Other cationic substances, such as *Poldine Methylsulphate*, which are administered orally to inhibit gastric secretion, are not appreciably absorbed and for this reason, rather than because of biliary secretion, are excreted mainly in the faeces (Langley, Lewis, Mansford and Smith, 1966). Organic bases of sufficiently high molecular weight, such as *Erythromycin* (MW 734; Piller and Bernstein, 1955) and *Rifamycin* (MW 698; Furesz and Scotti, 1961) are secreted in substantial amount unchanged in bile in man. Lower molecular weight bases (MW < 500) are not appreciably secreted in bile, but their anionic conjugates (glucuronides and sulphates) are excreted by this route if of sufficiently high molecular weight.

Enterohepatic circulation of drugs occurs in the same way as with the bile acids as a result of reabsorption from the intestines and transmission back to the liver. For most drugs which are secreted as water-soluble anionic conjugates, this involves microbiological deconjugation in the gut, releasing the parent drug or unconjugated metabolite for reabsorption. Because of the low molecular weight cut-off point for biliary secretion, enterohepatic circulation is a major factor in drug metabolism in the rat. For the same reason, it is much less extensive in man. It has, however, been clearly established for *Digitoxin* for which urinary excretion is prolonged due to this effect for up to 50 days (Katzung and Meyers, 1965). Other compounds, such as *Morphine* and *Stilboestrol* (Diethylstilbestrol), for which urinary excretion is also prolonged in man, though to a much lesser extent (*ca* 5 days), may also experience some enterohepatic circulation, though the low level of biliary secretion of morphine suggests that the reasons for its delayed excretion lie elsewhere.

URINARY EXCRETION OF DRUGS

Glomerular Filtration

The kidney, which controls urinary excretion, is composed of nephrons. Each nephron consists of three, linked, anatomically distinct parts, each with a distinct physiological role, namely:

(a) the glomerulus,
(b) the proximal tubule,
(c) the distal tubule.

The glomerular membrane permits the passage of almost all solutes by a process known as glomerular filtration. In effect, it produces an ultrafiltrate of blood plasma containing foreign compounds and their metabolites in much the same concentration as they are present in the plasma, but free of plasma proteins.

Reabsorption

The walls of the distal tubule behave as a lipid membrane, permitting reabsorption of lipid-soluble molecules by passive diffusion. Many natural metabolites are lipid-insoluble, and hence are readily excreted in urine, unless they are small enough to be reabsorbed by diffusion or are reabsorbed by active transport, as in the case of endogenous amino acids and sugars.

Urinary excretion rates are dependent on several factors. Some 137 litres of plasma are filtered by the glomeruli per day. Hence, the total plasma volume of *ca* 2.9 litres is filtered and recycled on average some 47 times per day. In consequence, only those compounds with very high diffusion coefficients for reabsorption are retained at significant plasma levels for substantial periods of time.

The pH–partition mechanism operates for tubular reabsorption of lipid-soluble drugs, the extent of absorption being dictated by the degree of ionisation (Figs. 29–31) at urinary pH (*ca* 6.0). Drugs with high lipid solubility and low water solubility, which are incapable of ionisation or which are not appreciably ionised at the pH of the tubular epithelium, favour the maintenance of high blood levels, and will only be excreted after metabolism to more highly water-soluble products. In contrast, low lipid solubility and high water solubility *per se*, as in quaternary salts, such as *Tubocurarine Chloride*, favour urinary elimination and rapid lowering of blood levels.

Active Secretion

A process of active secretion also occurs in the proximal tubule. Strong organic acids, such as *p*-aminohippuric acid, penicillins, thiazides and glucuronides, are secreted as anions. Quaternary ammonium salts, such as *Hexamethonium*, are also secreted by an active transport mechanism (p. 166), a process which contrasts with the passive diffusion of unionised structures which occurs at most other membranes. The nature of the carrier mechanism is not known, but individual substances may compete with each other for the secretory mechanism, so that administration of one drug may inhibit the excretion of another, as for example, in the slowing of *Benzylpenicillin* excretion by *Probenecid*. The latter also inhibits uric acid excretion from the proximal tubule. Its uricosuric action, and that of other uricosuric agents, such as *Sulphinpyrazone*, depends on its ability to suppress the reabsorption of uric acid in the distal tubule.

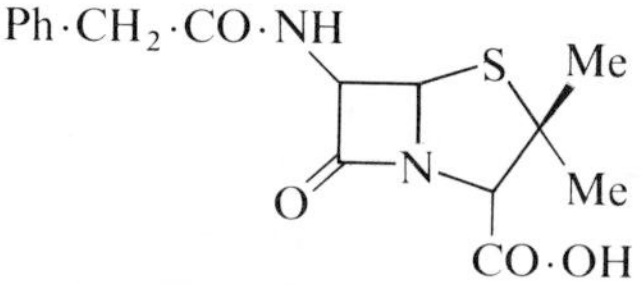

Ph·CH_2·CO·NH, S, Me, Me, N, O, CO·OH

Benzylpenicillin

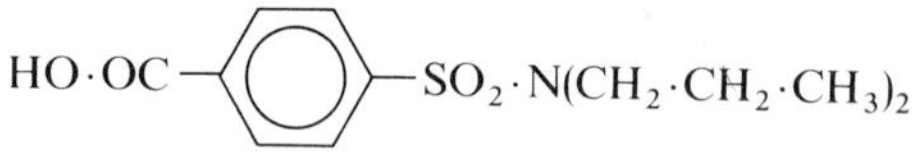

HO·OC—⟨benzene ring⟩—SO_2·$N(CH_2·CH_2·CH_3)_2$

Probenecid

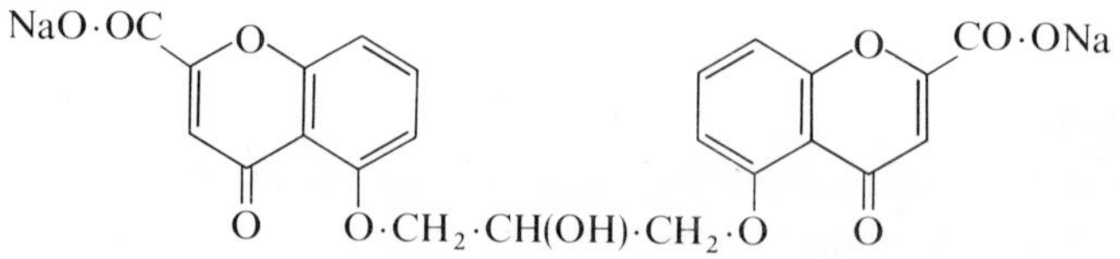

NaO·OC, O, O, O·CH_2·CH(OH)·CH_2·O, O, O, CO·ONa

Sodium Cromoglycinate

Active secretion usually leads to excretion rates which are considerably in excess of those which result solely from glomerular filtration (Smith, 1951). Thus, renal clearance of *Sodium Cromoglycate* in rabbits is reported to be equivalent to that of *p*-aminohippuric acid (p. 211) and some three times that calculated on the basis of inulin clearance (p. 211) for glomerular filtration (Cox *et al.*, 1970). Active secretion of high molecular weight compounds, such as *Sodium Cromoglycate* (MW 512), which also undergo extensive biliary secretion, leads to much higher concentration in the bile than in plasma (1379 and 0.07 μg/ml in bile and plasma respectively 2 hr after inhalation of the powder in the monkey).

Urinary pH and Excretion

Orloff and Berliner (1956) have shown that changes in tubular pH influence the excretion of weak electrolytes. Thus, administration of *Sodium Bicarbonate*

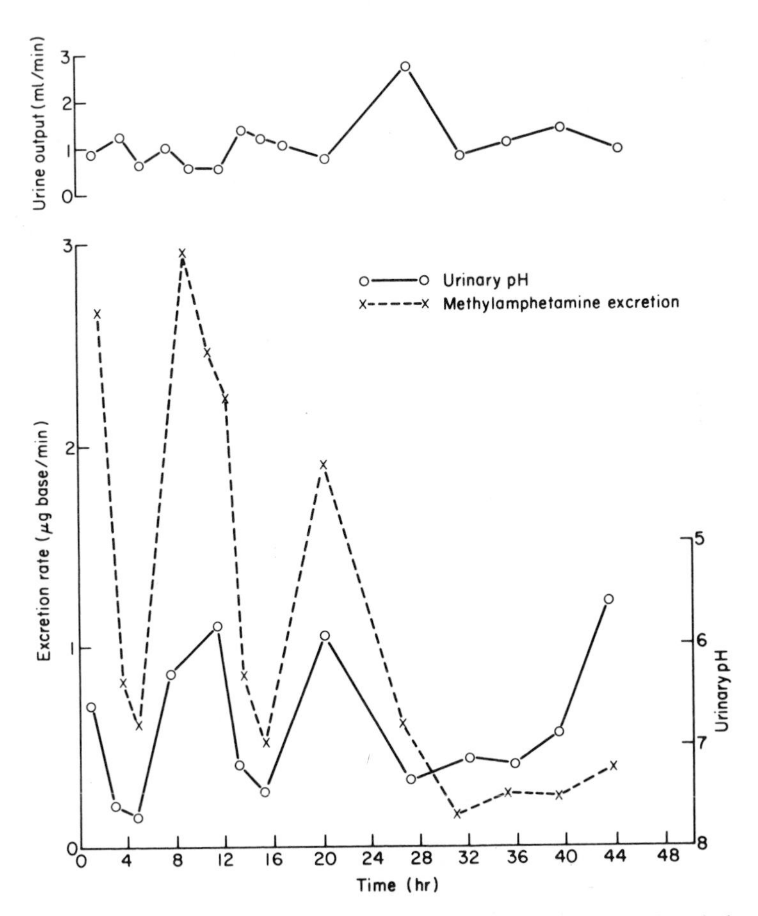

Fig. 44 The influence of urinary pH and urine output on the excretion of methylamphetamine (from Beckett and Rowland, 1965)

decreases the excretion of basic drugs as a result of the increase in concentration of unionised drug in the more alkaline tubular lumen. This observation is in conformity with the pH–partition hypothesis (p. 168), and Gutman, Yü and Sirota (1955) have similarly observed an increase in the excretion of acidic drugs under the same conditions. Conversely, lowering the tubular pH by oral administration of ammonium chloride increases the proportion of bases in the ionised form. Accordingly, reabsorption is retarded and excretion increased (Beckett and Rowland, 1964; 1965). Fig. 44 shows the effect of urinary pH on the excretion of *Methylamphetamine*.

In Caucasian subjects receiving a normal balanced western diet, urinary pH varies between subjects and within subjects throughout the day. The average pH is about 6.0 (Yarbro, 1956; Elliott, Sharp and Lewis, 1959). Recent observations by Hadzija (1969) on Negro subjects have shown that low protein diets lead to production of a more alkaline urine (pH 7.3–7.7) to an extent which could markedly decrease the rate of excretion of basic drugs due to suppression of ionisation (Fig. 45). The reverse, however, is true for acidic drugs.

In practice, the effect of diet is rather more complex, since oxidative metabolism, and hence the excretion rates of almost all drugs, whether basic, acidic or neutral, is lowered when protein intake is reduced. The retention of basic drugs is, therefore, doubly favoured by the combination of depressed metabolism and reduced excretion kinetics, whilst for acidic drugs, these are opposing factors and little change in the elimination rate results.

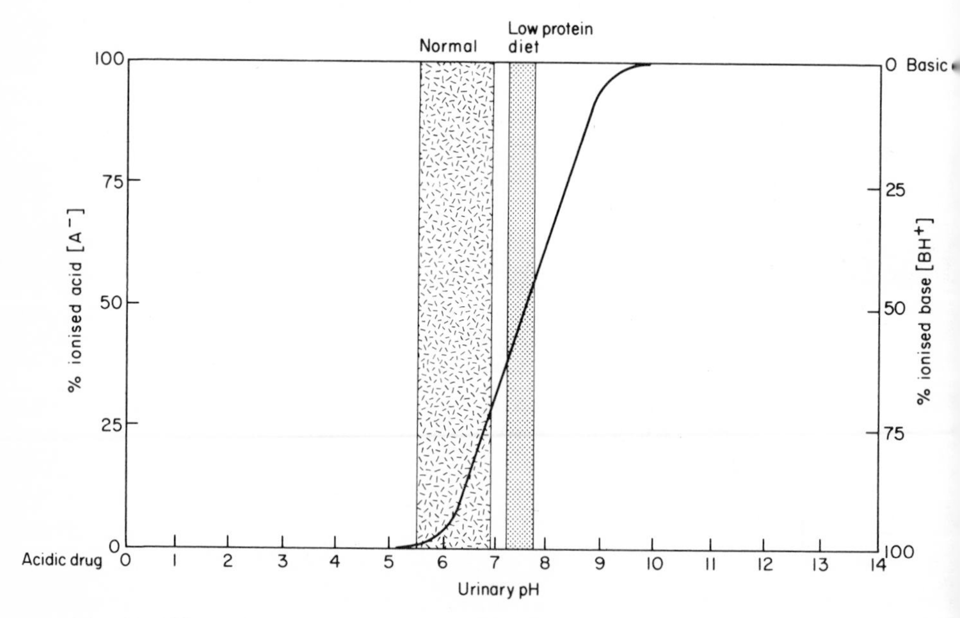

Fig. 45 Effect of low protein diets on urinary pH and on the excretion of ionised acids and bases

Excessive self-medication with alkalising antacids can also raise urinary pH. Likewise, patients consistently on large doses of acidic anti-rheumatic agents or taking massive doses of ascorbic acid are likely to excrete excessively acidic urine. Similarly, a low urinary pH is characteristic of patients suffering from gout, due to inhibition of ammonia production in the renal tubules (Gutman and Yü, 1963). Medication with uricosuric agents should, therefore, be accompanied by alkalisation of the urine to prevent deposition of uric acid crystals.

Urinary Clearance Rates

Renal function is normally assessed by the rate at which the kidneys are capable of excreting soluble waste products such as urea and uric acid. This is usually expressed in terms of **urinary clearance values**, which are defined as the volume of plasma in millilitres cleared by the substance in question by the kidneys per minute. The rate of elimination is, therefore, related to the plasma concentration by the expression:

$$\text{Clearance} = \frac{U \cdot V}{P}$$

where U = concentration of substance in urine in mg/ml
V = volume of urine excreted per min
P = concentration of substance in plasma in mg/ml

For example, if urea is eliminated in the urine at the rate of 18 mg/ml and plasma concentration is 0.25 mg/ml, then:

$$\text{Urea clearance} = \frac{18}{0.25}$$

$$= 72 \text{ ml/min}$$

Clearance values are particularly valuable for assessing the efficiency of kidney excretory function. Glomerular filtration is measured by determining the clearance of inulin, a high molecular weight plant polysaccharide (MW 5200) derived from fructose, which is non-toxic, and excreted quantitatively without tubular reabsorption and unchanged by metabolism. Clearance rates are measured following administration and normalised to a body surface area of 1.73 m^2 using the factor 1.73/patient's body surface in m^2. Normal values lie in the range 110–142 mg/ml for men and 101–133 mg/ml for women. Creatinine clearance values are usually somewhat lower, since some tubular reabsorption of creatine occurs in man (not in the dog). Normalised creatinine clearance values range from 97–135 for men and 85–120 for women.

Tubular secretion is similarly measured by clearance of *p*-aminohippuric acid (PAH). At concentrations less than that required to saturate the secretory mechanism, PAH is completely cleared at a single pass, and the clearance value is equivalent to the **renal plasma flow**. Normalised values usually lie in the range 491–817 ml/min for men and 439–745 ml/min for women. When the plasma

PAH concentration is raised to saturate the tubular secretion mechanism, any further increase in PAH plasma concentration due to glomerular filtration will be evident and measurable from the change in slope of the excretion rate plot.

Renal clearance values can be used as a rough guide to the excretion pathway. Normal values around 130 ml/min are indicative of glomerular filtration, while significantly higher values suggest active secretion, and lower values, extensive tubular reabsorption.

5 Drug Metabolism

FUNDAMENTAL ASPECTS

Introduction

The process of metabolism and excretion, together, are mainly responsible for terminating the action of drugs in the human body. The metabolic changes to which drugs are subject *in vivo* are directed primarily towards the formation of metabolites, which have physico-chemical properties favourable to their excretion. In practice, this means that they undergo transformation to more highly water-soluble derivatives usually by the unmasking or introduction of such polar groups as COOH, OH, and NH_2. These processes may be extended by subsequent formation of water-soluble conjugates with sulphuric acid, glucuronic acid or amino acids. They may also be accompanied by fragmentation leading to the formation of volatile small molecules, such as carbon dioxide and formaldehyde, capable of pulmonary excretion in exhaled air from the lungs.

Metabolic processes of this sort often result in detoxification of otherwise toxic substances as in the transformation of phenol to phenyl sulphate ($PhO \cdot SO_2OH$). Alternatively, they may lead to products of much greater toxicity as in the well-known examples of **lethal synthesis** from fluoroacetic acid and fluoroacetamide, in which they are converted metabolically to fluorocitrate, a specific inhibitor of the enzyme aconitase (Peters, Wakelin, Rivett and Thomas, 1953). Fluorocitrate, therefore, blocks the tricarboxylic acid cycle, an

HSCoA
F·CH₂·CO·**OH** → F·CH₂·CO·**SCoA** → F·CH(CO·O⁻)·C(CO·O⁻)(OH)·CH₂·CO·O⁻
H₂O

Fluoroacetic acid — Fluoroacetyl-coenzyme A — Fluorocitrate

CH₂·CO·O⁻ / C(CO·O⁻)(OH) / CH₂·CO·O⁻ —Aconitase→ ⁻O·OC–CH=C(CO·O⁻)–CH₂–CO·O⁻

Citrate — *cis*-Aconitate

essential feature of cellular energy supply, at the point where citrate is normally converted to *cis*-aconitate.

Another example of lethal synthesis is the metabolic conversion of the insecticide, *Parathion*, which is non-toxic *per se*, to its highly potent oxygen analogue, *Paroxon*, a powerful inhibitor of cholinesterase.

$$O_2N-C_6H_4-O\cdot P(=S)(OEt)\cdot OEt \longrightarrow O_2N-C_6H_4-O\cdot P(=O)(OEt)\cdot OEt$$

Parathion *Paroxon*

Frequently, however, metabolic transformation can lead to the unmasking of substances which are pharmacologically inactive, or of only low activity, to give highly active products. One of the earliest and best known examples of this form of unmasking is the transformation of the inactive drug, *Prontosil*, to the potent antibacterial, *Sulphanilamide* (Sulfanilamide), by liver azoreductase.

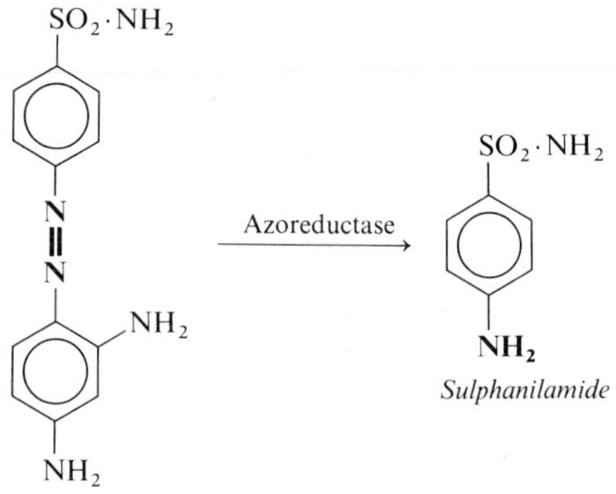

$$H_2NO_2S-C_6H_4-N{=}N-C_6H_3(NH_2)_2 \xrightarrow{\text{Azoreductase}} H_2NO_2S-C_6H_4-NH_2$$

Prontosil *Sulphanilamide*

The analogous oxidative cleavage of *Arsenphenamine* is similarly responsible for its conversion to *Oxophenarsine Hydrochloride* which is the effective anti-protozoal agent. An arsenoxide is also formed by metabolic reduction of the trypanocide, *Tryparsamide* ($H_2N\cdot CO\cdot CH_2\cdot NH\cdot C_6H_4\cdot AsO(OH)ONa$).

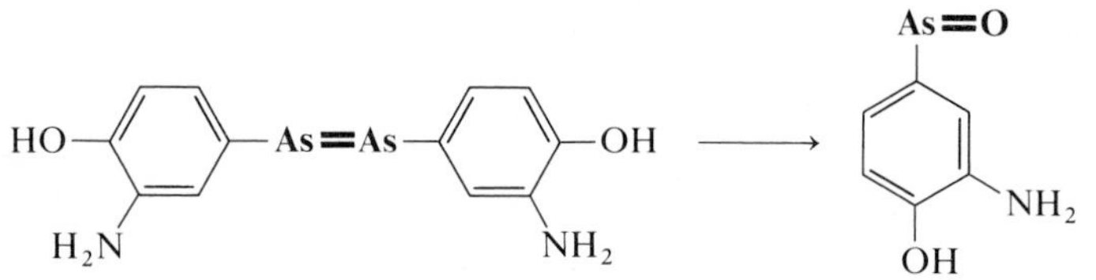

$$HO(H_2N)C_6H_3-As{=}As-C_6H_3(NH_2)OH \longrightarrow O{=}As-C_6H_3(NH_2)OH$$

Arsphenamine *Oxophenarsine Hydrochloride*

O-Demethylation of *Codeine* results in the formation of formaldehyde as a small volatile fragment, and release of *Morphine* as the effective analgesic (Adler, 1963). Unlike *Morphine*, however, *Codeine* is appreciably lipid-soluble, and hence is readily absorbed when administered orally. By virtue, therefore, of its demethylation to morphine, it is able to function as a weak orally-administered analgesic.

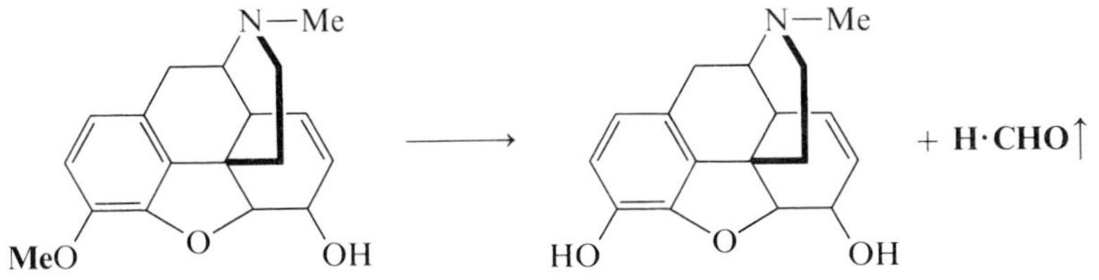

N—Me; MeO; O; OH → N—Me; HO; O; OH + H·CHO↑

Codeine *Morphine*

Similarly, the analgesic activity of *Phenacetin* is attributable to the formation of *Paracetamol* (Acetaminophen) by *O*-de-ethylation with release of acetaldehyde. *Paracetamol*, which is used in its own right as a mild pain killer, is also formed *in situ* from *Acetanilide* (Acetanilid) by ring hydroxylation.

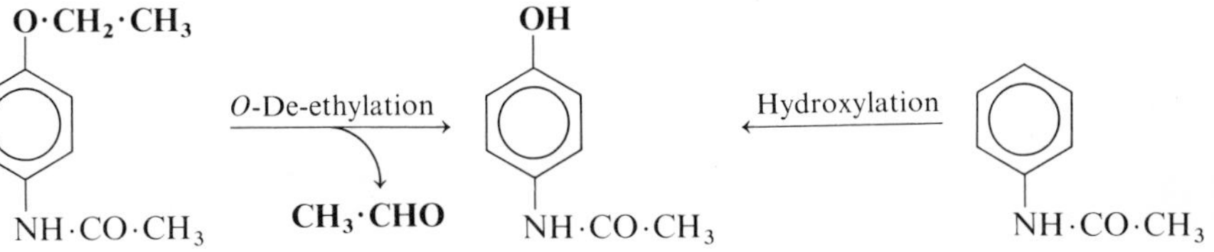

$O \cdot CH_2 \cdot CH_3$; $NH \cdot CO \cdot CH_3$ —*O*-De-ethylation→ ($CH_3 \cdot CHO$) OH; $NH \cdot CO \cdot CH_3$ ←Hydroxylation— $NH \cdot CO \cdot CH_3$

Phenacetin *Paracetamol* *Acetanilide*

Tertiary amines may also undergo *N*-demethylation, as in the case of the anti-depressant drug, *Imipramine*. The drug itself is pharmacologically inactive, and has a very slow onset of action (2 weeks), as a result of bio-demethylation to the active anti-depressant, *Desipramine*.

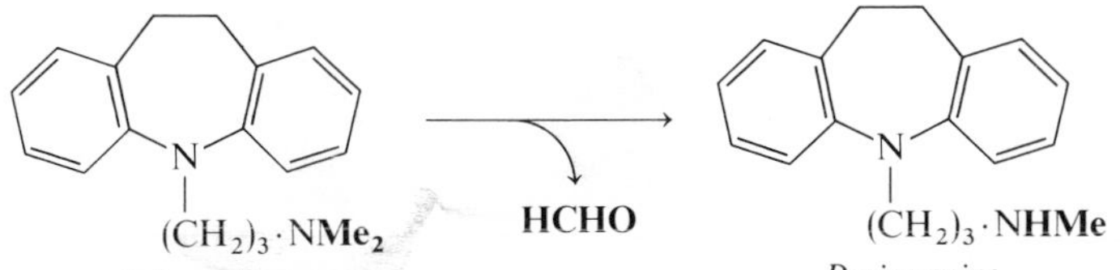

N; $(CH_2)_3 \cdot NMe_2$ → (HCHO) N; $(CH_2)_3 \cdot NHMe$

Imipramine *Desipramine*

Ring hydroxylation occurs in the metabolic oxidation of the anti-rheumatic drug, *Phenylbutazone* (Phenbutazone), with conversion to the more effective and less toxic *Oxyphenbutazone*.

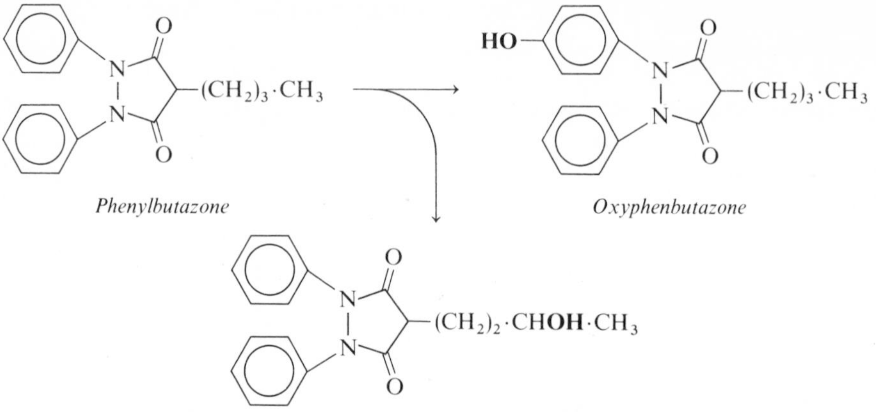

Phenylbutazone → *Oxyphenbutazone*

$(CH_2)_3 \cdot CH_3$ → $(CH_2)_2 \cdot CHOH \cdot CH_3$

Other examples of metabolic transformations leading to active products include reduction of *Chloral Hydrate* to the hypnotic trichloroethanol, and reductive ring closure of *Proguanil* (Chloroguanide) to the active antimalarial dihydrotriazine.

$$CCl_3 \cdot \mathbf{CHO} \xrightarrow[\text{dehydrogenase}]{\text{Liver alcohol}} CCl_3 \cdot \mathbf{CH_2 \cdot OH}$$

Chloral Hydrate *Trichloroethanol*

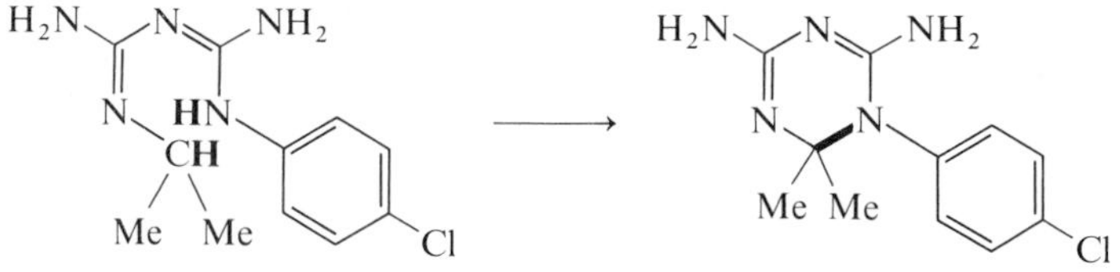

Proguanil

Evolutionary Aspects. Metabolism in the Foetus, Infants and the Aged

According to Brodie (1956), many of the drug metabolising enzymes are not the normal enzymes of intermediary metabolism, but an evolutionary development to permit higher animals to dispose of lipid-soluble substances, such as plant wax hydrocarbons, terpenes, alkaloids and other toxic substances ingested in foods. Thus fish, which for example cannot dealkylate amidopyrine, do not require oxidative metabolism, since they are able to dispose of lipid-soluble organic compounds through their gill membranes. Similarly, turtles, frogs and other aquatic amphibia are incapable of oxidising hexobarbitone, and oxidising enzymes only appear in land-based reptiles, birds and higher animals which have developed water-impermeable skins to prevent moisture loss, and in consequence have more complex excretory systems.

In keeping with this concept of evolutionary development of drug metabolising enzymes, the mammalian foetus, which develops in an embyro sac cushioned by the aqueous environment of amniotic fluids, may be considered to be at the fish stage of evolutionary development. Not surprisingly, the foetus is found to be almost entirely lacking in oxidising enzymes capable of metabolising foreign compounds, and these only appear in appreciable amount some weeks or, in some cases, months after birth (Jondorf, Maickel and Brodie, 1958; Fouts and Adamson, 1959). The human foetus, however, even at twelve weeks is capable of hydrolysing *Aspirin* (Ahmed, 1972). The ability to activate latent oxidising enzymes, which are suppressed at birth, has been demonstrated in newborn rats (Iba, Soyka and Schulman, 1975). The phospholipid components of oxidase enzymes in newborn rats contain twice as many unsaturated bonds as those from adults, and treatment which reduces these double bonds activates the enzyme.

Whilst the placenta appears to act as a barrier to the transfer of larger molecules of molecular weight greater than about 1000, compounds of lower molecular weight not only diffuse across the placental barrier, but do so for the most part without undergoing metabolic change (**2**, 4). Studies of human placental homogenates (Juchau, Niswander and Yaffe, 1968) obtained during therapeutic abortion at 9–12 weeks gestation have shown the absence of nitroreductase, oxidases, and acetylases. Glucuronyltransferase is also lacking (Schmid, Buckingham, Mendilla and Hammaker, 1959), but there is some evidence of hydrolysing capability (Petten, Hirsch and Cherrington, 1968; Ahmed, 1972).

Similarly, whilst enzymes which hydrolyse esters and glucuronides are not deficient in the newborn infant, ester and glucuronide synthesis is deficient, and there is a deficiency of glucuronyltransferase (Dutton, 1959). Thus, *Chloramphenicol*, which is normally excreted as the glucuronide, is highly toxic in premature infants; mortality with such treatment can be as high as 85% compared with 25% or less in infants at a later stage of development when glucuronide synthesis has developed, as it does within the first 30 days of life. Reduction of azo and nitro compounds is also deficient in newborn mammals (Jondorf, Maichel and Brodie, 1958).

Very little is known of the state of drug metabolising capability in old age. Geriatric patients, however, not infrequently suffer from massive hypothermias. Others on low incomes may be on protein-deficient diets, whilst yet others (males) may suffer a fall in testosterone production, all factors which could lead to a diminution in drug metabolising capability. Indeed, it has been shown that the mean plasma half-lives of orally-administered *Phenazone* (Antipyrene; 18 mg/kg in aqueous solution) and *Phenylbutazone* (6 mg/kg in tablets) are 45% and 29% respectively greater than in young adults. The half-life of *Phenazone* in older women alone was 78% greater than in the younger age group, a result which may merely reflect the longer survival of women to an age where drug metabolising ability rapidly declines (O'Malley, Crooks, Duke and Stevenson, 1971). There are also reports that *Nitrazepam* is a particularly unsuitable hypnotic for the aged, who not infrequently show considerable signs of toxification due to impaired metabolising ability (Evans and Jarvis, 1972).

Species Differences

Significant differences are known to exist in the metabolism of drugs in different animal species. Such differences are especially important in the extrapolation of results from animal metabolism studies to man. In some instances, the differences are merely quantitative differences in the rate of metabolism, since in general the smaller the animal the higher the metabolic rate. Thus, a dose of 50 mg/kg of *Hexobarbitone* (Hexobarbital), which will maintain anaesthesia for several hours in man, produces at the same dose level a sleeping time of only minutes in the mouse. Similarly, whilst more than 80% of a single dose of *Pentobarbitone* (Pentobarbital; p. 232) is excreted within 24 hr in dogs, in man only 45% is excreted in 48 hr (Maynert and Dawson, 1952). Pharmacokinetic factors other than excretion rates, which may account for such quantitative differences, include fat deposition, protein binding and body weight.

Much more significant differences in the pattern of drug metabolism may be observed when there are differences in the rate of metabolism, or even marked differences in the actual metabolic pathway. Thus, rabbits, in distinction to man, mice, rats, dogs and monkeys, remain relatively free from the toxic effects of *Methotrexate* (**1**, 23), owing to its more rapid metabolism in the rabbit to 4,7-dihydroxymethotrexate and excretion (Johns, Ianotti, Sartorelli, Booth and Bertino, 1964; Redetzki, Redetzki and Elias, 1966). Similarly, differences in the metabolism of *Imipramine* in various species readily account for the different pharmacological responses seen in the rat and man on the one hand and in the rabbit on the other. Thus, metabolism of *Imipramine* in the rat and man gives rise mainly to desipramine (Dingell, Sulser and Gillette, 1964), which has a potent anti-reserpine action, whilst the principal metabolite in the rabbit is 2-hydroxy-imipramine, which has negligible anti-reserpine activity. Similarly, the marked species difference between the metabolic pathways of *Malathion* in insects and man accounts for its effectiveness, as an insecticide, and its safety to man (**2**, 2).

Even more clear-cut species differences are shown in conjugating mechanisms (p. 268). Thus, insects, bacteria, and plants conjugate alcohols, thiols and phenols with glucose (and other sugars) to form glycosides utilising uridine diphosphate glucose (UDPG) and glycosyltransferase.

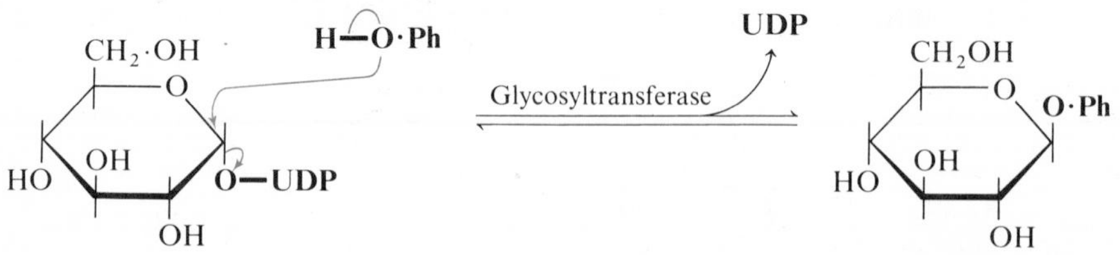

$CH_2{\cdot}OH$
H—O·Ph
O
HO
OH
O—UDP
OH
UDP
Glycosyltransferase
CH_2OH
O
O·Ph
HO
OH
OH

In contrast, all mammals except the cat (Robinson and Williams, 1958), which is deficient in glucuronyltransferase (Dutton and Greig, 1957), marsupials, amphibia, reptiles and birds form glucuronides (p. 273). Some well-defined species ideosyncracies in conjugating mechanisms (Smith, 1969), including the

Table 39. Some Species Differences in Conjugation Mechanisms

Metabolic Pathway	Cat	Dog	Rhesus Monkey	Man
N-Acetylation of aromatic acids	++	−[1]	++	++ (or +)
Glucuronidation	−	+	++	++
Glutamine conjugation of aromatic acids	−	−	++	++
Glycine conjugation of aromatic acids	++	++[2]	−	+ (or −)
Mercapturic acid conjugation	++	++	+ (or −)	+

− absent or very ineffective
+ slow
++ fast
[1] Sulphonamides undergo acetylation of the sulphonamido group (N^1-acetylation)
[2] Fenclofenac conjugates almost exclusively with taurine ($H_2N \cdot CH_2 \cdot SO_2OH$; Jordan and Rance, 1974)

inability of dogs to acetylate aromatic amines (Marshall, Cutting and Emerson, 1937) are listed in Table 39.

Sex Differences

A number of important sex differences in response to drugs have been observed in the rat, which relate to differences in metabolism. Thus, the sleeping time with barbiturates is longer in female rats than in males (Holck, Kanan, Mills and Smith, 1937), due to more rapid metabolism in the male. Production of androgen has been shown to be the key factor in these differences, since treatment of female rats with testosterone enhances the rate of drug metabolism (Quinn, Axelrod and Brodie, 1958), whilst the feeding of oestradiol to male rats produces the opposite response (Brodie, Maichel and Jondorf, 1958). In contrast, sex differences in the metabolism of drugs appear to occur only rarely in guinea pigs, rabbits, dogs and man (Quinn, Axelrod and Brodie, 1958).

Some metabolic retardation has, however, been observed in pregnancy, when oestrogen and progestogen levels are abnormally high, particularly in the third trimester. Thus, oxidative *N*-demethylation of *Pethidine* (Meperidine) is reduced in pregnancy, as also is the metabolism of *Chlorpromazine* (Rudofsky and Crawford, 1966). Similar depression of metabolising capability also occurs on administration of *Stilboestrol* (Diethylstilbestrol) and oral contraceptives (O'Malley, Stevenson and Crooks, 1972). *Amylobarbitone* (Amobarbital) also appears to stimulate the metabolism of the antithyroid drug, *Methimazole*, in patients suffering from hyperthyroidism (Stenlake and Skellern, 1972).

Genetic Effects

Studies of *Dicoumarol* (Dicumarol), *Phenazone*, *Desipramine* and *Nortriptyline* in sets of identical and non-identical twins have quite clearly established that drug metabolism is subject to genetic variation. The plasma half-life of *Dicoumarol*, which is excreted only after metabolic hydroxylation, was almost identical after a single oral therapeutic dose in identical twins, but showed significant differences in most pairs of fraternal twins (Vesell and Page, 1968). Similar results have been obtained by Hammer and Sjöqvist (1967) with *Desipramine* and *Nortriptyline*.

Two clinically important drugs, which show some quite startling differences in drug metabolism due to genetic factors, are *Suxamethonium* (Succinylcholine Chloride) and *Isoniazid*. The former is a short-acting neuromuscular blocking agent, which owes its short action to metabolic hydrolysis and consequent disruption of the molecule by plasma pseudocholinesterases. This mechanism is deficient in a small proportion of the population. Subjects with this deficiency are, therefore, at risk if this potent and otherwise non-reversible neuromuscular blocking agent is used on them (Lehman and Ryan, 1956). Kalow and Staron (1957) have shown this to be due to a genetically-determined deficiency, by typing human serum cholinesterase activity in some 1700 subjects using an inhibition test with *Cinchocaine* (Dibucaine). The percent enzymic inhibition achieved in this test, termed the **Dibucaine Number** (DN), was found to lie in one of three categories described as usual (high; DN 78 $\pm$ 7, intermediate (DN 62 $\pm$ 4) or atypical (DN $<$ 20). Such a classification implies the existence of two types of enzyme, which occur either singly or in admixture. The proportion of the two enzymes present is genetically determined and, therefore, varies from person to person. The majority of subjects (*ca* 97 %) have a high DN with the two enzymes present in equal proportions. About one in 30 persons has an intermediate DN with some deficiency of one of the enzymes, whilst an atypical DN occurs with a frequency of one in 3000 to one in 10 000. Subjects in this latter category who appear to possess only one atypical pseudocholinesterase with a low affinity for *Suxamethonium*, are particularly at risk with this drug.

The rate at which the anti-tubercular drug, *Isoniazid*, is acetylated shows quite large individual variation, and patients can be classified as either **slow** or **fast acetylators** (Peters, Miller and Brown, 1965; Jenne, 1965).

$CO{\cdot}NH{\cdot}NH_2$ $\quad CH_3{\cdot}Co{\cdot}S{\cdot}\overline{CoA}$ $\longrightarrow$ $CO{\cdot}NH{\cdot}NH{\cdot}CO{\cdot}CH_3$

N $\qquad HS{\cdot}\overline{CoA}$ $\qquad$ N

Isoniazid $\qquad$ Acetylisoniazid

Circulating levels of *Isoniazid* are much higher in slow acetylators, and such patients are more prone to show toxic side-reactions (Wade, 1970). Failures in the treatment of pulmonary tuberculosis with *Streptomycin* and *Isoniazid* in single large doses weekly are also higher amongst rapid than amongst slow acetylators.

Slow acetylation of *Isoniazid* is due to a deficiency of the appropriate hepatic *N*-acetyltransferase. The same enzyme is also involved in the acetylation of *Sulphadimidine* (Sulfamethazine) and *Hydrallazine* (Hydralazine), since slow

Isoniazid acetylators are unable to metabolise these drugs efficiently (White and Evans, 1968; Van Oudtshoorn and Potgieter, 1972). In contrast, however, the acetylation of *Sulphanilamide* and *Aminosalicylic Acid* is monomorphic in humans, so that a different acetyltransferase must be involved in the metabolism of these particular compounds (Peters, Gordon and Brown, 1965).

The distribution of slow and fast acetylators shows some interesting variations in particular ethnic groups. About 75% of Caucasians and Negroes are slow acetylators, whereas Japanese, and Eskimos, who are also of Mongolian descent, are largely (90% or more) fast acetylators.

Classification of Metabolic Transformations

Much, but not all metabolism of drugs is catalysed by microsomal enzymes of the endoplasmic reticulum of the liver, and to a lesser extent, kidney, lung and some other tissues. Non-microsomal enzymic metabolism takes place in a number of other specialised tissues. Sub-classification is based on the type of reaction involved, so that main classes of metabolic process may be listed as follows:

(a) microsomal oxidation,
(b) microsomal reduction,
(c) non-microsomal oxidation,
(d) non-microsomal reduction,
(e) hydrolysis (microsomal and non-microsomal),
(f) conjugation.

The Endoplasmic Reticulum and Microsomal Enzymes

The endoplasmic reticulum forms a tubular tissue network consisting of two types of tissue, known as **rough** and **smooth** endoplasmic reticulum. The rough-surfaced form consists of tubules studded with large numbers of small dense bodies known as ribosomes, which play an essential rôle in protein synthesis. In contrast, smooth endoplasmic reticulum, as its name implies, has a smooth surface and is devoid of ribosomes. Oxidase enzyme activity is associated mainly with smooth endoplasmic reticulum, although enzyme synthesis appears to be concentrated in the rough form (Ernster and Orrenius, 1965).

The endoplasmic reticulum is broken down to form small vesicles, known as microsomes when liver tissue is homogenised. Microsomal preparations are obtained by centrifuging the homogenate at 10 000 g (10 min) to deposit cell debris, and then centrifuging again at 100 000 g (1 hr) when the microsomes, which are deposited as a pellet, separate from the so-called supernatant soluble fraction.

There is substantial evidence that microsomal enzymes are associated with a lipid membrane and are lipid-dependent. Thus, sonic vibration or the use of hypotonic solutions fail to solubilise them, whilst treatment of microsomes with desoxycholic acid, which solubilises lipid membranes, destroys the activity of the oxidative enzymes. Also lauric acid, which undergoes ω-hydroxylation, inhibits *C*-oxidation of other compounds (Orrenius and Thor, 1969) due to solubilisation and hence denaturation of microsomal enzymes. Oxidative *O*-dealkylation of alkoxy-*p*-nitrophenols, similarly, decreases with increasing alkyl chain length (McMahon, Culp, Mills and Marshall, 1963).

The membrane-dependent UDP-glucuronyltransferase (p. 273) has been shown to be inactivated in microsomal preparations by both phospholipase A and phospholipase C (Graham and Wood, 1969). These enzymes, which are Ca^{2+}-dependent, degrade phosphatidylcholine and related membrane phospholipids, lysolecithins and glycerophosphates respectively.

$$\begin{array}{l} CH_2\cdot O\cdot CO\cdot R^1 \\ | \\ CH\cdot \mathbf{O\cdot CO\cdot R^2} \\ |\qquad\ \ O \\ |\qquad\ \ \| \\ CH_2\cdot O\cdot P\cdot \mathbf{O\cdot CH_2\cdot CH_2\cdot \overset{+}{N}Me_3} \\ \qquad\quad\ | \\ \qquad\quad\ O^- \end{array}$$

Phospholipase A $\rightarrow$

$$\begin{array}{l} CH_2\cdot O\cdot CO\cdot R^1 \\ | \\ CH\cdot \mathbf{OH} \\ |\qquad\ \ O \\ |\qquad\ \ \| \\ CH_2\cdot O\cdot P\cdot \mathbf{O\cdot CH_2\cdot CH_2\cdot \overset{+}{N}Me_2} \\ \qquad\quad\ | \\ \qquad\quad\ O^- \end{array} \quad + \mathbf{R^2\cdot CO\cdot OH}$$

Phospholipase C $\downarrow$

$$\begin{array}{l} CH_2\cdot O\cdot CO\cdot R^1 \\ | \\ CH\cdot \mathbf{O\cdot COR^2} \quad + \mathbf{HO\cdot CH_2\cdot CH_2\cdot \overset{+}{N}Me_3} \\ |\qquad\ \ O \\ |\qquad\ \ \| \\ CH_2\cdot O\cdot P\cdot \mathbf{OH} \\ \qquad\quad\ | \\ \qquad\quad\ O^- \end{array}$$

The fall in microsomal glucuronyltransferase activity was shown to be in parallel with the release of these hydrolysis products, though the products themselves were not inhibitors. More significantly, glucuronyltransferase activity was almost completely restored after treatment with phospholipase A by addition of a phospholipid dispersion. Activity, however, was only partially restored after treatment with phospholipase C which suggests that the system is able to restore the fatty acid ester link, but not the phosphorylcholine link. This technique has been applied by Bock and Fleischer (1975) to obtain membrane-dependent bovine heart mitochondrial D-β-hydroxybutyrate apodehydrogenase in a homogeneous soluble but inactive form, which is devoid of lipid. This apoenzyme, which is stable (half-life *ca* 450 hr at 0°), can be reactivated by aqueous microdispersions of lecithin (Gazzotti, Bock and Fleischer, 1975).

Rogers and Strittmatter (1973) have similarly shown that two associated membrane-dependent enzymes, cytochrome b_5 and NADH-cytochrome b_5 reductase, which catalyse oxidation–reduction processes in the endoplasmic reticulum, bind the vesicles prepared from purified microsomal lipids. Model studies show that these two enzymes, which are attached to the surface of the endoplasmic reticulum by an apparently hydrophobic protein, are able to catalyse the reduction of cytochrome *c* by NADH (Rogers and Strittmatter, 1975) when bound simultaneously to egg lecithin liposomes. This complex enzyme interaction is temperature-dependent, occurring readily at 37°, but becoming frozen at 24°, not through inactivation of the individual enzymes, but by physical immobilisation of the process of alignment in the lipid membrane without which interaction is impossible (Fig. 46).

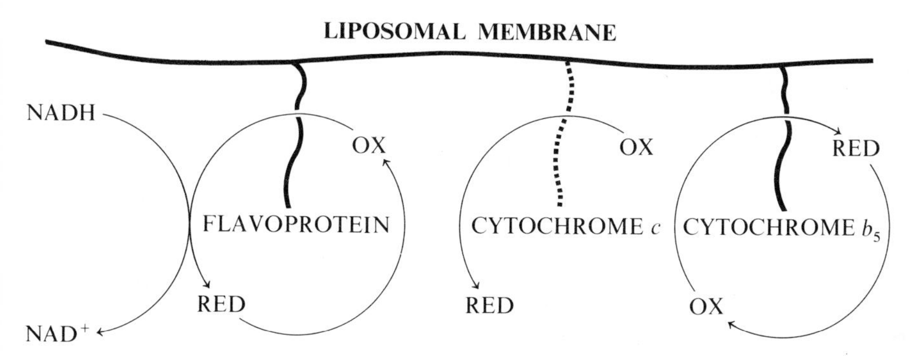

Randomised attachment of enzymes immobilised at 24°

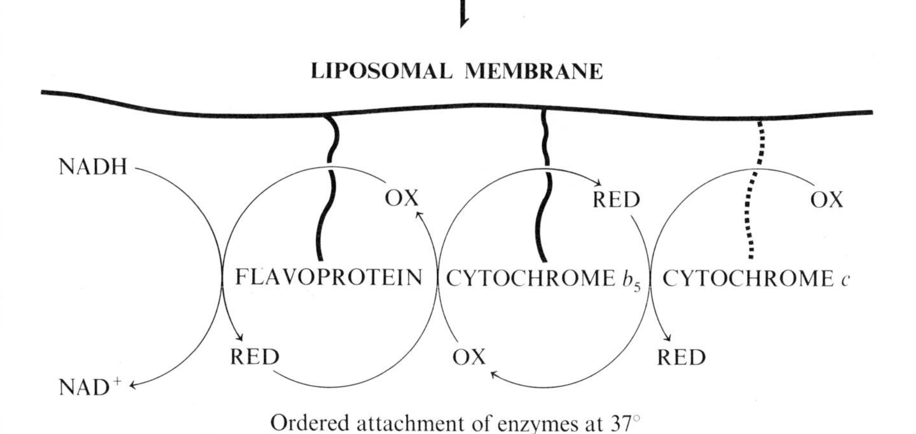

Ordered attachment of enzymes at 37°

Fig. 46 Interaction of membrane-dependent enzymes in egg-lecithin lyposomes

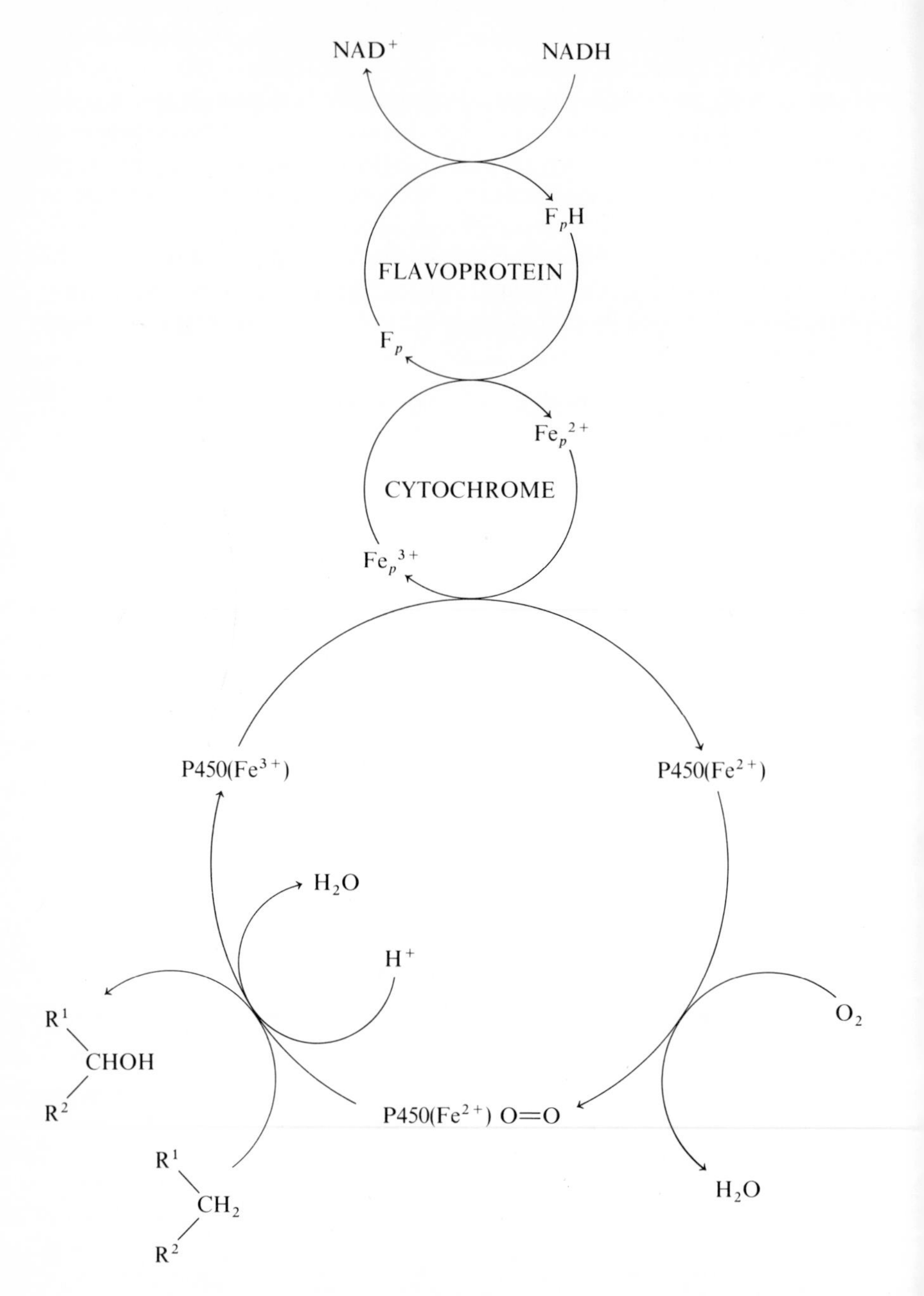

Fig. 47 Microsomal oxidation

Microsomal oxidation requires both nicotinamide–adenine dinucleotide phosphate in its reduced form (NADPH) and molecular oxygen (Brodie, Gillette and La Du, 1958). Experiments with $^{18}O_2$ and $H_2{}^{18}O$ show that the oxygen in the oxidised compound originates in molecular oxygen, and not water, and that the second oxygen from each oxygen molecule is reduced to water by NADPH. The enzymes responsible are, therefore, classified (Mason, 1957) as mixed function oxidases.

The specific oxidising enzyme is a particular cytochrome, known as cytochrome P450 (Estabrook, Cooper and Rosenthal, 1963) on account of the characteristic peak at 450 nm in its carbon monoxide adduct difference spectrum. This enzyme requires an electron transport chain for its reduction, and this consists of a flavoprotein enzyme, cytochrome *c* reductase and cytochrome P450 reductase, which transfers the electron from the flavine to cytochrome P450 (Fig 47). Cytochrome *c* reductase in turn requires NADP as coenzyme. Carbon monoxide (CO) acts as an inhibitor of reduced cytochrome P450 by competing with molecular oxygen. It is the CO adduct which has the characteristic absorption maximum at 450 nm, and a trough at 410 nm, as opposed to the CO-difference spectrum of haemoglobin, which shows a peak at 420 nm and a trough at 450 nm. Oxidised cytochrome P450 has an absorption maximum at 422 nm which shifts to 407 nm on reduction.

Substrates of cytochrome P450 combine with the oxidised form of the enzyme (Gillette, 1969) with characteristic spectral shifts. These are of two types. Type I spectral shifts, which are seen with *Hexobarbitone* and *Chlorpromazine*, show a peak at 385 nm and a trough at 420 nm. Type II shifts exhibit a peak at 430 nm and a trough at 395 nm and are characteristic of pyridine and aniline (Fig. 48; Remmer, Schenkman, Estabrook, Sasame, Gillette, Narasimhulu, Cooper and Rosenthal, 1966).

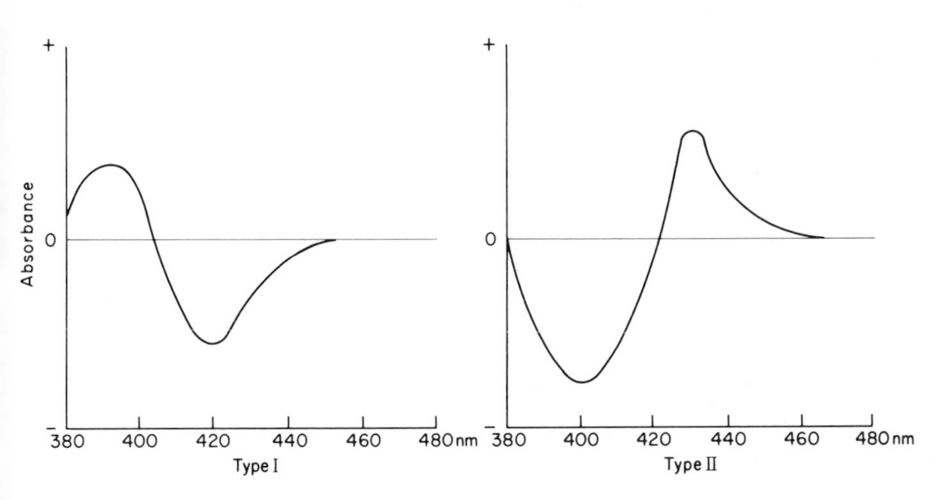

Fig. 48 Cytochrome P450—substrate difference spectra

Enzyme Induction

Oxidative drug metabolism is stimulated in animals by prior or co-administration of polycyclic aromatic hydrocarbons (Brown, Miller and Miller, 1954) and various drugs, notably *Phenobarbitone* (Phenobarbital; Remmer, 1959), but also for example by *Tolbutamide* and *Nikethamide* (Remmer and Merker, 1965). This drug-induced increase in drug metabolising activity is accompanied by an increase in liver weight (Goldberg, 1966) and particularly by proliferation of the smooth endoplasmic reticulum (Remmer and Merker, 1965; Ernster and Orrenius, 1965), following an increase in RNA polymerase. Less marked changes occur with benzpyrene and methylcholanthrene than with *Phenobarbitone* and other inducer drugs. Polycyclic aromatic hydrocarbons also enhance the metabolism of only a small number of drugs, whilst enhancement by *Phenobarbitone* is more widespread in its effect.

Liver changes, following *Phenobarbitone* induction, are marked by an increase in the associated cytochrome P450 and NADPH-cytochrome *c* reductase (Orrenius, 1965). Parallel increases in phospholipid biosynthesis and incorporation of ^{32}P into endoplasmic reticulum phospholipids (Ernster and Orrenius, 1964), bilirubin conjugation, vitamin D metabolism (Dent, Richens, Rowe and Stamp, 1970) and γ-glutamyltranspeptidase activity (Idéo, de Franchis, Del Ninno and Dioguardi, 1971; Martin, Martin and Goldberg, 1975) are also observed. Oxidative metabolism, including both hydroxylation and oxidative *N*-dealkylation, is correspondingly enhanced. Non-oxidative microsomal enzymic action, e.g. *Chloramphenicol* nitroreductase activity, but not cortisol Δ^3-ketoreductase activity, is similarly increased (Remmer and Merker, 1965). Liver changes are reversed on withdrawal of the inducing drug. They are also inhibited by *Puromycin* and other protein synthesis inhibitors (Conney and Gilman, 1963).

Barbiturates, *Glutethimide*, *Phenylbutazone* (Phenbutazone), *Chloral Hydrate* and ethanol, have all been shown to increase the rate of metabolism of other drugs in human subjects. Thus anticoagulant activity and plasma levels of *Dicoumarol* (Cucinell, Conney, Sansur and Burns, 1965) and *Warfarin* (MacDonald, Robinson, Sylwester and Jaffé, 1969) are reduced in patients also receiving *Phenobarbitone* (Phenobarbital), but pre-treatment plasma levels of the drugs and anticoagulant activity were restored on withdrawal of *Phenobarbitone* medication. Metabolism of *Dicoumarol* and *Warfarin* is similarly enhanced by *Chloral Hydrate* (Cucinell, O'Dessky, Weiss and Dayton, 1966) and by *Glutethimide* (Schmid, Cornu, Inhof and Keberle, 1964) respectively. Plasma levels of the anticonvulsant, *Phenytoin*, are also reduced on simultaneous administration with *Phenobarbitone* to about one-third of those obtained with therapeutic doses of *Phenytoin* alone. The intrinsic anticonvulsant action of *Phenobarbitone*, however, compensates for the loss due to *Phenytoin* when combined therapy is used in the management of epilepsy (Conney, 1969).

Phenylbutazone enhances the rate of oxidative *N*-dealkylation of *Amidopyrine* (Chen, Vrindten, Dayton and Burns, 1962), and also stimulates endogenous hydroxylation of cortisol (*Hydrocortisone*) to 6β-hydroxycortisol which is normally a minor metabolite (Kuntzman, Jacobson, Levin and Conney, 1968). Hydroxylation of *Hydrocortisone* which is also enhanced by *Phenobarbitone* (Bledsoe, Island, Ney and Liddle, 1964) and *Phenytoin* (Werk, MacGee and Sholiton, 1964), is now frequently used as an index of metabolism induction (Stevenson, Browning, Crooks and O'Malley, 1972). It has been used to confirm that Mandrax, a proprietary medicament containing *Methaqualone* and *Diphenhydramine*, stimulates *Antipyrine* metabolism, whereas neither component alone is active in this respect.

A consistently high consumption of alcohol not only enhances ethanol oxidation, but also markedly increases oxidative metabolism of *Pentobarbitone* (Rubin and Lieber, 1968) and *Tolbutamide* (Kater, Tobon and Iber, 1969). In contrast, however, it is perhaps significant that metabolic activity of liver homogenates *in vitro* is markedly reduced in the presence of ethanol. According to Conney (1969), this may explain both the increased tolerance of alcoholics to barbiturates when sober, and their enhanced response to the same hypnotics when inebriated.

Some drugs are able to induce their own metabolism. Thus, glucuronide formation is markedly enhanced in patients stabilised on *Glutethimide*, but shows a corresponding decline on re-administration following a 20-day withdrawal period (Schmid, Cornu, Inhof and Keberle, 1964). Hydroxylation of *Meprobamate* (Douglas, Ludwig and Smith, 1963) and of *Guanoxan* (Jack, Stenlake and Templeton, 1971) is also markedly enhanced on repeated administration to human subjects. Various indicators of metabolism induction are used. These include urinary excretion of 6β-hydroxycortisol (Poland, Smith, Kuntzman, Jacobson and Conney, 1970) and D-glucaric acid ($HO \cdot OC \cdot (CHOH)_4 \cdot COOH$), and plasma levels of γ-glutamyltranspeptidase (Idéo, de Franchis, Del Ninno and Dioguardi, 1971). Thus, a survey of plasma γ-glutamyltranspeptidase (GGT) activity in 100 patients (Martin, Martin and Goldberg, 1975) shows a positive correlation between GGT activity and both serum triglyceride and serum pre-lipoprotein levels. This correlation reflects hepatic microsomal induction in hyperlipidaemic patients, which increases their hepatic content of the enzymes for triglyceride synthesis. Notwithstanding such observations, GGT is also raised in patients with liver damage, kidney and prostate disease, and even after administration of some drugs which do not increase liver microsomal activity. For these reasons, plasma GGT levels are not a good indicator of microsomal induction.

Diet

The rate of drug metabolism can be considerably influenced by diet. Starvation inhibits the oxidative metabolism of *Hexobarbitone* and *Chlorpromazine* (Dixon, Shultice and Fouts, 1960; Kato and Takanaka, 1967), whilst increase in dietary protein in rats from the normal level of about 20% to around 50%, increases metabolic oxidation (Kato, Chiesara and Vassanelli, 1962; McLean and

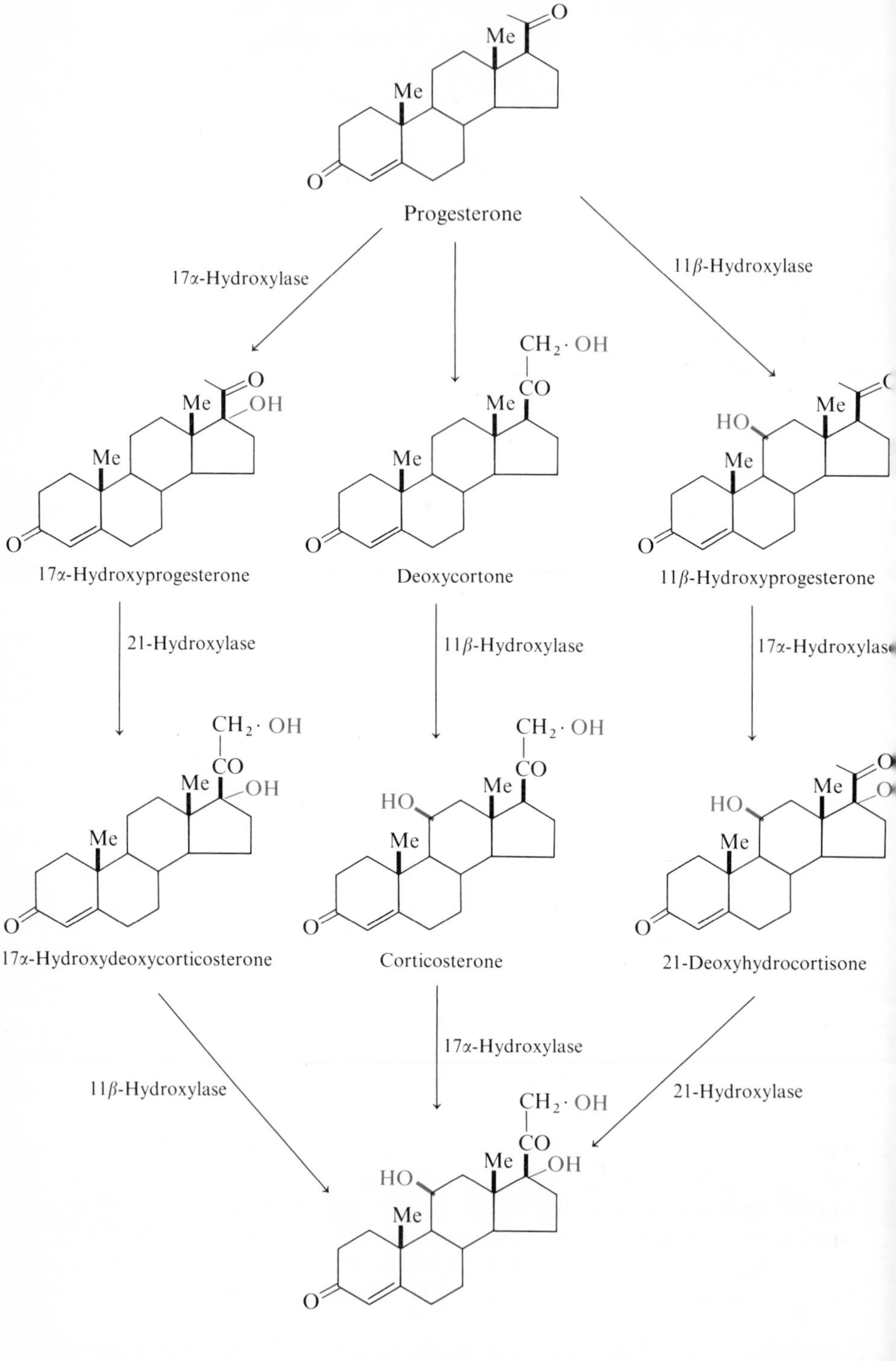

Me
Me
O
O
Progesterone

17α-Hydroxylase

11β-Hydroxylase

$CH_2 \cdot OH$
CO
Me
OH
Me
O
O
Me
Me
HO
Me
Me
O
O
O

17α-Hydroxyprogesterone

Deoxycortone

11β-Hydroxyprogesterone

21-Hydroxylase

11β-Hydroxylase

17α-Hydroxylase

$CH_2 \cdot OH$
CO
Me
OH
Me
O
$CH_2 \cdot OH$
CO
Me
HO
Me
O
Me
HO
Me
O
O
O

17α-Hydroxydeoxycorticosterone

Corticosterone

21-Deoxyhydrocortisone

17α-Hydroxylase

11β-Hydroxylase

21-Hydroxylase

$CH_2 \cdot OH$
CO
Me
OH
HO
Me
O

Verschuuren, 1969), due to increase in microsomal enzymic activity (Kato, Oshima and Tomizawa, 1968). This effect is greater in males than in females. The carbohydrate content of the diet is also an important factor in controlling the effect of oxidising enzymes (Kato, 1953).

In contrast to the inhibitory effect of protein deficiency on oxidative metabolism, glucuronide conjugation, but not sulphate conjugation, is markedly increased in rats and particularly in immature animals of the species (Wood and Woodcock, 1970; Woodcock and Wood, 1971; Graham, Wood and Woodcock, 1972). An important factor in this situation is the increase in glycogen content of the liver in protein deficiency, increasing carbohydrate availability for glucuronide synthesis. Administration of glucose and fructose in humans is likewise followed by an increase in glucuronide conjugation (Beck and Richter, 1962).

Low protein diets usually result in the excretion of alkaline urine (pH 7.3–7.7), which can also indirectly affect urinary excretion of ionised drugs and their metabolites (Hadzija, 1969; **2**, 4).

OXIDATION BY MICROSOMAL ENZYMES

C-Oxidation

Hydroxylation of Saturated Hydrocarbons

C-Oxidation forms one of the most important pathways for the metabolism of organic compounds, and is characteristic of the metabolism of paraffin hydrocarbons, saturated hydrocarbon moieties and aromatic ring systems. Steroid hydroxylases in the adrenal cortex normally play a vital rôle in the biosynthesis of endogenous hydroxy-steroids. There are a number of such steroid hydroxylases, each specific for the oxidation of a specific position in the steroid nucleus. This specificity is seen in the biosynthesis of *Hydrocortisone* from *Progesterone*, which involves successive oxidation at 11β, 17α and 21-positions, though not necessarily in that order, so that the complete synthesis may be achieved by three alternative pathways (Usui and Yamasaki, 1960; Halkerston, Eichhorn and Hechter, 1961). 17α-Hydroxylation of *Progesterone* occurs in *in vitro* cultures of *Rhizopus nigricans*, and is used in the manufacture of *Hydrocortisone* from *Progesterone* (Hayano, Gut, Dorfman, Sebek and Peterson, 1958). (See opposite.)

McCarthy (1964) showed that when [^{14}C]-hexadecane and [^{14}C]-octadecane are administered orally to goats, rats and chickens, ω-oxidation occurs, and that the radioactivity is recovered in fatty acids of the same chain length. Oxidation proceeds via the alcohol, which is then further oxidised to the carboxylic acid (Kusunose, Ichihara and Kusunose, 1969).

Microsomal ω-hydroxylation of long-chain hydrocarbons is centred on the liver, but shorter chain hydrocarbons, e.g. n-decane, are also actively oxidised

$$CH_3(CH_2)_{14}\cdot CH_3 \xrightarrow{\text{Oxidation}} CH_3(CH_2)_{14}\cdot \mathbf{CH_2\cdot OH}$$

Hexadecane → Hexadecanol

$$\downarrow \text{Oxidation}$$

$$CH_3(CH_2)_{14}\cdot \mathbf{CO\cdot OH}$$

Palmitic acid

by both lung and kidney (Ichihara, Kusunose and Kusunose, 1969a). ω-Hydroxylation is also typical of decanoic, lauric and other long-chain fatty acids (Ichihara, Kusunose and Kusunose, 1969b; Das, Orrenius and Ernster, 1968).

$$CH_3(CH_2)_8\cdot CO\cdot OH \xrightarrow{\text{Oxidation}} \mathbf{HO\cdot CH_2}(CH_2)_8\cdot CO\cdot OH$$

Decanoic acid → 10-Hydroxydecanoic acid

$$\swarrow \text{Oxidation}$$

$$\mathbf{HO\cdot OC}(CH_2)_8\cdot CO\cdot OH$$

Sebacic acid

Small alkyl substituents are also subject to ω-hydroxylation, as for example in *Barbitone* (Barbital), which although not appreciably metabilised, forms small amounts (3 %) of 5-ethyl-5β-hydroxyethylbarbituric acid.

Barbitone

5-Ethyl-5-hydroxyethylbarbituric acid (3 %)

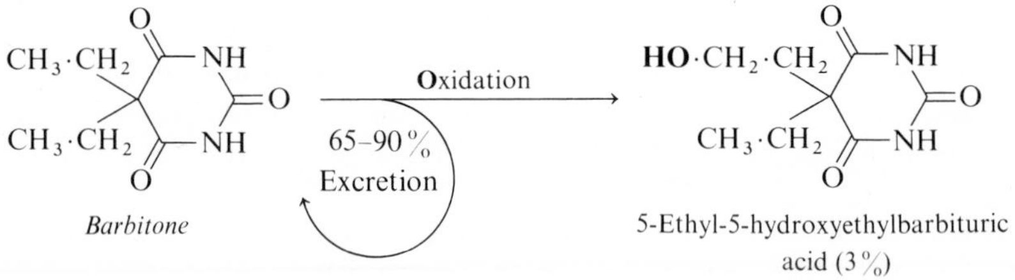

Metabolic oxidation of alkyl substituents in aromatic rings occurs much more readily, particularly at C—H bonds α to the ring, which are activated by resonance effects. Typical examples of this type of ω-oxidation are seen in the metabolism of *Butylated Hydroxytoluene* (BHT) (Ladomery, Ryan and Wright, 1967; Holder, Ryan, Watson and Wiebe, 1970) and *Tolbutamide* (Thomas and Ikeda, 1966).

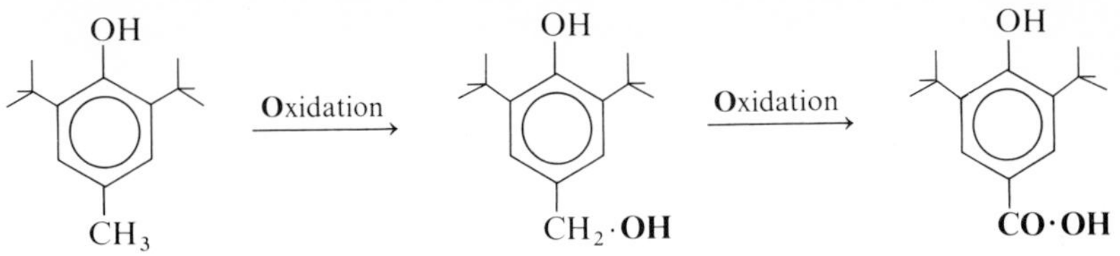

OH OH OH
Oxidation Oxidation
CH_3 $CH_2 \cdot$**OH** **CO·OH**

Butylated Hydroxytoluene

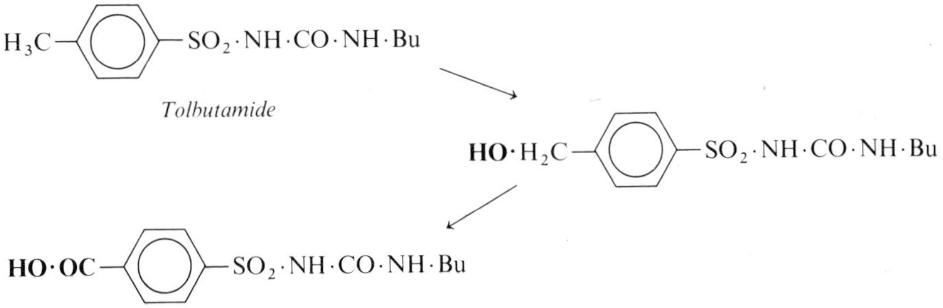

H_3C—⬡—$SO_2 \cdot NH \cdot CO \cdot NH \cdot Bu$

Tolbutamide

HO·H_2C—⬡—$SO_2 \cdot NH \cdot CO \cdot NH \cdot Bu$

HO·OC—⬡—$SO_2 \cdot NH \cdot CO \cdot NH \cdot Bu$

Metabolism of tetralin and fluorene similarly gives rise mainly to tetral-1-ol and fluoren-9-ol respectively. In both cases there is some evidence of an intermediate hydroperoxide (Elliott, Hanam, Parke and Williams, 1964; Elliott and Hanam, 1968; Chen and Lin, 1968, 1969).

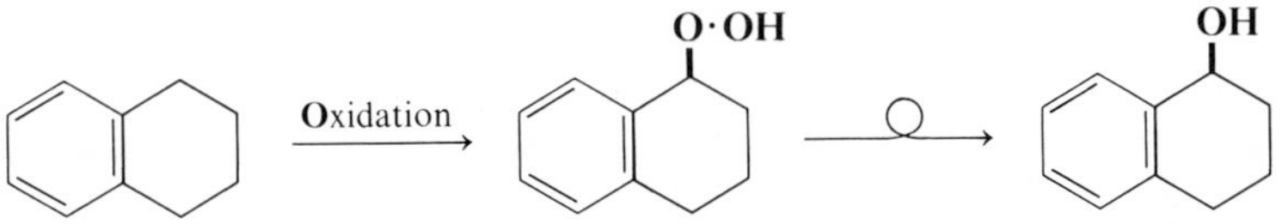

O·OH **OH**
Oxidation
Tetralin Tetralin-1-hydroperoxide Tetral-1-ol

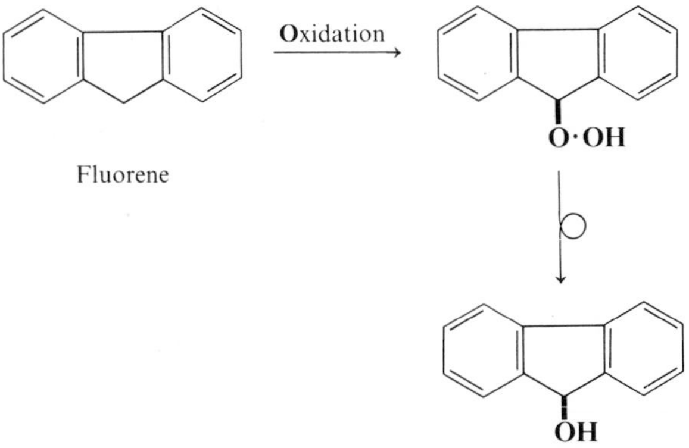

Oxidation
O·OH
Fluorene

OH
Fluoren-9-ol

It is perhaps significant that whilst long-chain fatty acids undergo ω-oxidation, longer chain alkyl substituents in otherwise lipophilic molecules are characteristically metabolised in the sub-terminal $(\omega - 1)$ position, as for example in *Pentobarbitone* (Cooper and Brodie, 1955a), *Meprobamate* and *Ethosuximide*.

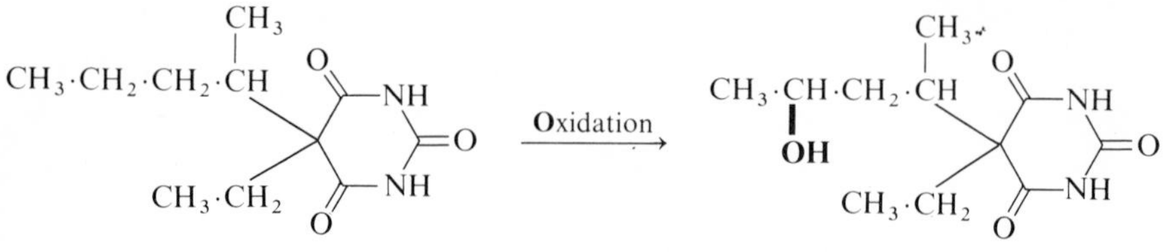

$CH_3\cdot CH_2\cdot CH_2\cdot CH(CH_3)$ / $CH_3\cdot CH_2$ (barbiturate ring) $\xrightarrow{\text{Oxidation}}$ $CH_3\cdot CH(\mathbf{OH})\cdot CH_2\cdot CH(CH_3)$ / $CH_3\cdot CH_2$ (barbiturate ring)

Pentobarbitone

Me, $CH_2\cdot CH_3$ (succinimide ring) $\longrightarrow$ Me, $CH(\mathbf{OH})\cdot CH_3$ $\longrightarrow$ Me, $\mathbf{CO}\cdot CH_3$

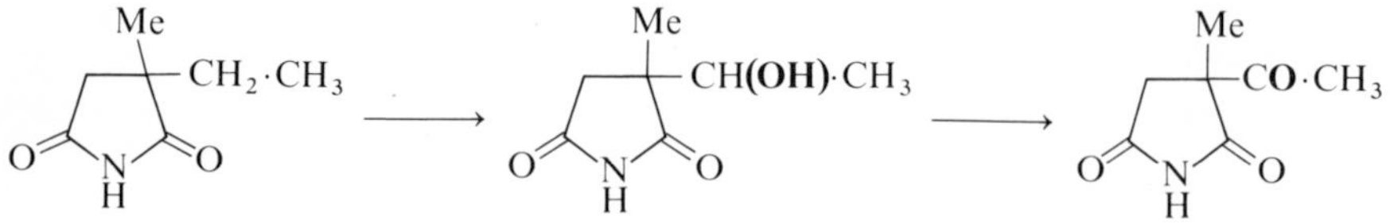

Ethosuximide

This seems to be a metabolic device to avoid the formation (by ω-oxidation) of a lipophilic carboxylic acid which like the long-chain fatty acids themselves would inhibit cytochrome P450 by detergent action on the lipophilic membrane to which it is attached and on which it is dependent (p. 222). In this connection, it is interesting to note that the *O*-dealkylation of alkoxy *p*-nitrophenols is not only increasingly inhibited with increasing length of the alkyl chain (McMahon, Culp, Mills and Marshall, 1963), but that the longer chain compounds are increasingly metabolised by *C*-oxidation (Yoshimura, Tsuji and Tsukamoto, 1966). Significantly, whilst the butyl and iso-amyl *p*-nitrophenyl ethers are subject to ($\omega - 1$) oxidation, the intermediate propyl ether undergoes ω-oxidation. Thus, only those hydrocarbon chains capable of producing acids, of sufficient chain length to have detergent properties, undergo ($\omega - 1$) oxidation.

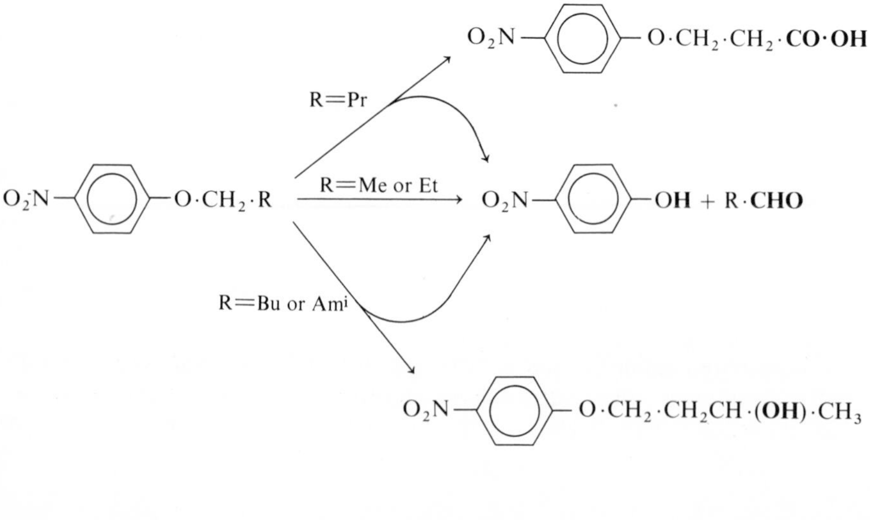

$O_2N-C_6H_4-O\cdot CH_2\cdot R$

R=Pr → $O_2N-C_6H_4-O\cdot CH_2\cdot CH_2\cdot \mathbf{CO\cdot OH}$

R=Me or Et → $O_2N-C_6H_4-\mathbf{OH} + R\cdot\mathbf{CHO}$

R=Bu or Am[i] → $O_2N-C_6H_4-O\cdot CH_2\cdot CH_2CH\cdot(\mathbf{OH})\cdot CH_3$

Meprobamate, which also embodies a n-propyl substituent, actually shows both ω- and $(\omega - 1)$-oxidation (Douglas, Ludwig and Smith, 1963).

Alicylic hydrocarbons are also metabolised by microsomal *C*-oxidation. Cyclohexane gives cyclohexanol (Ullrich, 1969). C—H bond cleavage, which is clearly involved, however, is not rate-limiting, since no kinetic isotope effect is seen in the metabolism of dodecadeuterocyclohexane. Methylcyclohexane is similarly hydroxylated (Elliott, Tao and Williams, 1965), but gives rise to a mixture of 2-, 3- and 4-hydroxy-1-methylcyclohexanes in which the latter predominate. The flexibility of the cyclohexane ring no doubt accounts for the formation of both *cis*- and *trans*-3- and 4-hydroxy compounds, though the conformationally more stable 3-*cis*- and 4-*trans*-compounds predominate.

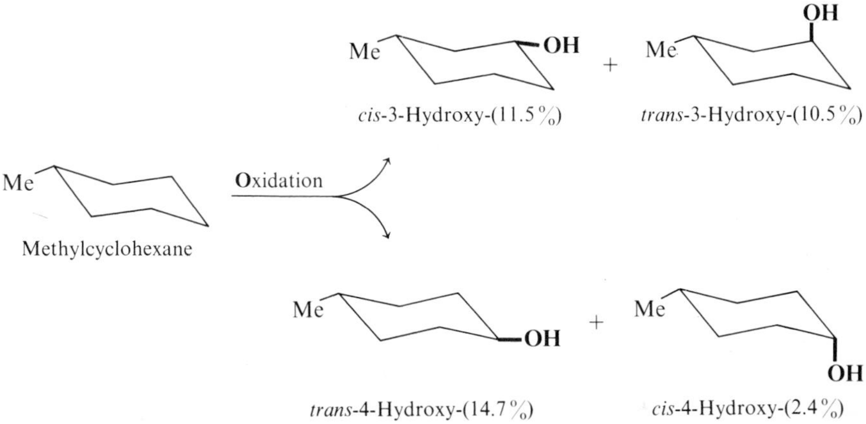

Me OH + Me OH

cis-3-Hydroxy-(11.5%) *trans*-3-Hydroxy-(10.5%)

Me Oxidation

Methylcyclohexane

Me OH + Me OH

trans-4-Hydroxy-(14.7%) *cis*-4-Hydroxy-(2.4%)

Stereochemical effects are much more readily discernible in *cis*- and *trans*-decalins, which are metabolised in substantial amount to *cis*-*cis*-2-decalol and *trans*-*cis*-2-decalol respectively (Elliott, Robertson and Williams, 1966). The entering hydroxyl group, which is *cis* to the ring junction C-10 hydrogen and fixed in a 1,3-bi-axial arrangement in the *trans*-decalin derivative (though not necessarily so in the *cis*-decalin compound because of conformational flexibility), appears to be directed in this way by the same stereochemical constraints that apply in the metabolic hydroxylation of *Testosterone* and *Hydrocortisone* to $6\beta,7\alpha,11\beta$- and 16α-hydroxytestosterones and 6β-hydroxycortisol respectively.

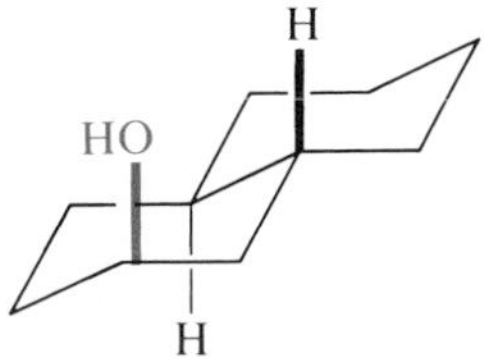

H HO H

trans-*cis*-2-decalol

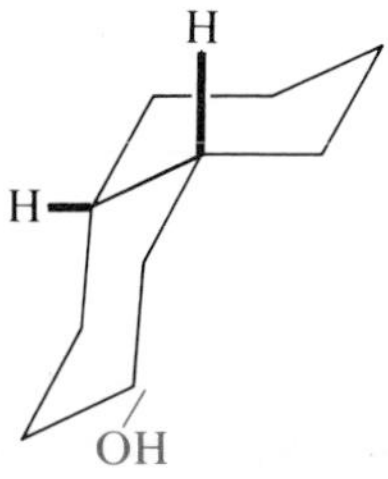

H H OH

cis-*cis*-2-decalol

The stereospecificity of these hydroxylations is still further underlined by the observations that 7α-[^{3}H]-cholesterol is metabolised in the rat to 7α-hydroxycholesterol with only 7% retention of tritium, whereas 7β-[^{3}H]-cholesterol is oxidised to 7α-hydroxy-7β-[^{3}H]-cholesterol (Bergström, Lindstedt and Samuelsson, 1958).

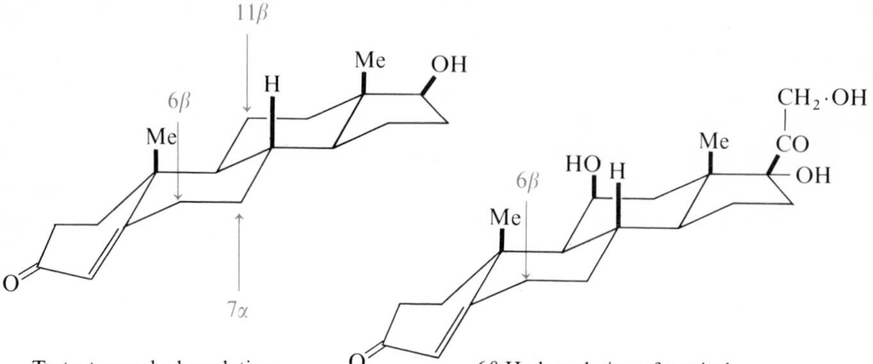

Testosterone hydroxylation

6β-Hydroxylation of cortisol

Similar considerations apply in the metabolic hydroxylation of non-steroid alicylic compounds. Both *Cyclobarbitone* (Cyclobarbital) and *Hexobarbitone* are metabolised in the allylic position, which is also the 3′-position, to give the 3′-hydroxy compound, and this is further oxidised to the 3′-oxo-derivative. *Hexobarbitone* also undergoes *N*-demethylation (Cooper and Brodie, 1955b).

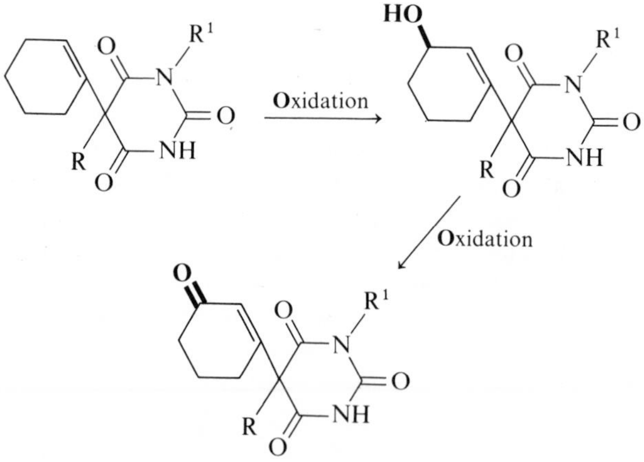

Cyclobarbitone (R = Et; R′ = H)
Hexobarbitone (R = Me; R′ = Me)

The oxidative metabolism of *Glutethimide*, however, provides a rare example of metabolic *C*-oxidation which proceeds by an entirely different pathway in the two optical enantiomers (Keberle, Riess, Schmidt and Hofmann, 1963).

The absolute stereochemistry of these isomers appears not to have been established, but it would seem that one enantiomer is stereochemically capable of coupling with the enzyme and the other not, possibly as shown.

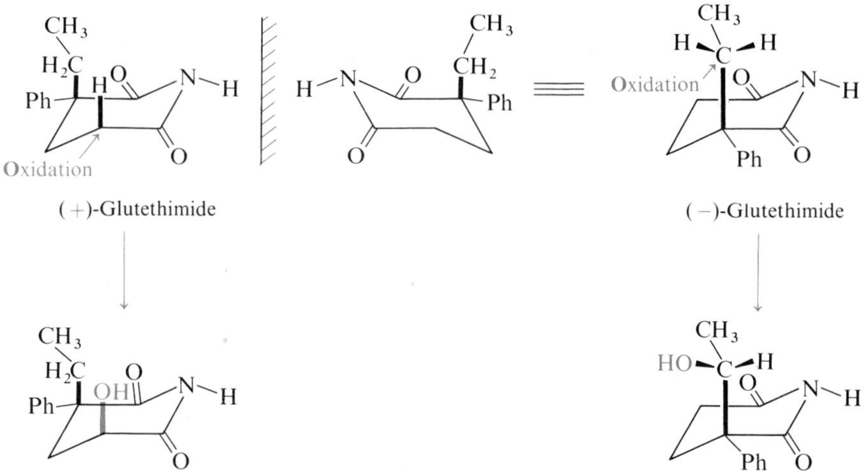

(+)-Glutethimide

(−)-Glutethimide

3-Ethyl-3-phenyl-5-hydroxy glutethimide (45%)

Hydroxylation of Aromatic Hydrocarbons

Aromatic compounds are also metabolised by *C*-oxidation. Thus, benzene undergoes detoxification by this route to phenol and catechol, both of which are excreted as the glucuronide and sulphate ester. Oxidation appears to take place via the 1,2-epoxide and the 1,2-dihydrodiol (Sato, Fukuyama, Suzuki and Yoshikawa, 1963). There is, however, no direct evidence of free (as opposed to enzyme-bound) epoxide being formed, though liver microsomes are able to utilise benzene epoxide as substrate (Witkop, 1969). (See structure at the top of p. 236.)

Similar hydroxylations occur in the metabolism of most substituted aromatics including, for example, chlorobenzene, *Phenobarbitone*, *Phenylbutazone* (p. 216) and *Guanoxan* (Jack, Stenlake and Templeton, 1971).

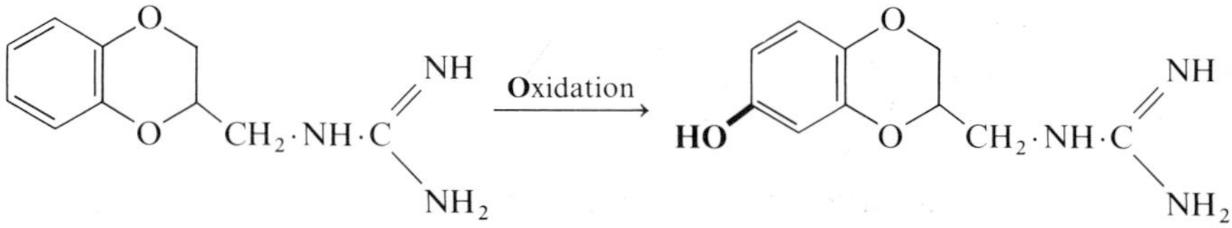

Guanoxan → Oxidation → 7-Hydroxyguanoxan

The NIH Shift

The NIH shift is a particular intramolecular rearrangement which occurs in the course of aromatic hydroxylation. The rearrangement is a hydride shift which normally is not apparent, but is observable in the metabolic oxidation of some appropriately tritiated and deuterated compounds. It was first observed in the enzymic hydroxylation of 4-tritiophenylalanine, which unexpectedly gave 3-tritiotyrosine with 90% retention of the tritium label (Guroff, Daly, Jerina, Renson, Witkop and Udenfriend, 1967).

The reactivity of substrates towards microsomal hydroxylation correlates well with their reactivity in electrophilic substitution. Rings which are activated for

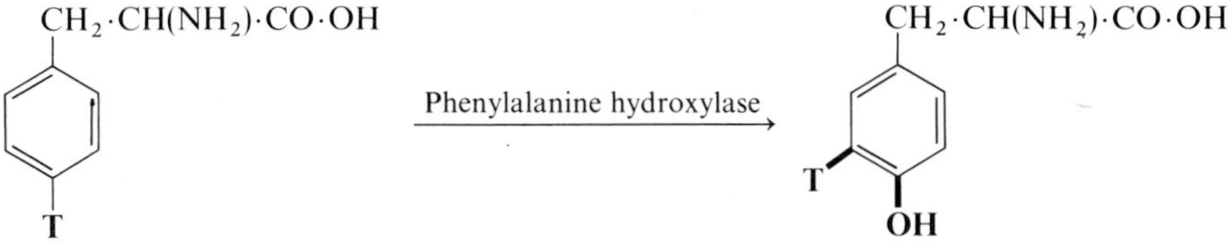

$CH_2 \cdot CH(NH_2) \cdot CO \cdot OH$ (4-Tritiophenylalanine) —Phenylalanine hydroxylase→ $CH_2 \cdot CH(NH_2) \cdot CO \cdot OH$ (3-Tritiotyrosine)

4-Tritiophenylalanine

3-Tritiotyrosine

electrophilic substitution (i.e. electron-rich) are the best substrates. The microsomal hydroxylating system, therefore, acts as a weak selective electrophile (Daly, Jerina and Witkop, 1968). The extent of migration and retention of the labelled atom likewise depends on the nature of the ring substituent. Substrates in which the ring substituent cannot readily lose a proton, such as 4-[^{2}H]-anisole, 4-[^{2}H]-diphenyl ether, 4-[^{2}H]-toluene, 4-[^{2}H]-chlorobenzene, 4-[^{2}H]-benzonitrile, 4-[^{2}H]-benzamide and 4-[^{2}H]-nitrobenzene show 40–60% deuterium retention with label shift according to the following rearrangement.

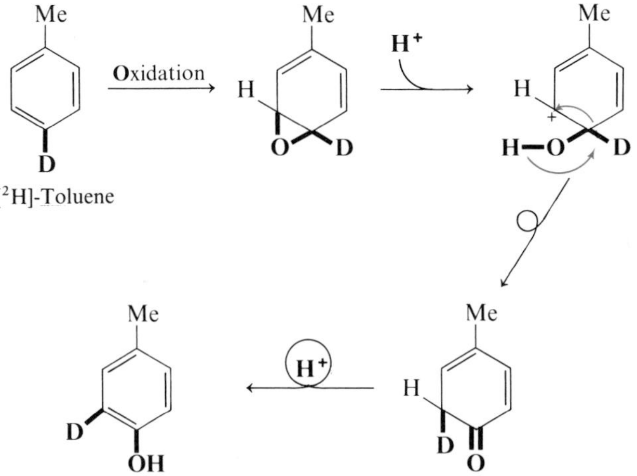

Me, D —Oxidation→ Me, H, O, D —H^+→ Me, H, +, H—O, D → Me, H, D, O —H^+→ Me, D, OH

4-[^{2}H]-Toluene

In contrast, substrates capable of ionisation by loss of a proton from the ring substituent, such as 5-[^{2}H]-salicylic acid, 4-[^{2}H]-aniline, 4-[^{2}H]-*N*-phenylbenzenesulphonamide and various 4-[^{2}H]-*N*-acylanilines, show a much lower label retention and shift (0–30% at pH 8.0) which, however, is pH dependent. Thus, the degree of label retention with 4-[^{2}H]-acetanilide ranges from 21% in an incubating medium at pH 10 to 44% at pH 6.0. Similarly, as the acidity of the amide hydrogen increases in a series of *N*-acylanilines, migration and retention of deuterium decreases, in accord with the following hydroxylation mechanism.

4-[^{2}H]-Acetanilide

It seems likely that an analogous alkyl shift would explain the rearrangement involved in the enzymic conversion of *p*-hydroxyphenylpyruvate to homogentisate (Schepartz and Gurin, 1949).

p-Hydroxyphenylpyruvate

Homogentisate

O-, S- and N-Dealkylation

Metabolic *O*-dealkylation of alkyl aryl ethers (Axelrod, 1956) takes place by breaking of the O—C bond, since studies with liver homogenates and $^{18}O_2$ show no incorporation of the heavy oxygen isotope into the products (Renson, Weissbach and Udenfriend, 1965). Typical *O*-dealkylations give rise to a phenol, and an aldehyde or ketone which results from *C*-oxidation at the carbon adjacent to the ether oxygen. Deuteration of the methyl group in *o*-nitroanisole inhibits oxidative *O*-demethylation by as much as 50% (Mitoma, Yasuda, Tagg and Tanabe, 1967) in accord with oxidation of a methoxy C—H bond as the rate-limiting step giving rise to a hemiacetal (Brodie, Gillette and La Du, 1958). *Phenacetin* gives rise to paracetamol and *Codeine* to *Morphine* by *O*-dealkylation. The volatile carbonyl compound is usually excreted via the lungs or further oxidised.

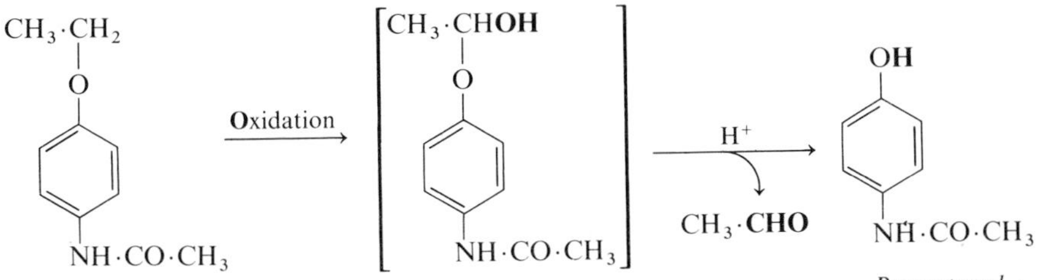

Phenacetin *Paracetamol*

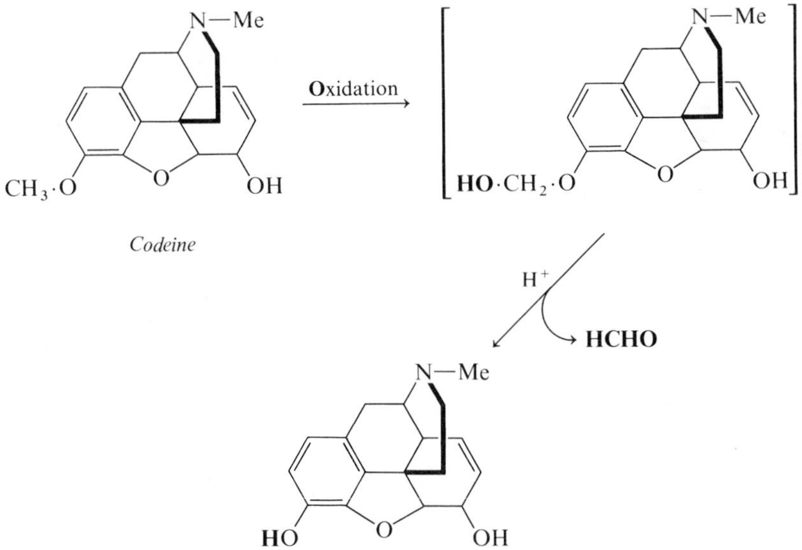

Codeine

Morphine

S-Demethylation of methanethiol, methylisothiourea and 6-methylthiopurine also occurs by a similar mechanism with release of formaldehyde (Mazel,

Henderson and Axelrod, 1964). Methionine and dimethyl sulphide, however, are not dealkylated by this pathway.

N-Dealkylation of tertiary amines is one of two alternative pathways by which tertiary amines are metabolised. Like *O*- and *S*-dealkylation, it too involves *C*-oxidation at the α-carbon atom (Bickel, 1971) catalysed by microsomal cytochrome P450, as shown by its suppression in the presence of the specific inhibitors, carbon monoxide (CO) and SKF-525A. In contrast, the alternative *N*-oxidation of tertiary amines is not inhibited by either CO or SKF-525A, in accord with its catalysis by an entirely different and independent cytochrome-free flavoprotein microsomal enzyme system (Pettit, Orme–Johnson and Ziegler, 1964; Ziegler, Mitchell and Jollon, 1969; Masters and Ziegler, 1971). The relative importance of these two pathways is subject to considerable species variation. Both *in vivo* and *in vitro* studies can be markedly influenced by a number of important experimental factors, including substrate concentrations, co-factors and pre-treatment with metabolism inducers (Beckett and Hewick, 1967; Arrhenius, 1970; Gorrod, Jenner, Keysell and Beckett, 1971; Gigon and Bickel, 1971). For example, enzyme induction in female rats, following pre-treatment with *Phenobarbitone* does not influence *N*-oxidation, but increases *N*-demethylation (Beckett, 1971).

Typical examples of *N*-demethylation include *Imipramine* (p. 215), *Propoxyphene* (McMahon and Sullivan, 1964) and *Methadone*. Studies of *N*-demethylation rates of some typical compounds (Hansch, Steward and Iwasa, 1965b) have shown that they correlate well with the following equation.

$$\log(\text{Demethylation rate}) = 0.470 \log P - 0.268\,(\mathrm{p}K_a - 9.5) - 1.305 \qquad (r = 0.890)$$

in which P is the calculated partition coefficient of the amine between octanol and water

and K_a is the ionisation constant of the amine.

The $(\mathrm{p}K_a - 9.5)$ term relates the difference between the ionisation constant of the amine and that of the base, $C_6H_5\cdot CH_2\cdot CH_2\cdot CH_2N\cdot(CH_3)_2$, and provides a measure of the electron density on nitrogen. This equation demonstrates the dependence of the process both on the lipophilicity of the amine (log P term), and the availability of electrons on nitrogen. The negative coefficient of the $(\mathrm{p}K_a - 9.5)$ term implies that low electron density favours high demethylation rates.

Demethylation rates are not significantly influenced by steric effects (McMahon and Easton, 1961), and the steric requirements for the protonation of bases are low compared with those for bulkier substituents. Hansch and his collaborators concluded, therefore, that *N*-demethylation could involve direct interaction between the enzyme and the *N*-lone pair of electrons. Whether or not this is correct, studies of metabolic *N*-debenzylation of *N*-benzyl*nor*pethidine in the presence of $^{18}O_2$ have shown that the heavy oxygen isotope is incorporated in the resulting benzaldehyde (McMahon, Culp and Occolowitz, 1969) in keeping with the following metabolic pathway.

$$>N\cdot CH_2\cdot Ph \xrightarrow{^{18}O_2} >N\cdot CH(^{18}OH)\cdot Ph \longrightarrow >N\text{—}H + Ph\cdot CH\cdot {}^{18}O$$

Oxidative *N*-dealkylation of tertiary amines is favoured with increase in alkyl group size, no doubt as a function of increasing lipophilicity. Thus, whereas *NN*-dimethylaminoalkylphenothiazines are metabolised both by *N*-oxidation and *N*-dealkylation, the latter represents the sole pathway for the metabolism of *NN*-diethylaminoethylphenothiazines (Beckett, 1971).

Phenothiazine–$CH_2\cdot CH_2\cdot CH_2\cdot NR_2$ —Oxidation (R = Me)→ Phenothiazine–$CH_2\cdot CH_2\cdot CH_2\cdot \overset{+}{N}(O^-)Me_2$

Phenothiazine–$CH_2\cdot CH_2\cdot CH_2\cdot NR_2$ —(R = Me), – HCHO→ Phenothiazine–$CH_2\cdot CH_2\cdot CH_2\cdot N(H)\cdot Me$

Phenothiazine–$CH_2\cdot CH_2\cdot CH_2\cdot NR_2$ —Oxidation (R = Et), – $CH_3\cdot$**CHO**→ Phenothiazine–$CH_2\cdot CH_2\cdot CH_2\cdot N(H)\cdot Et$

N-De-ethylation is also the major metabolic route of *Lignocaine* (Lidocaine) in man (Beckett, Boyes and Appleton, 1966).

2,6-Me$_2C_6H_3$–$NH\cdot CO\cdot CH_2\cdot NEt_2$ (*Lignocaine*) —Oxidation, – CH_3**CHO**→ 2,6-Me$_2C_6H_3$–$NH\cdot CO\cdot CH_2\cdot$**NH**$\cdot Et$

Excretion

N-Dealkylation also occurs as the major pathway in the metabolism of some secondary amines. The extent of *N*-dealkylation is also influenced by stereochemical parameters. Thus, the *dextro*(2*S*)-isomers of *N*-alkylamphetamines show decreasing excretion and increasing *N*-dealkylation as the alkyl group size increases from methyl, through ethyl to isopropyl (Vree, Muskens and van Rossum, 1971). The *laevo* isomers, however, show no such effect, indicating that the asymmetric α C—H bond is implicated in binding to the dealkylating enzyme as part of the rate-limiting step and, in accord with this conclusion, *dextro*-α-deutero-*N*-isopropylamphetamine shows a kinetic isotope effect, whereas the *laevo* isomer does not (Vree, Gorgels, Muskens and van Rossum, 1971). Similar steric effects are shown in the *N*-dealkylation of the (1*S*, 2*S*)-, and (1*R*, 2*S*)-alkyl*nor*ephedrines.

N-Oxidation

Tertiary Amines

N-Oxidation provides an alternative pathway for the metabolism of tertiary amines, which although dependent on liver microsomes, NADPH and molecular oxygen, is independent of cytochrome P450. For this reason, *N*-oxidation is not induced by *Phenobarbitone* (Uehlake, 1971) and is not inhibited by CO or SKF-525A (Bickel, 1971). It is, however, inhibited by dithiothreitol and cysteamine (Gorrod, Jenner, Keysell and Beckett, 1971), and accordingly is considered to be catalysed by a cytochrome-free flavoprotein enzyme system (Ziegler, Mitchell and Jollon, 1969; Masters and Ziegler, 1971).

N-Oxidation is typical of the metabolism of both aliphatic and aromatic tertiary amines, for example trimethylamine, *Chorpromazine* (Beckett and Hewick, 1967), *NN*-dimethylaniline, *Nicotinamide*, *Nikethamide* (Niacinamide) and *Tripelennamine* (Gorrod, 1971).

$CH_2{\cdot}CH_2{\cdot}CH_2{\cdot}NMe_2$
Chlorpromazine

$CH_2{\cdot}CH_2{\cdot}CH_2{\cdot}\overset{+}{N}{\cdot}Me_2$ (O^-)
Chlorpromazine-*N*-oxide (34 %)

HCHO

$CH_2{\cdot}CH_2{\cdot}CH_2{\cdot}NH{\cdot}Me$
Norchlorpromazine (12 %)

$CH_2{\cdot}CH_2{\cdot}CH_2{\cdot}NMe_2$
Chlorpromazine-*S*-oxide (10 %)

N-Oxidation is remarkably sensitive to steric factors. Thus, introduction of a methyl substituent into the phenothiazine basic side-chain reduces *N*-oxidation (Hewick, 1967). Similarly, *N*-demethylation progressively increases with increasing steric hindrance about nitrogen in the series normethadone, isomethadone, *Methadone*, at the expense of *N*-oxidation (Beckett, 1971), as shown by the increase in excretion of cyclic metabonate.

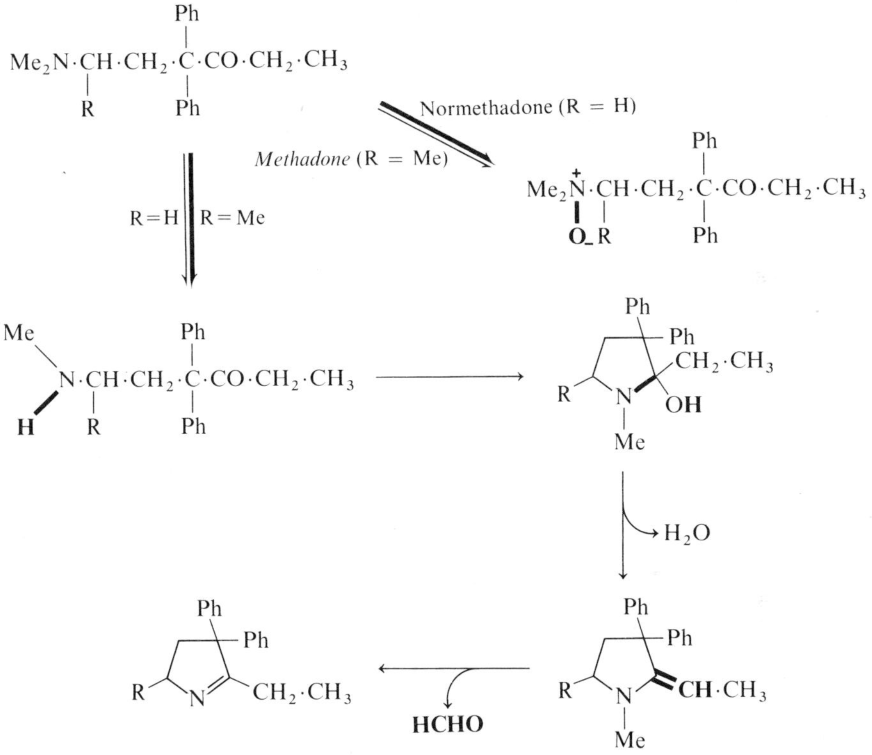

$Me_2N \cdot CH(R) \cdot CH_2 \cdot C(Ph)_2 \cdot CO \cdot CH_2 \cdot CH_3$

Normethadone (R = H)

Methadone (R = Me)

$Me_2\overset{+}{N}(O^-) \cdot CH(R) \cdot CH_2 \cdot C(Ph)_2 \cdot CO \cdot CH_2 \cdot CH_3$

R=H R=Me

$Me(H)N \cdot CH(R) \cdot CH_2 \cdot C(Ph)_2 \cdot CO \cdot CH_2 \cdot CH_3$

Ph, Ph, $CH_2 \cdot CH_3$, OH, R, N, Me

H_2O

Ph, Ph, R, N, $CH_2 \cdot CH_3$

HCHO

Ph, Ph, R, N, Me, $CH \cdot CH_3$

N-Hydroxylation of Primary and Secondary Amines

Both primary and secondary amines undergo *N*-hydroxylation by microsomal enzymes of liver and other extra-hepatic tissues such as lung and kidney. The hydroxylation of primary and secondary amines is, however, influenced by different factors. Hydroxylation of secondary amines, in contrast to primary amines, resembles *N*-oxidation of tertiary amines in that it does not require cytochrome P450 and is not subject to stimulation by enzyme induction (Uehlake, 1971). Thus, although metabolism of *N*-methylaniline is stimulated by *Phenobarbitone* pre-treatment, enhancement is confined to the *N*-demethylation and *p*-hydroxylation routes. Similarly, *N*-hydroxylation of *N*-benzylamphetamine is inhibited by dithiothreitol, whereas the alternative *N*-dealkylation pathways are inhibited by SKF-525A (Beckett, 1971). The formation of hydroxylamines from secondary amines appears to take place via the corresponding *N*-oxide. The detection of urinary metabolites which might help to confirm this view, however, is complicated by their extensive chemical degradation which occurs under the acidic and alkaline conditions of the usual extraction procedures. This is demonstrated in the study of *N*-benzylamphetamine which is metabolised by the following pathways.

$$C_6H_5\text{—}CH_2\cdot C(H)(Me)\cdot NH\cdot CH_2\cdot Ph$$

Oxidation

$$Ph\cdot CH_2\cdot C(\mathbf{OH})(Me)\cdot NH\cdot CH_2\cdot Ph \quad\quad Ph\cdot CH_2\cdot C(H)(Me)\cdot NH\cdot CH(\mathbf{OH})\cdot Ph \quad\quad Ph\cdot CH_2\cdot C(H)(Me)\cdot \overset{+}{N}H(\mathbf{O^-})\cdot CH_2\cdot Ph$$

$$\downarrow$$

$$Ph\cdot CH_2\cdot C(H)(Me)\cdot N(\mathbf{OH})\cdot CH_2\cdot Ph$$

$$\downarrow \; + Ph\cdot\mathbf{CHO}$$

$$Ph\cdot CH_2\cdot\mathbf{CO}\cdot CH_3 + H_2N\cdot CH_2\cdot Ph \quad\quad Ph\cdot CH_2\cdot C(H)(Me)\cdot NH_2 + Ph\cdot\mathbf{CHO} \quad\quad (Ph\cdot CH_2)(Me)C{=}N\cdot\mathbf{OH}$$

syn- and *anti*-benzylmethyl ketoxime

Metabolism of primary amines, including aniline, 2-naphthylamine, *Sulphanilamide*, *Dapsone*, phenetidine (Uehlake, 1971) and *Amphetamine* (Beckett, 1971) gives rise to the corresponding *N*-hydroxy compounds, of which some have been shown to be carcinogenic. The products are considered to be formed via the *N*-oxide, but the mechanism of oxidation is different from that of secondary amines, since it requires cytochrome P450 and is subject to enzyme induction by *Phenobarbitone*, suggesting that it may, perhaps, occur by direct oxidation of the N—H bond.

Some light is shed on the complexities of the process by the $^{18}O_2$ incorporation studies of Parli, Wang and McMahon (1971a) which provide evidence for the following pathways.

$$R_2{\cdot}CH{\cdot}NH_2 \xrightarrow{^{18}O_2} \left[R_2{\cdot}C(^{18}OH){\cdot}NH_2\right] \xrightarrow{-H_2{}^{18}O} R_2C{=}NH \xrightarrow{^{18}O_2} R_2C{=}N{\cdot}{}^{18}OH$$

$$\left[R_2{\cdot}C(^{18}OH){\cdot}NH_2\right] \xrightarrow{-NH_3} R_2C{=}{}^{18}O \xleftarrow{\text{Chemical hydrolysis}} R_2C{=}N{\cdot}{}^{18}OH$$

$R_2C{=}{}^{18}O$: 25% Incorporation of $^{18}O_2$

$R_2C{=}N{\cdot}{}^{18}OH$: 95% Incorporation of $^{18}O_2$

Further evidence for the imine pathway comes from the cytochrome P450-dependent microsomal *N*-hydroxylation of 2,4,6-trimethylacetophenonimine (Parli, Wang and McMahon, 1971b) to 2,4,6-trimethylacetophenone oxime.

$$2,4,6\text{-}(CH_3)_3C_6H_2{-}C(CH_3){=}NH \xrightarrow{\text{Oxidation}} 2,4,6\text{-}(CH_3)_3C_6H_2{-}C(CH_3){=}N{\cdot}OH$$

N-Hydroxylation of Amides

Both primary and secondary amides are known to be subject to metabolic *N*-hydroxylation. *Urethane* (Boyland and Nery, 1965) and *N*-acetylaminofluorene (Cramer, Miller and Miller, 1960) are both converted to highly carcinogenic *N*-hydroxyamides.

$CH_3{\cdot}CH_2{\cdot}O{\cdot}CO{\cdot}NH{\cdot}OH$

N-Hydroxyurethane

$C_{13}H_9{-}N(OH){\cdot}CO{\cdot}CH_3$

N-Hydroxyacetylaminofluorene

The harmful side-effects of *Phenacetin* have also been attributed (Nery, 1971a) to the formation of an *N*-hydroxy metabolite which functions as a reactive intermediate for rearrangement to 2-hydroxyphenacetin, and for acetylation and alkylation of proteins and nucleic acids (Nery, 1971b). There is some evidence, too, that the reactivity of the hydroxyamines is considerably enhanced by conjugation as glucuronides and sulphate esters, and that these are important intermediates in their further reaction (Irving, 1971).

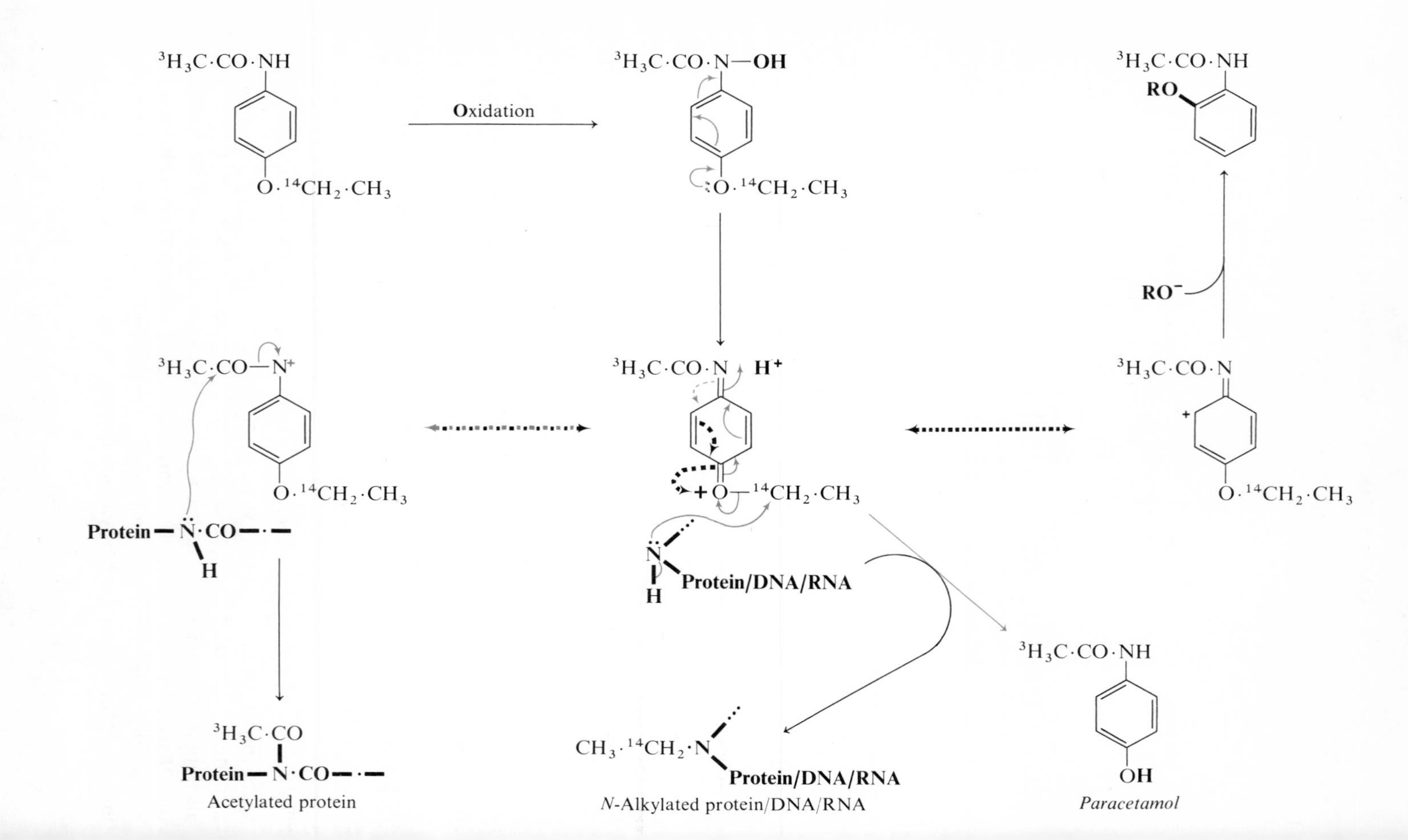

Oxidation
RO⁻
Protein/DNA/RNA
Acetylated protein
N-Alkylated protein/DNA/RNA
Paracetamol

The *N*-hydroxylation of natural peptide N—H bonds occurs in the biosynthesis of the *N*-hydroxypeptide, *Desferrioxamine* (**1**, 11) by certain microorganisms. *N*-Hydroxypeptides have also been reported to be present in spontaneous human tumour tissue, in mouse virus tumours, and in chemically-induced rat tumours (Neunhoeffer, 1970).

REDUCTION BY MICROSOMAL ENZYMES

Microsomal enzymes of the endoplasmic reticulum are also concerned with reduction, catalysing the reduction of azo and nitro compounds and the reductive dechlorination of alkyl halides. All three processes are complex, involving more than one pathway. All, however, are dependent on NADPH–cytochrome *c* reductase and almost certainly mediated by flavoprotein enzymes such as FADH (Mueller and Miller, 1950).

Azo Reduction

Azo compounds undergo reductive cleavage to primary aromatic amines by azoreductase. Thus, *Prontosil* owes its *in vivo* antibacterial activity to its reduction to sulphanilamide (p. 214). Similarly, the carcinogenic activity of *p*-dimethylaminoazobenzene is suppressed by dietary riboflavine which stimulates the formation of FAD, and hence its reduction to *p*-dimethylaminoaniline. The reductions are thought to parallel the corresponding chemical reductions proceeding via the hydrazo compounds (Bray, Hybs, Clowes, Lake and Thorpe, 1951).

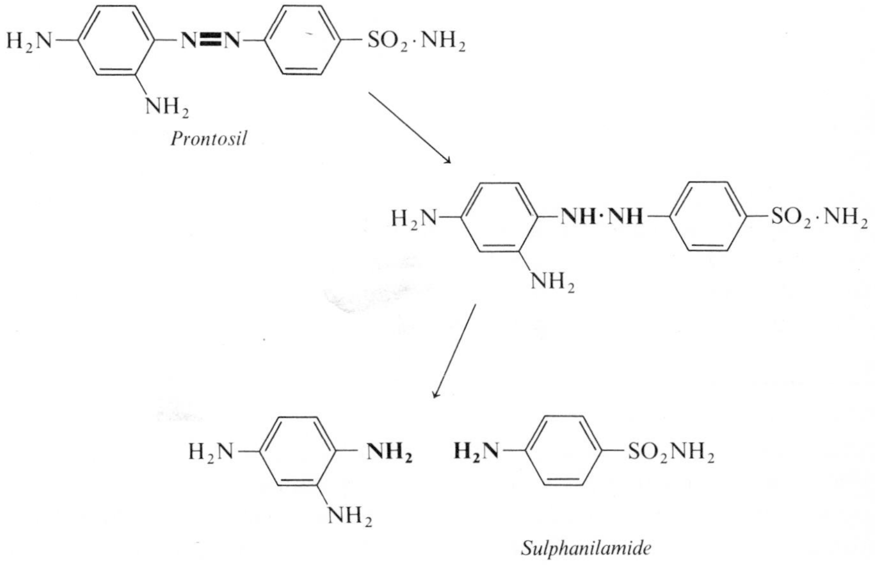

H_2N—$C_6H_3(NH_2)$—N=N—C_6H_4—$SO_2 \cdot NH_2$

Prontosil

↓

H_2N—$C_6H_3(NH_2)$—**NH·NH**—C_6H_4—$SO_2 \cdot NH_2$

↓

H_2N—$C_6H_3(NH_2)$—**NH₂** **H₂N**—C_6H_4—SO_2NH_2

Sulphanilamide

According to Fouts, Kamm and Brodie (1957), azo reductase in rabbits is present mainly in the liver, but also in kidney, lung and other tissues, and although a proportion of the enzyme activity is found in the microsomes, it is mainly in the soluble fraction from liver tissue. There is a distinct similarity between liver azo and nitro reductases. Both are markedly stimulated by the addition of excess FAD, FMN and riboflavine. Azo reductase can, however, be separated from nitro reductase by ammonium sulphate precipitation.

Azo dyes used as pharmaceutical colours, such as *Tartrazine* (FD & C Yellow No. 5), are reduced by azo reductases and excreted as the corresponding sulphanalic acids when administered orally (Jones et al., 1964, Ryan et al., 1968).

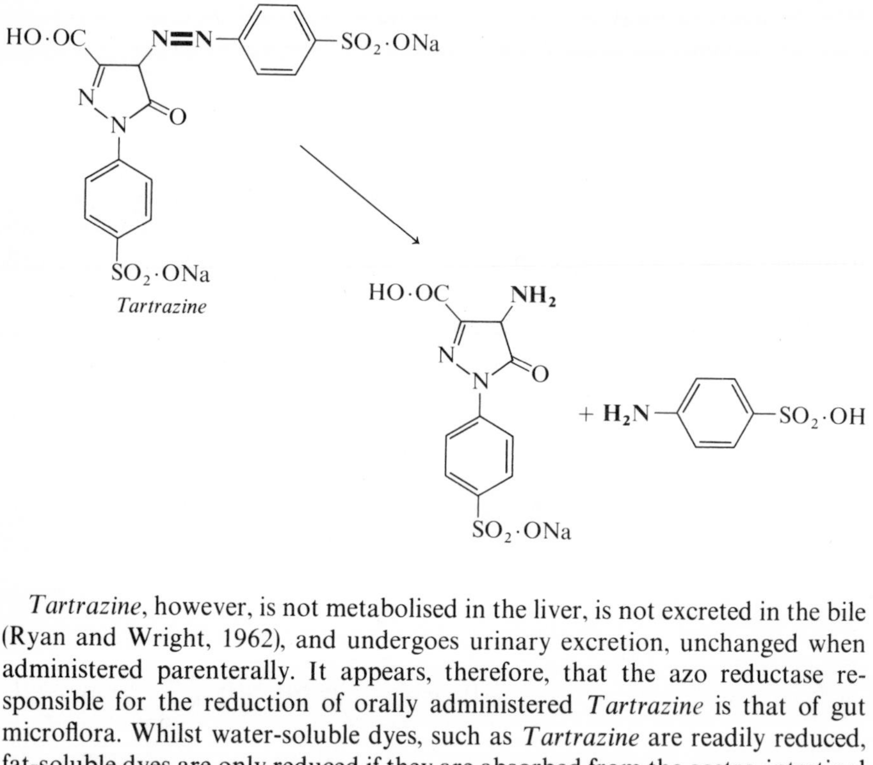

HO·OC N=N— $SO_2 \cdot ONa$

N N O

$SO_2 \cdot ONa$

Tartrazine

HO·OC NH_2

N N O

+ H_2N— $SO_2 \cdot OH$

$SO_2 \cdot ONa$

Tartrazine, however, is not metabolised in the liver, is not excreted in the bile (Ryan and Wright, 1962), and undergoes urinary excretion, unchanged when administered parenterally. It appears, therefore, that the azo reductase responsible for the reduction of orally administered *Tartrazine* is that of gut microflora. Whilst water-soluble dyes, such as *Tartrazine* are readily reduced, fat-soluble dyes are only reduced if they are absorbed from the gastro-intestinal tract and excreted in the bile as water-soluble conjugates.

Nitro Reduction

The reduction of aromatic nitro compounds to the corresponding amines by nitro reductase occurs in most species (Adamson, Dixon, Francis and Rall, 1965), and is thought to proceed via the corresponding nitroso and hydroxylamino compounds (*cf* chemical reduction, **1**, 15). Thus, nitro reductase reduces nitrosobenzene and phenylhydroxylamine to aniline more rapidly than it reduces nitrobenzene (Fouts and Brodie, 1957), and *m*-dinitrobenzene is metabolised in the rabbit to *m*-nitrosonitrobenzene and *m*-nitrophenylhydroxylamine in addition to *m*-nitroaniline (Parke, 1961).

$$m\text{-}C_6H_4(NO_2)NO_2 \longrightarrow m\text{-}C_6H_4(NO_2)NO \longrightarrow m\text{-}C_6H_4(NO_2)NHOH \longrightarrow m\text{-}C_6H_4(NO_2)NH_2$$

Nitro reductase is capable of utilising either NADPH or NADH as hydrogen donor. It is, however, inactive in air, due possibly to depletion of the hydrogen donor following air-oxidation of the hydroxylamino intermediate. Activity is also inhibited by carbon monoxide which, like inhibition in air, is characteristic of cytochrome P450 involvement, whilst reductase activity is also enhanced by *Phenobarbitone* administration (Gillette, Kamm and Sasame, 1968; Sasame and Gillette, 1969). According to Kato, Oshima and Tanaka (1969), however, two separate reductases are involved in the reduction of nitro compounds, one oxygen-sensitive and inducible which reduces the nitro compound to the hydroxylamine and the other, oxygen-insensitive and non-inducible which reduces the hydroxylamino metabolite.

Most benzenoid aromatic nitro compounds including the antibacterial *Chloramphenicol* (**1**, 15) and *p*-nitrophenylarsonic acid (Moody and Williams, 1962), which is used in poultry and pig food to reduce gastro-intestinal parasites, are metabolised by nitro reductases. The heteroaromatic urinary antiseptic, *Nitrofurazone*, however, is not metabolised by azo reductase (Umar and Mitchard, 1968).

Reductive Dehalogenation

A number of important polyhalogen compounds, including the general anaesthetics, *Halothane*, *Methoxyfluorane*, and the insecticide DDT (*Dicophane*, Chlorophenothane) are metabolised by reductive dechlorination. The carbon fluorine bond is metabolically stable, but other carbon–halogen bonds are susceptible to reductive dechlorination (van Dyke and Chenoweth, 1965).

$$\underset{\textit{Halothane}}{CF_3\cdot CH\mathbf{BrCl}} \longrightarrow CF_3\cdot \mathbf{CH_3}$$

$$\underset{\textit{Methoxyfluorane}}{CH\mathbf{Cl_2}\cdot CF_2\cdot O\cdot CH_3} \longrightarrow \mathbf{CH_3}\cdot CF_2\cdot O\cdot CH_3$$

The reductive enzyme utilises NADPH and molecular oxygen, and is induced by *Phenobarbitone* and by sub-anaesthetic doses of *Methoxyfluorane* (van Dyke, 1966). The corresponding reductive dechlorination of DDT to DDD forms the main metabolic pathway in fish (Datta, Laug and Klein, 1964). DDD is an intermediate in the formation of DDA, the major metabolic product in mammals, which is excreted in urine as amino acid conjugates. The reductive dechlorination of DDT represents an alternative to the metabolic dehydrochlorination to DDE, which occurs in the liver. DDE, like DDT, accumulates in body fat. Mean concentrations of 3 ppm DDT and 7.5 ppm DDE have been reported as normal in human body fat (Hoffman, Fishbein and Andelman, 1964).

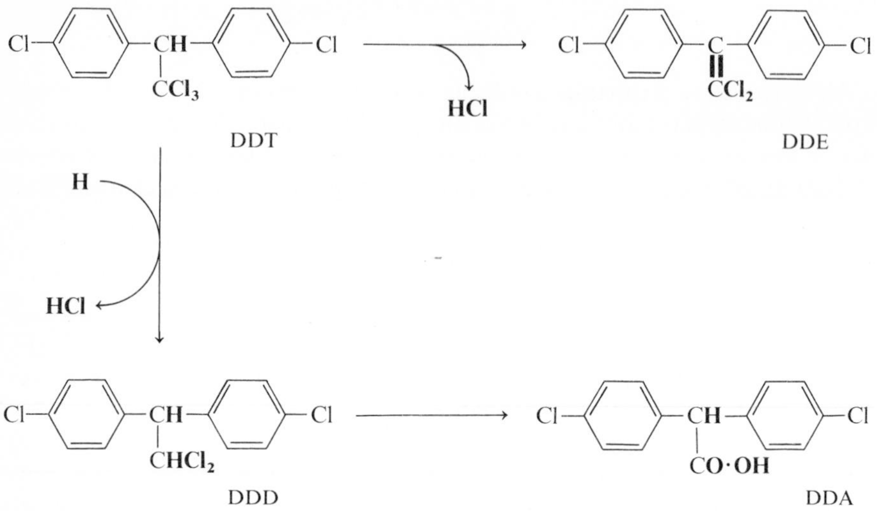

Cl–C₆H₄–CH(CCl₃)–C₆H₄–Cl → (–HCl) → Cl–C₆H₄–C(=CCl₂)–C₆H₄–Cl

DDT → DDE

DDT ↓ (H, –HCl)

Cl–C₆H₄–CH(CHCl₂)–C₆H₄–Cl → Cl–C₆H₄–CH(CO·OH)–C₆H₄–Cl

DDD → DDA

NON-MICROSOMAL OXIDATIONS

Oxidation of Alcohols

Primary and secondary alcohols are oxidised enzymatically in mammalian systems to aldehydes and ketones respectively by pathways which resemble those of chemical oxidation. Liver alcohol dehydrogenase (LAD), which is present in the soluble fraction of liver, kidney and lung tissues of most mammalian species, catalyses the oxidation of primary alcohols ranging from ethanol (Theorell and Bonnichsen, 1951) to retinol, including such alcohols as fluoroethanol (Merritt and Tomkins, 1959), phenylethanol (Gillette, 1959), benzyl alcohol (Bliss, 1951) and geraniol (Donninger and Ryback, 1964). Methanol, however, is a poor substrate and is only slowly oxidised at high concentration (Kini and Cooper, 1961).

The oxidation of secondary alcohols by LAD takes place much more slowly than that of primary alcohols due to their lower acidic strength (Dalziel and Dickinson, 1967) and increased steric hindrance, which inhibit enzyme–substrate formation. Enzymic oxidation of tertiary alcohols is similarly inhibited, though not simply because they are even weaker acids, but rather because the adjacent tertiary carbon atom by its very nature lacks the α-C—H bond essential for direct oxidation. It is this resistance to oxidation, combined with high lipid solubility which accounts for the greater effectiveness of such tertiary alcohols as *Methylpentynol* (Meparfynol; **1**, 7), *Amylene Hydrate* and *Trichloroethanol* as hypnotics. Resistance to oxidation, however, is to some extent counteracted metabolically by conjugation, which serves as an alternative detoxification pathway for all alcohols, but increasingly in the order primary < secondary < tertiary alcohols respectively (p. 275).

Liver alcohol dehydrogenase-catalysed oxidation of alcohols is reversible, and requires the presence of NAD^+ as coenzyme.

$$CH_3 \cdot \mathbf{CH_2 \cdot OH} + NAD^+ \xrightleftharpoons{\text{LAD}} CH_3 \cdot \mathbf{CHO} + NADH + \mathbf{H^+}$$

Oxidation occurs in a ternary complex formed of enzyme, alcohol and hydrogen acceptor (NAD^+). The rate-determining step, however, is not formation, but the dissociation of the resultant NADH–enzyme complex (Dalziel and Dickinson, 1966) formed in the oxidation.

Kini and Cooper (1961) have shown that $NADP^+$ can also function as hydrogen acceptor. This coenzyme, however, is unimportant in the oxidation of alcohols *in vivo*, since it is present in the liver mainly in its reduced form, NADPH. Since, however, the reverse reaction, the oxidation of aldehydes and ketones, is also catalysed by LAD, it seems probable that NADPH functions as coenzyme in that reaction.

$$\underset{\text{Chloral Hydrate}}{Cl_3C \cdot \mathbf{CH(OH)_2}} + NADPH + \mathbf{H^+} \rightleftharpoons \underset{\text{Trichloroethanol}}{Cl_3C \cdot \mathbf{CH_2 \cdot OH}} + NADP + \mathbf{H_2O}$$

The H^+ term in the above equations implies that the action of the enzyme is pH-dependent. Its effectiveness, therefore, increases with proton availability, and hence in parallel with the acidic strength of the alcoholic —OH function. This increases in the order $3° < 2° < 1°$ < methanol. The inability of methanol to function as an efficient substrate for LAD cannot, therefore, be due to this factor, and must arise from the lack of a suitable hydrophobic group to bind with the enzyme.

Experiments with 1-[^{2}H]-deuteroethanol clearly demonstrate that as in chemical oxidation of alcohols (**1**, 7), severance of an α-C—H bond with, in this case, transfer of the departing hydride ion (H^-) to NAD^+, is also an important step in the reaction. The enzyme (LAD), therefore, co-ordinates two functions, the withdrawal of the acidic proton from the hydroxyl group and the alignment of the coenzyme (NAD^+) to facilitate hydride ion acceptance. These two requirements for effective oxidation may, however, conflict so far as the substrate itself is concerned, since electron-withdrawing substituents, which enhance acidic strength, will at the same time oppose the loss of a hydride ion from the α-carbon atom. The ability of the oxidation to proceed would then be determined by the relative electron attracting powers of the substituent and NAD^+. This would explain why reduction of *Chloral Hydrate* to *Trichloroethanol* is favoured rather than the reverse reaction.

Pyridine nucleotide-linked enzymes fall into two groups according to which side of the NAD^+ nucleus the transferred hydrogen is added (Levy, Talalay and Vennesland, 1962). Liver alcohol dehydrogenase belongs to class A, in which the hydrogen is added to that side of the nicotinamide ring on which the ring atoms 1 to 6 appear in an anti-clockwise order. Enzymes of class B add hydrogen to the opposite side of NAD^+.

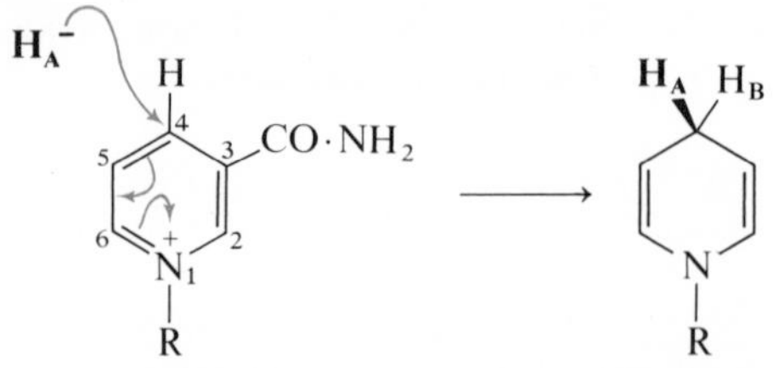

NAD^+ $\longrightarrow$ NADH

Accordingly, degradation of 4-[^{2}H]-NADH formed in a class A transfer from 1,1-[^{2}H]-3-methyl-2-butenol gives (2*R*)-monodeuterosuccinic acid by the following route (Cornforth, Ryback, Popják, Donninger and Schroepfer, 1962).

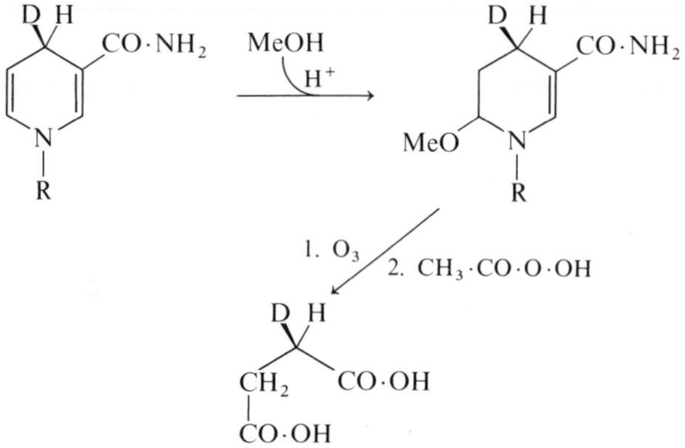

MeOH, H^+; 1. O_3; 2. $CH_3{\cdot}CO{\cdot}O{\cdot}OH$

(2*R*)-Monodeuterosuccinic acid

Liver alcohol dehydrogenase is also **stereoselective** in respect of its substrate. Thus, the heavy isotope in 1-[^{3}H]-geraniol is transferred to NAD^+ in preference to the non-isotopic H-1 group in the stereoisomer of the (*R*)-configuration (Donninger and Ryback, 1964). Yeast alcohol dehydrogenase (YAD) also detaches the heavy isotope in (1*R*)-1-[^{2}H]-ethanol in a similar oxidation which, however, is completely **stereospecific** (Levy, Loewus and Vennesland, 1957). LAD and YAD also have similar stereoselectivity for (2*S*)-(+)-isomers in A-specific oxidations of butan-2-ol and octan-2-ol (Dalziel and Dickinson, 1967). The authors postulate that enzyme–substrate binding is promoted by hydrophobic interaction to the n-alkyl group, and this controls the orientation of the substrate. This hypothesis accounts for the observed enzyme stereospecificity both towards the primary alcohols (1*R*)-1-[^{3}H]-geraniol and (1*R*)-1-[^{2}H]-ethanol, and the secondary alcohols (2*S*)-butan-2-ol and (2*S*)-octan-2-ol, which have similar orientations about the carbinol carbon.

Hydrophobic interactions

1*R*-1-[^{2}H]-ethanol   2*S*-1-[^{2}H]-butan-2-ol

Taking all the above factors into account, the enzyme–substrate–coenzyme complex, and its oxidation to enzyme–NADH complex and aldehyde may be represented diagrammatically, as shown in Fig. 49.

Consumption of alcohol during treatment with *Levodopa* has been shown to lead to urinary excretion of salsolinol due to condensation of acetaldehyde formed by metabolic oxidation with dopamine.

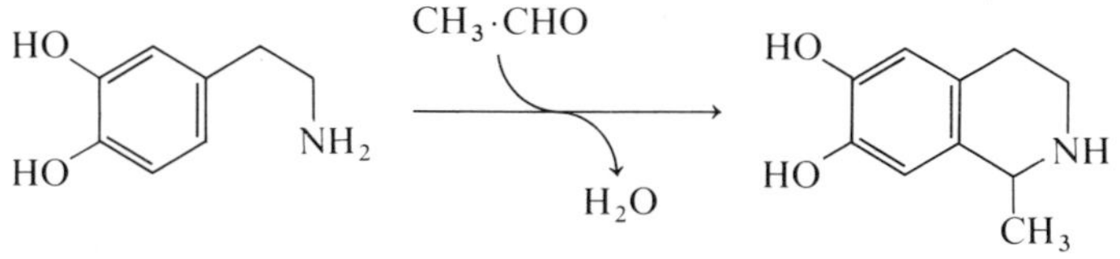

Dopamine   Salsolinol

Oxidation of Aldehydes

Aldehydes, whether of metabolic origin, or otherwise, are oxidised in the liver to the corresponding carboxylic acids. The oxidation is catalysed mainly by liver aldehyde dehydrogenase (Lundquist, Fugmann, Rasmussen and Svendsen, 1962), though the molybdo-flavoprotein enzymes, xanthine dehydrogenase (xanthine oxidase) and hepatic aldehyde oxidase, may also be involved. Liver aldehyde dehydrogenase, like LAD, is NAD^+-dependent and present in the soluble-fraction of liver tissue (Kraemer and Dietrich, 1968). Although the enzyme is effective under anaerobic conditions, its efficiency is almost doubled in air, which promotes re-oxidation of NADH formed in the reaction. Oxidation of NADH is inhibited by azide, so the most likely pathway for the aerobic re-oxidation is by way of NADH–microsomal cytochrome *c* reductase.

Cleavage of the α-C—H bond with hydride transfer to NAD^+ is a key step in aldehyde oxidation, just as in the enzymic oxidation of alcohols. The oxidation of aldehyde to acid, however, implies a second oxidative step involving, in

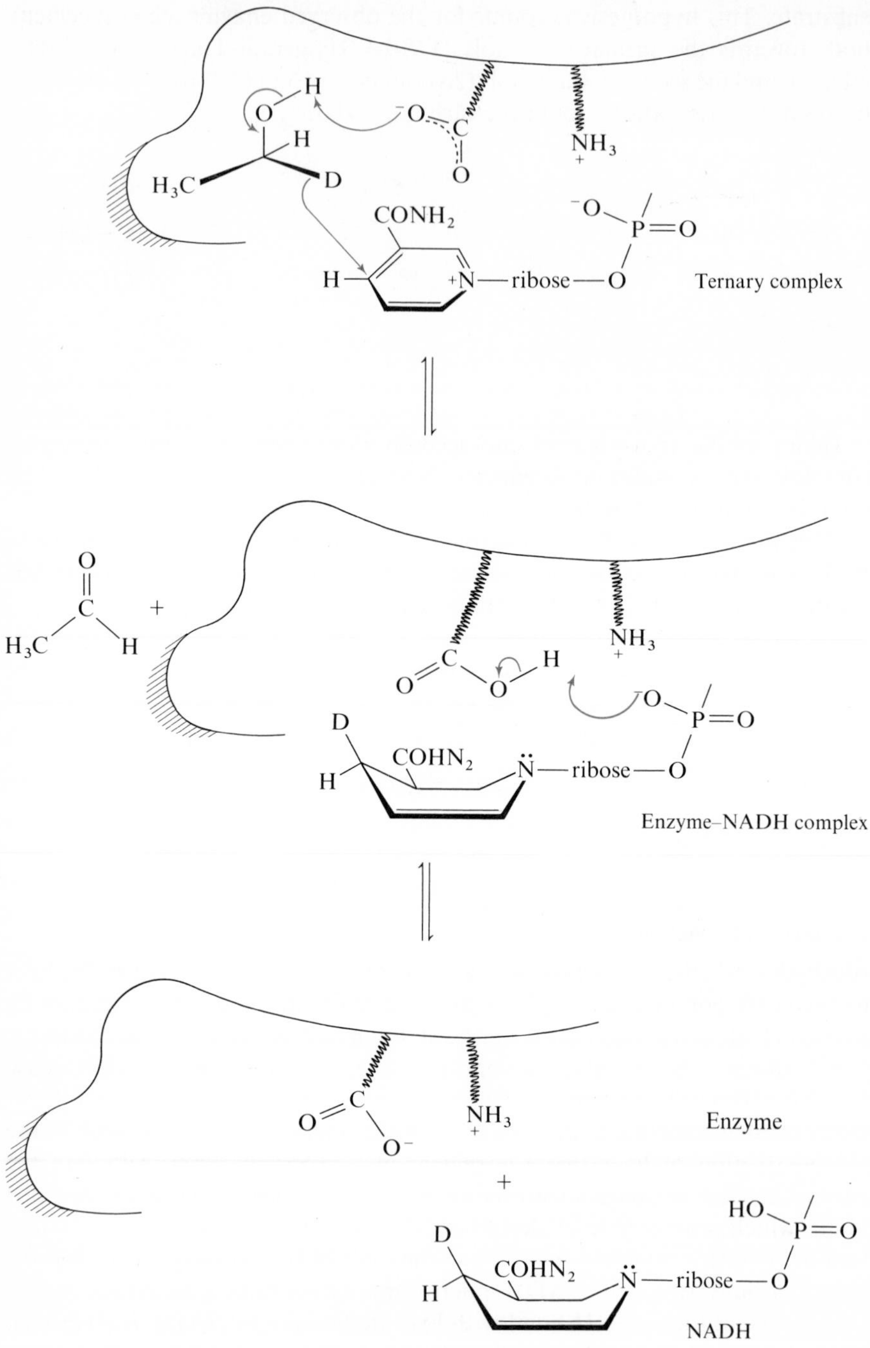

Ternary complex

Enzyme–NADH complex

Enzyme

NADH

Fig. 49 Alcohol oxidation by liver alcohol dehydrogenase

all probability, a nucleophilic oxygen function (H_2O or HO^-). In accord with this concept, enzymic activity increases with increasing pH up to pH 10, above which reversible inactivation occurs. Hydroxyl ions, therefore, appear to be implicated in the oxidation mechanism.

^-OH

$CH_3—C=O$

H

H

$CO \cdot NH_2$

N

R

NAD^+

OH

$CH_3—C$

O

+

H H

$CO \cdot NH_2$

N

R

NADH

The enzyme is capable of oxidising a wide variety of aldehyde substrates, including acetaldehyde, formaldehyde, benzaldehyde, indole-3-aldehyde and glyoxal, but not glycollic acid, *Chloral Hydrate* or strophanthin. Ketones lack the essential α-C—H bond and hence are resistant to metabolic oxidation.

Deamination by Monoamine Oxidase (MAO)

Monoamine oxidase (Hare, 1928), the physiological enzyme responsible for the oxidation of biogenic and exogenous amines, is widely distributed in the mitochondria of a variety of mammalian tissues including liver, kidney, intestinal mucosa, lung, blood vessels and plasma, heart muscle and nerves. The enzyme, a flavoprotein located in the outer membrane of the mitochondria (Tipton, 1967; Waite, 1969) is capable of oxidising primary amines to the corresponding aldehyde with loss of ammonia in the reaction:

$$R \cdot CH_2 \cdot NH_2 + \tfrac{1}{2}O_2 + H_2O \longrightarrow R \cdot CHO + NH_3 + H_2O$$

Its tissue distribution is such that it is readily available to provide protection from the otherwise toxic effects of biogenetic amines, whether these are released by nervous stimulation or as products of bacterial metabolism in the gut. Typical substrates which are readily deaminated by monoamine oxidase include benzylamine, phenylethylamine, *Noradrenaline* (Norepinephrine), dopamine, tyramine and 5-hydroxytryptamine. According to Sandler and Youdim (1972), there are several different forms of MAO, depending on the species and tissue in which it occurs, with varying substrate specificities and catalytic ability. All, however, appear to bring about oxidation *via* the corresponding imine, and it is by this group that the benzylamine oxidase from pig plasma attaches its substrate (Buffoni, Della Corte and Knowles, 1968). This enzyme, which has copper(II)-pyridoxal as coenzyme, is alkali-sensitive, stable and active at acid pH, but inhibited competitively in strongly acidic media (Hare, 1928). This activation in mildly acidic conditions accords with a mechanism

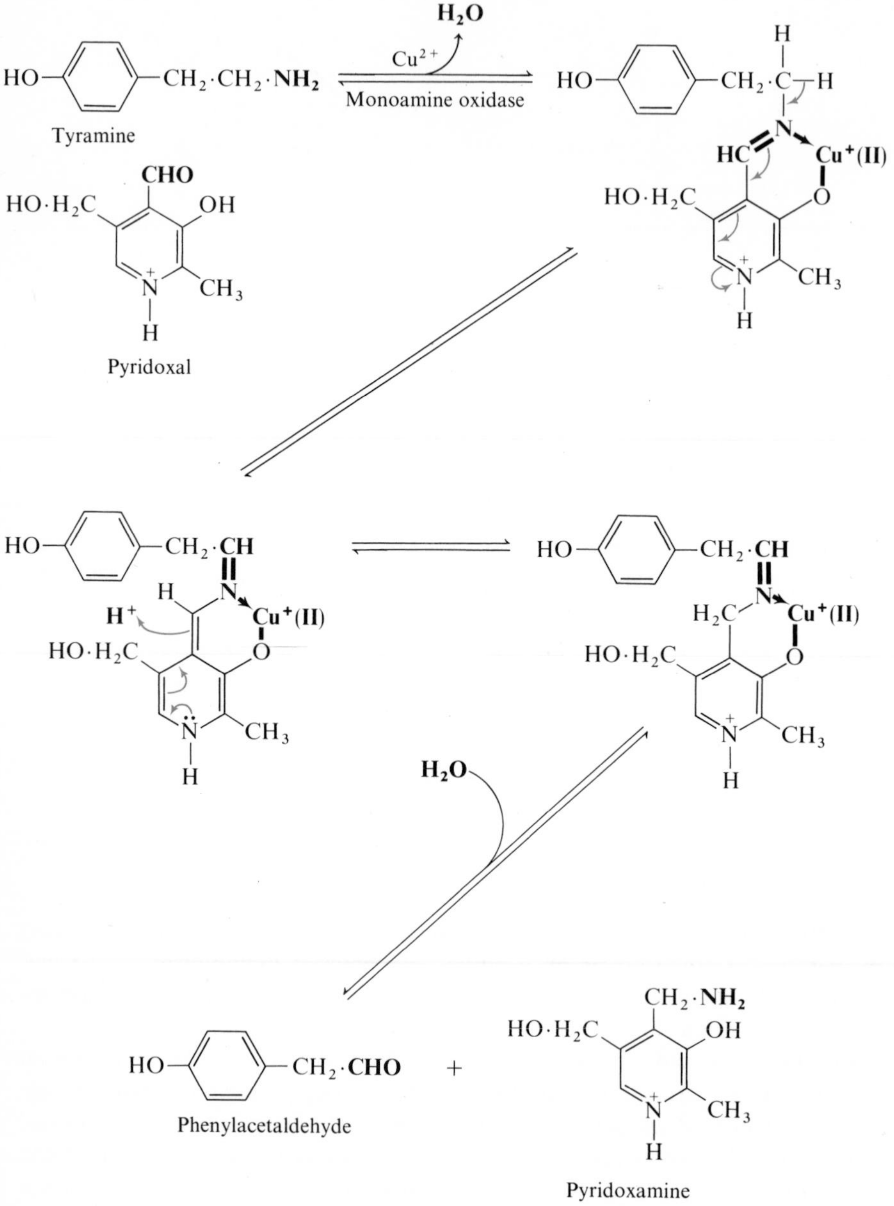

HO–C6H4–$CH_2 \cdot CH_2 \cdot$ **NH_2**

Tyramine

CHO

$HO \cdot H_2C$ OH

CH_3

N⁺–H

Pyridoxal

H_2O

Cu^{2+}

Monoamine oxidase

HO–C6H4–$CH_2 \cdot C$–H (H)

N → **$Cu^+(II)$**

HC

$HO \cdot H_2C$ O

CH_3

N⁺–H

HO–C6H4–$CH_2 \cdot$ **CH**

N → **$Cu^+(II)$**

H

H^+

$HO \cdot H_2C$ O

CH_3

N–H

HO–C6H4–$CH_2 \cdot$ **CH**

N → **$Cu^+(II)$**

H_2C

$HO \cdot H_2C$ O

CH_3

N⁺–H

H_2O

HO–C6H4–$CH_2 \cdot$ **CHO** +

Phenylacetaldehyde

$CH_2 \cdot$ **NH_2**

$HO \cdot H_2C$ OH

CH_3

N⁺–H

Pyridoxamine

involving a pyridoxal copper chelated Schiff's base which undergoes acid-catalysed rearrangement and hydrolysis, as in typical pyridoxal mediated transaminations (**1**, 10). It should be noted, however, that it is by no means clear that all other monoamine oxidases act by a similar mechanism. Thus, the activity of some appears to be linked to the uptake of dietary iron (Symes, Missala and Sourkes, 1971).

Ammonia, formed by hydrolysis of the imine, is swept via pyridoxamine into the normal biosynthetic pathway for urea synthesis in the liver (**1**, 17), whilst the aldehydes are oxidised further to the corresponding carboxylic acid. Both these on-going metabolic transformations are dependent on cytochrome enzyme systems, which are also conveniently located in the mitochondria.

Monoamine oxidase is capable of oxidising alkylamines as well as aryl-alkylamines. Substrate specificity increases with chain length reaching an optimum in butylamine (Blaschko, 1952). Branched-chain primary amines, such as isobutylamine, are readily oxidised, but branching at the α-carbon atom, as in sec-butylamine, *Amphetamine* and *Tranylcypromine*, inhibits oxidation. The latter, a competitive and reversible inhibitor of MAO, demonstrates an important stereochemical feature of such monoamine oxidase inhibitors. The corresponding *cis*-isomer in which the amino and phenyl groups are eclipsed is not an effective inhibitor, presumably because the phenol group impedes imine formation with pyridoxal.

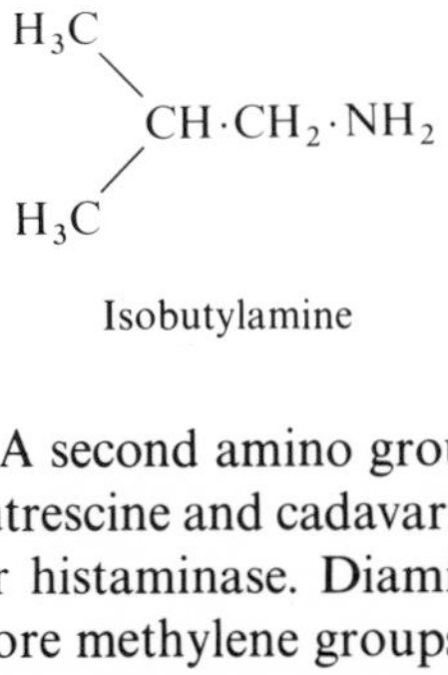

$$(H_3C)_2CH{\cdot}CH_2{\cdot}NH_2$$

Isobutylamine

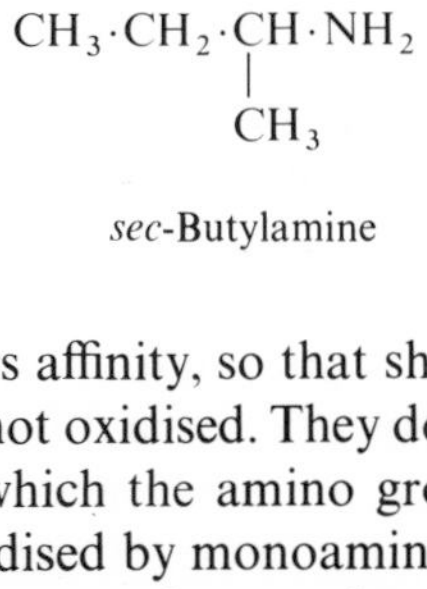

$$CH_3{\cdot}CH_2{\cdot}CH(CH_3){\cdot}NH_2$$

sec-Butylamine

H, Ph, NH_2, H

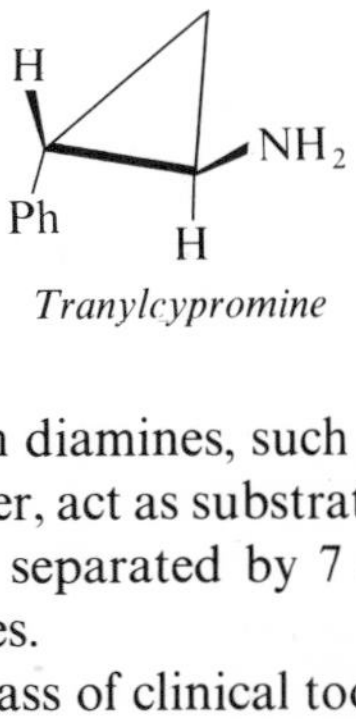

Tranylcypromine

A second amino group lowers affinity, so that short-chain diamines, such as putrescine and cadavarine, are not oxidised. They do, however, act as substrates for histaminase. Diamines in which the amino groups are separated by 7 or more methylene groups are oxidised by monoamine oxidases.

Monoamine oxidase inhibitors now form an important class of clinical tools for potentiating the effects on the central nervous system of biogenetic amines, which otherwise would normally have very limited duration of action when released in response to nervous stimulation. As a result, they heighten the effect of physiological amines, such as catecholamine, dopamine and 5-hydroxytryptamine, producing central nervous stimulation with powerful antidepressive effects. The most important group of monoamine oxidase inhibitors are hydrazines, such as *Iproniazid*, *Nialamide*, *Phenelzine* and *Isocarboxazid*. These are also non-competitive inhibitors and only slowly reversible. Their precise mechanism of action is not fully established, but there is considerable evidence that they also block a number of other important enzymes including some hepatic microsomal drug metabolising enzymes. They, therefore, often indirectly potentiate the effect of other drugs, including barbiturates, *Pethidine*, *Imipramine*, *Cortisone*, hypertensives such as *Ephedrine*, amphetamines, and the anti-hypertensive, *Methyldopa*.

$(H_3C)_2CH \cdot NH \cdot NH \cdot CO$–(4-pyridyl)

Iproniazid

$Ph \cdot CH_2 \cdot CH_2 \cdot NH \cdot NH_2$

Phenelzine

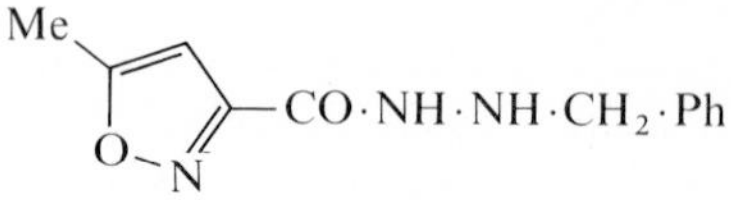

(5-Me-isoxazol-3-yl)–$CO \cdot NH \cdot NH \cdot CH_2 \cdot Ph$

Isocarboxazid

$Ph \cdot CH_2 \cdot CO \cdot CH_2 \cdot CH_2 \cdot NH \cdot NH \cdot CO$–(4-pyridyl)

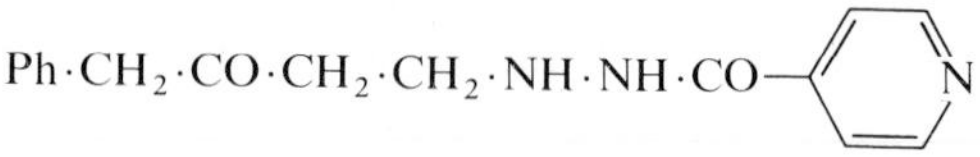

Nialamide

Aromatisation of Hydroaromatic Compounds

A few species, notably guinea pig and rabbit, but not man, cat, dog or mouse, are capable of metabolising hydroaromatic compounds of structure $C_6H_{11}(CH_2)_nCOOH$ (where n is an even number) to cyclohexanecarboxylic acid (as its *S*-coenzyme A ester) and thence to benzoic acid. This latter ring oxidation is catalysed by a mitochondrial enzyme present in the liver and kidney of guinea pig and rabbit (Babior and Bloch, 1966), which utilises molecular oxygen, ATP, coenzyme A and FAD, and is quite distinct from the enzyme responsible for the aromatisation of ring A in the biosynthesis of steroids.

The normal aromatisation of androgenic steroids in the biosynthesis of oestrogens in man is catalysed by an $NADPH_2$-dependent enzyme, which is present in human placenta and other tissues, and also requires oxygen (Ryan, 1959). Aromatisation is linked with the simultaneous removal of the C-19 methyl group, since 19-nortestosterone and related 19-nor-steroids are not readily converted to the corresponding oestrogens. The complete pathway involves both 19-hydroxymethyl and 19-formyl derivatives as intermediates (Longchampt, Gual, Ehrenstein and Dorfman, 1960; Morato, Hayano, Dorfman and Axelrod, 1961), with the latter undergoing aromatisation following oxidative attack at the aldehydic C—H bond, possibly by the following reaction pathway.

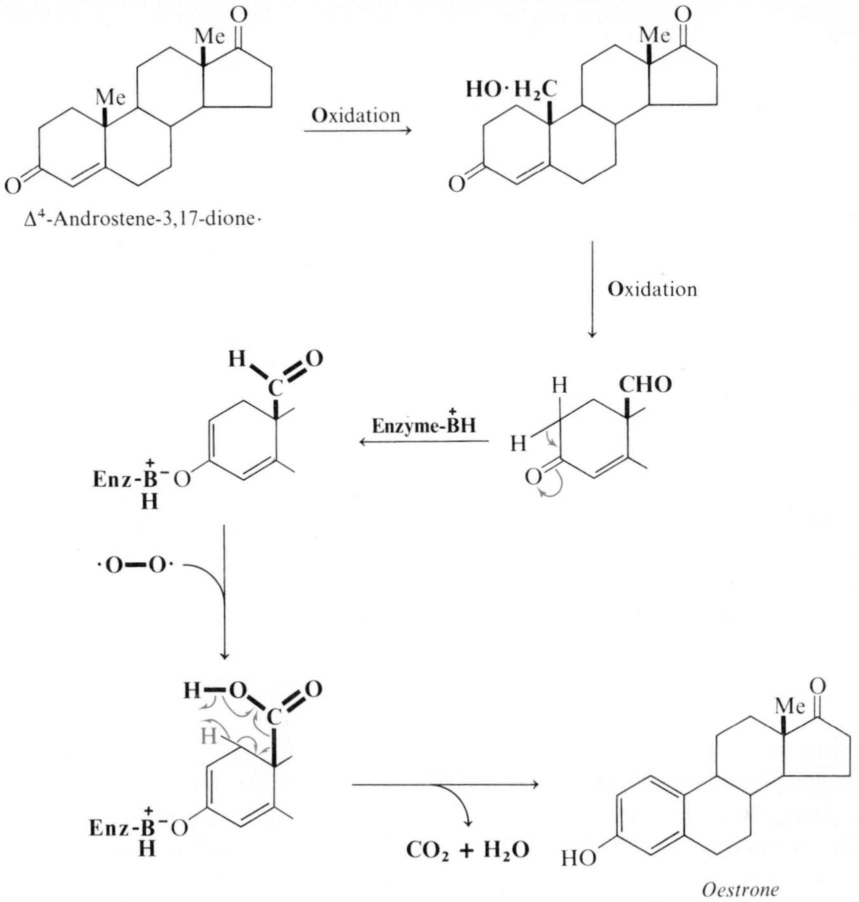

Me
O
Δ⁴-Androstene-3,17-dione·
Oxidation
HO·H₂C
Oxidation
H
CHO
Enzyme-B̊H
Enz-B⁻O
H
·O—O·
H—O
C
CO₂ + H₂O
HO
Oestrone

NON-MICROSOMAL REDUCTIONS

Reduction of Ketones

Metabolic interconversions of ketones and secondary alcohols are catalysed by a group of soluble enzymes, which are widely distributed in a variety of tissues. The liver alcohol dehydrogenase–NADH catalysed equilibrium favours the reduction of ketones to alcohols, rather than the reverse reaction in mammalian liver at pH 7.4. Metabolism by this pathway with or without conjugation of the product alcohol to glucuronide has been reported for *Acetohexamide* (Welles, Root and Anderson, 1961), cyclohexanone (Merritt and Tomkins, 1959), *Warfarin* (Lewis and Trager, 1970) and *Chloral Hydrate* (Friedman and Cooper, 1960).

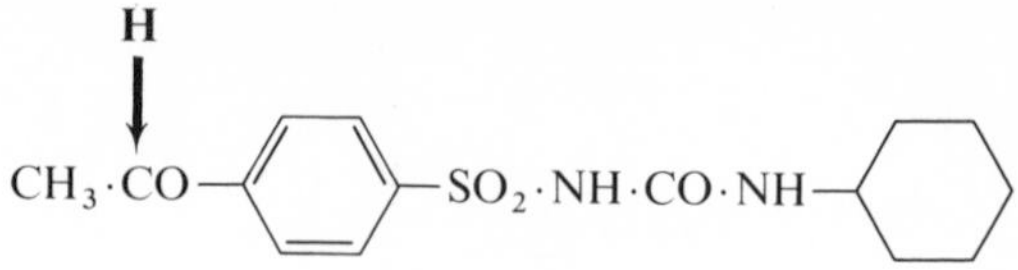

Acetohexamide

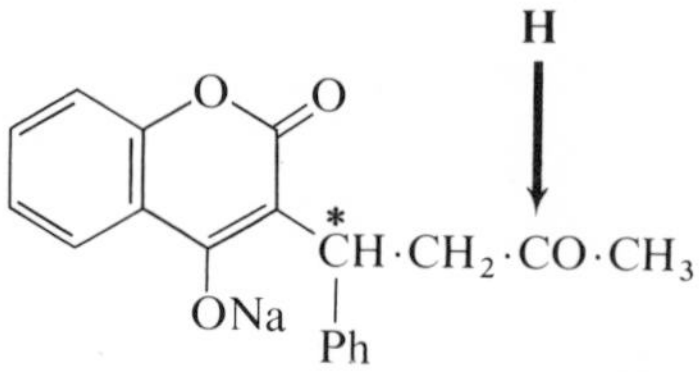

Warfarin Sodium

Metabolic reduction of racemic *Warfarin* gives rise to all four possible isomers. Studies with 2-t-butylcyclohexanones, however, show that this reduction is stereospecific, with the cyclic ketone reacting in its most stable conformation in which the t-butyl group is equatorial (Cheo, Elliott and Tao, 1967). The reduction is A-specific leading to the transfer of the C-4 H_A from NADH in a face-to-face interaction of ketone and NADH. The stereochemistry of the product is largely determined by adoption of that face-to-face alignment which avoids steric interactions between the C-2 substituent of the ketone and the 3-aminocarbonyl substituent of NADH.

The presence or absence of a 2-substituent, therefore, determines whether (+), (−), or both enantiomers are formed. Thus, whereas (±)-2-methylcyclohexanone gives a mixture of the optically active compounds (+)-*cis*-2-methylcyclohexanol and (+)-*trans*-2-methylcyclohexanol, (±)-3-methylcyclohexanone gives only the racemic (±)-*cis*- and (±)-*trans*-3-methylcyclohexanones (Elliott, Jacob and Tao, 1969). The relative orientation of the reactants in the enzyme–substrate complex determines whether hydrogen is transferred to either the **normal** or **reverse** side of the ketone carbonyl. Transfer to the **normal** side results in a product alcohol with an equatorial hydroxyl, whilst transfer to the **reverse** side yields the corresponding alcohol with an axially-orientated hydroxyl. In practice, minor non-bonded interactions favour reverse face interactions with the substrate in what the authors describe as a **perpendicular** orientation. The alignments shown on p. 261, therefore, apply in the reduction of (+)- and (−)-2-methylcyclohexanones.

In contrast, 3- and 4-alkylcyclohexanones are free from the steric control imposed on the reduction of the 2-alkyl ketones by the NADH–3-carboxyamido group. Both **normal** and **reverse** coenzyme–substrate interactions are, therefore, possible with each enantiomer of 3-methylcyclohexanone giving rise to a mixture of optically active *cis*- and *trans*-3-methylcyclohexanols.

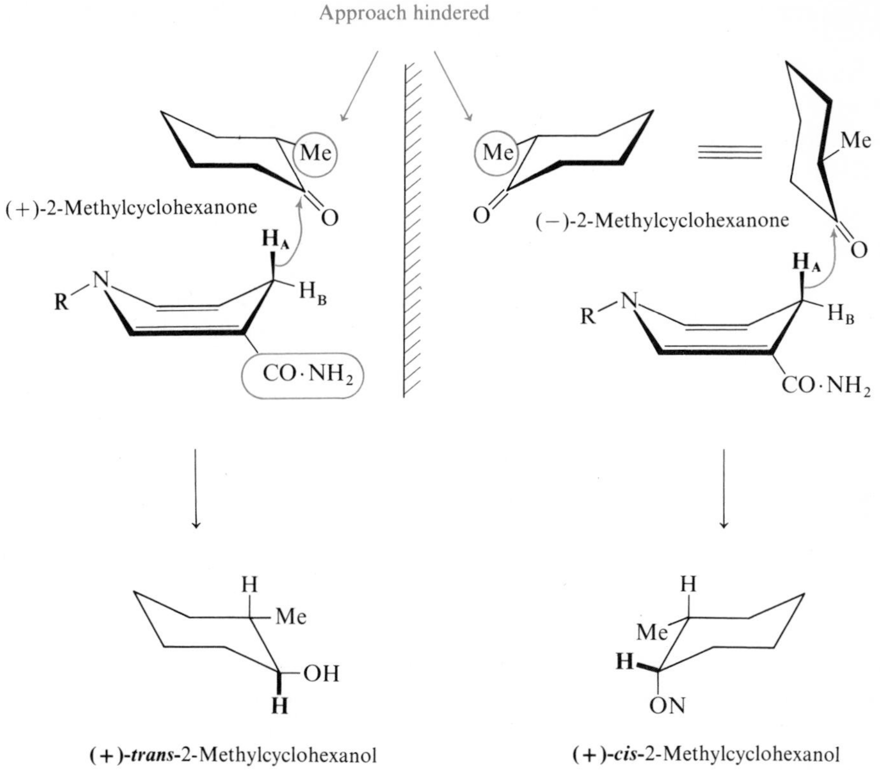
Approach hindered
Me
Me
Me
(+)-2-Methylcyclohexanone
(−)-2-Methylcyclohexanone
O
O
O
H_A
H_B
R
N
$CO \cdot NH_2$
H
Me
OH
ON
(+)-*trans*-2-Methylcyclohexanol
(+)-*cis*-2-Methylcyclohexanol

Aromatic ketones and aldehydes are reduced by a separate enzyme aldehyde–ketone reductase (A–K reductase) which is present in rabbit kidney and liver (Culp and McMahon, 1968). It catalyses the NADPH-dependent reduction of a wide spectrum of aromatic carbonyl compounds. Reduction is facilitated by

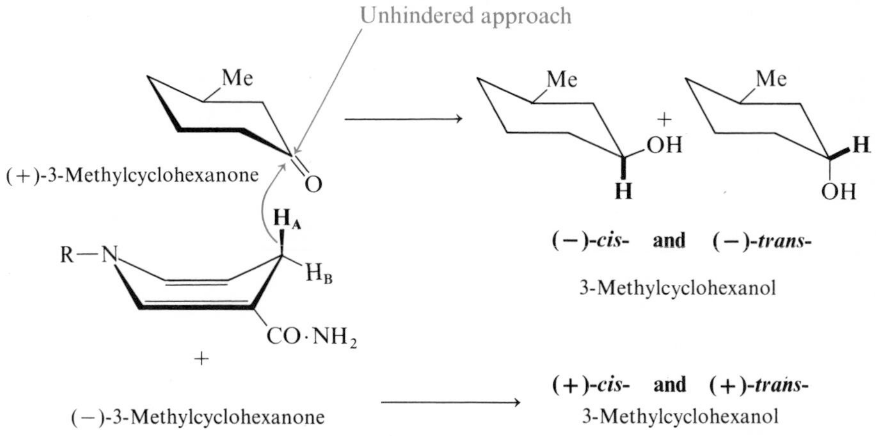
Unhindered approach
Me
(+)-3-Methylcyclohexanone
O
H_A
H_B
R—N
$CO \cdot NH_2$
+
(−)-3-Methylcyclohexanone
OH
H
(−)-*cis*- and (−)-*trans*-
3-Methylcyclohexanol
(+)-*cis*- and (+)-*trans*-
3-Methylcyclohexanol

the presence of electron-withdrawing substituents in the ring, and inhibited by electron-donating groups. However, the correlation of maximum reduction rates (v_{max}) in a series of substituted acetophenones with the Hammett substituent constant, σ, is poor, and significant correlations between reduction rate and structure are only obtained by Hansch analysis according to the expression (Hansch and Fujita, 1964):

$$\log v_{max} = K_\pi + \rho\sigma + k'$$

Positive K_π and $\rho\sigma$ terms are in accord with a reduction which is favoured by increased lipophilicity and the presence of electron-attracting substituents.

In contrast to LAD, aromatic A—K reductase is B-specific with respect to the C-4 hydrogen of NADPH (Culp and McMahon, 1968). Like LAD, however, it is stereospecific, reducing acetophenone to a mixture of 76% (*S*)-(−)- and 24% (*R*)-(+)-1-phenylethanol.

Reduction of Double Bonds

The metabolic reduction of ethylenic compounds to alkanes occurs in the reduction of the C24—C25 ethylenic bond in the biosynthesis of cholesterol from desmosterol by liver ethylenic reductase (Akhtar, Munday, Rahimula, Watkinson and Wilton, 1969). The enzyme is NADPH-dependent. Experiments with 4-[^{3}H]-NADPH show that reduction occurs by a Markownikoff-type electrophilic addition of a proton(enzyme-bound) to the more electron-rich terminus of the olefine to give a carbonium ion, followed by addition of a hydride ion from the 4-position of NADPH.

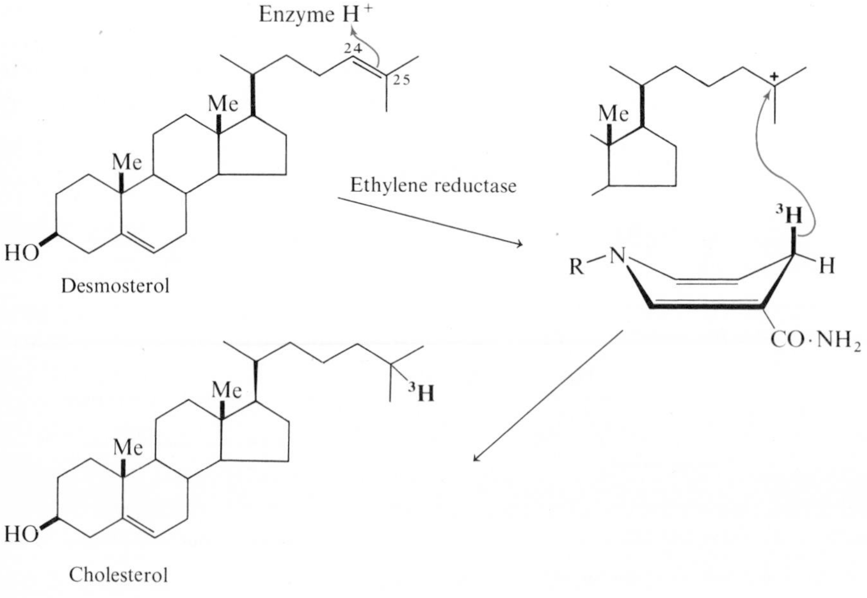

Enzyme H^+
24
25
Me
Me
HO
Desmosterol
Ethylene reductase
Me
^{3}H
R—N
H
CO·NH$_2$
Me
Me
^{3}H
HO
Cholesterol

Reduction of $\alpha\beta$-unsaturated ketones is also characteristic of the metabolism of most Δ^4-3-ketosteroids, such as *Progesterone* (Atherden, 1959), and 16α-ethylprogesterone (Stenlake, Templeton and Taylor, 1968). Both 5α- and 5β-pregnan-3,20-dione have been identified as metabolites of *Progesterone*, indicating that the double bond is the first point of attack; 3α- and 3β-hydroxy-5α- and 5β-pregnan-20-ones are also important metabolites resulting from subsequent reduction of the 3-keto group. A specific NADPH-dependent $\alpha\beta$-unsaturated ketone reductase, which catalyses the reduction of $\alpha\beta$-unsaturated ketones to saturated ketones, has been isolated from dog erythrocytes and human liver (Fraser, Peters and Hardinge, 1967).

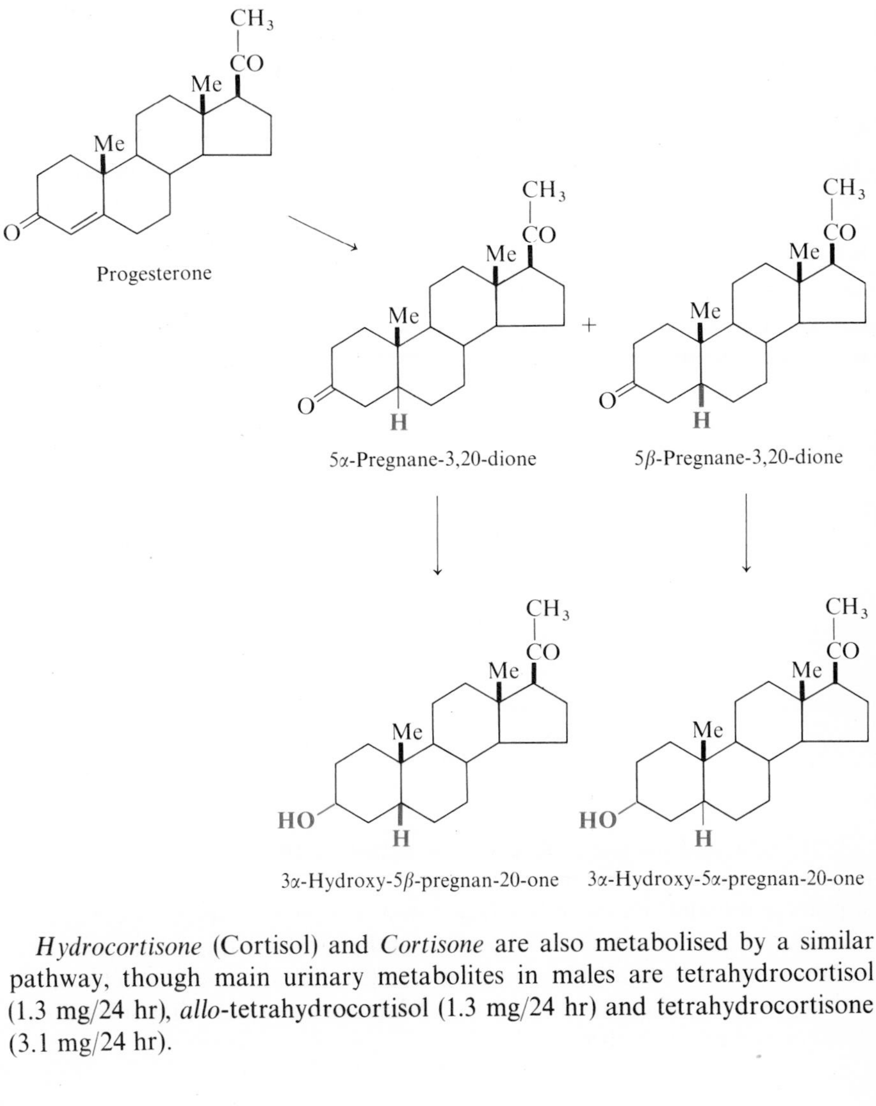

Progesterone

5α-Pregnane-3,20-dione + 5β-Pregnane-3,20-dione

3α-Hydroxy-5β-pregnan-20-one   3α-Hydroxy-5α-pregnan-20-one

Hydrocortisone (Cortisol) and *Cortisone* are also metabolised by a similar pathway, though main urinary metabolites in males are tetrahydrocortisol (1.3 mg/24 hr), *allo*-tetrahydrocortisol (1.3 mg/24 hr) and tetrahydrocortisone (3.1 mg/24 hr).

HYDROLYSIS

Carboxylic Acid Esters

Hydrolysis of esters, amides and related carboxylic acid derivatives, such as hydroxamic acids, hydrazides and nitriles, is readily catalysed by a variety of hydrolytic enzymes present in blood plasma, and in microsomal fractions of liver and kidney. Esterases and peptidases are also present in the alimentary canal essentially to facilitate digestion of food, but the same enzymes can be important in the metabolism of esters and amides used in medicine. Hydrolysis of esters and amides can also occur in the lower intestinal tract due to enzymes of the gut microflora.

Carboxylic ester hydrolases (esterases) show wide variation in substrate and inhibitor specificities. Table 40 lists the more important esterase enzymes with their International Enzyme Commission classification.

The range of substrate specificity of esterase enzymes ensures the rapid hydrolysis of most ester-based compounds, but the widespread distribution of the various enzymes in the body makes it difficult to predict the fate of particular compounds. Thus, acetylcholinesterase is specifically located in nervous tissues at synapses and nerve endings close to the sites of acetylcholine release. It is, however, also present in substantial concentration in erythrocytes, though not in blood plasma. Human blood plasma, however, does contain substantial but varying amounts of two non-specific esterases (pseudocholinesterases; p. 220) capable of splitting, aliphatic esters, particularly acetates and aromatic esters such as *Procaine* and *Cinchocaine* (Butacaine).

$H_2N-C_6H_4-\mathbf{CO{\cdot}O}{\cdot}CH_2{\cdot}CH_2{\cdot}NEt_2$

Procaine

$\downarrow$ $\mathbf{H_2O}$

$H_2N-C_6H_4-\mathbf{CO{\cdot}OH}$ $\quad$ $\mathbf{HO}{\cdot}CH_2{\cdot}CH_2{\cdot}NEt_2$

Both steric and electronic effects in the substrate affect the rate of hydrolysis by plasma esterases (Levine and Clarke, 1955). Enzyme-catalysed hydrolysis appears to resemble base-catalysed hydrolysis in that it is subject to substituent effects which influence nucleophilic attack at the carbonyl atom. Thus, electron-withdrawing substituents (4-fluoro) increase the double bond character of the carbonyl carbon and assist hydrolysis, whilst electron-donating substituents (4-amino) inhibit hydrolysis.

Table 40. Esterase Enzymes

EC Number	Systematic Name	Trivial Name	Typical Substrates
3.1.1.1	Carboxylic ester hydrolase	Carboxylesterase	aliphatic esters
3.1.1.2	Aryl ester hydrolase	Arylesterase	aromatic esters
3.1.1.3	Glycerol ester hydrolase	Lipase	triglycerides
3.1.1.6	Acetic ester acetyl-hydrolase	Acetylesterase	acetic esters
3.1.1.7	Acetylcholine acetyl-hydrolase	Acetylcholinesterase	acetylcholine
3.1.1.8	Acylcholine acyl-hydrolase	Cholinesterase (pseudo-cholinesterase)	acylcholines

R—C₆H₄—C(=O)—$O \cdot CH_2 \cdot CH_2 \cdot NEt_2$ + HO^- → R—C₆H₄—C(O^-)(OH)—$O \cdot CH_2 \cdot CH_2 \cdot NEt_2$ → (H^+) → R—C₆H₄—C(=O)OH + $HO \cdot CH_2 \cdot CH_2 \cdot NEt_2$

R	*Rate of hydrolysis* (μg/ml serum/hr; nucleophile HO^-)
F	3000
NH_2	500

Steric effects also affect the rate of enzymic hydrolysis, as shown by the fall in the rate of splitting as bulky substituents are introduced adjacent to the ester link.

H_2N—C₆H₄—$CO \cdot O \cdot C(R_1)(R_2) \cdot CH_2 \cdot NEt_2$

R_1	R_2	Rate of hydrolysis (μg/ml serum/hr)
H	H	500
H	CH_3	15
CH_3	CH_3	0

Many steroid esters, such as *Hydrocortisone Acetate* and *Oestradiol* (Estradiol) *Benzoate*, are used as such because they are absorbed and distributed in the body more efficiently than the parent steroid, but are readily metabolised and dependent for their activity upon enzymic hydrolysis. The topical steroid, *Betamethasone Valerate*, although not intended for absorption through the skin, likewise depends for its effect upon its ability to release the parent alcohol, *Betamethasone*, by hydrolysis in the skin (**1**, 22).

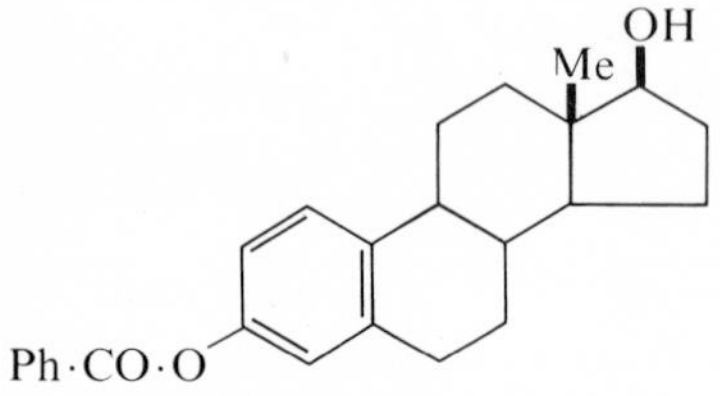

Ph·CO·O — Me — OH

Oestradiol Benzoate

$CH_2·O·CO·CH_3$ — CO — Me — OH — HO — Me — O

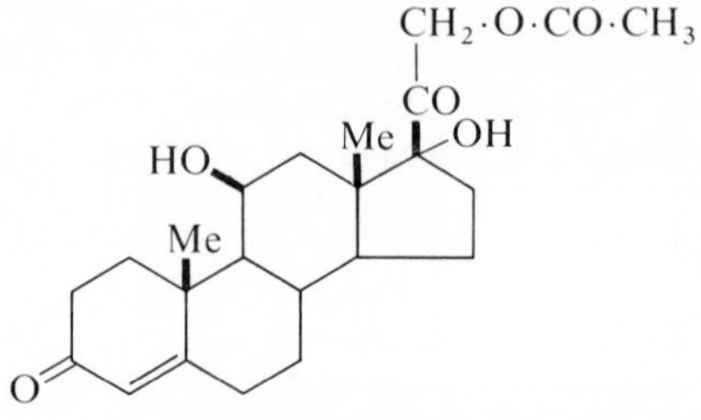

Hydrocortisone Acetate

Certain esters, however, show a high degree of stability in human plasma, which may be due to steric effects. Thus, both *Atropine* (Godeaux and Tonnesen, 1949) and *Cocaine* (Blaschko, Himms and Strömbled, 1955) are comparatively stable in human plasma, but readily hydrolysed in rabbit plasma. Other esters, such as *Pethidine*, which are stable to plasma esterases, are hydrolysed by liver esterases (Way, Swanson and Gimble, 1947).

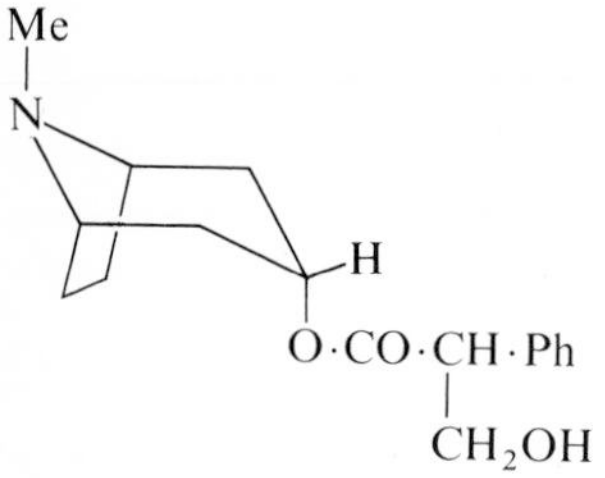

Me — N — H — O·CO·CH·Ph — CH_2OH

Atropine

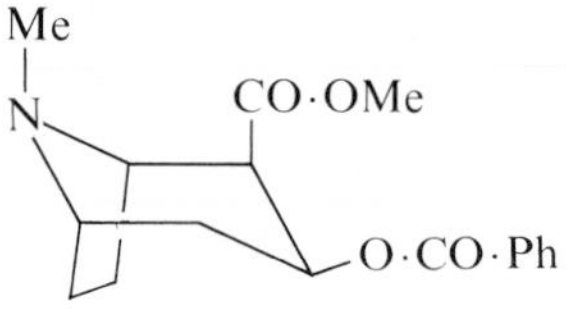

Me — N — CO·OMe — O·CO·Ph

Cocaine

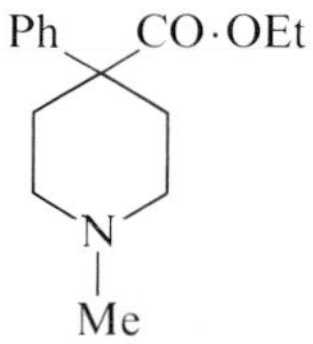

Ph — CO·OEt — N — Me

Pethidine

Carbamates

Urethanes (carbamate esters) are also hydrolysed by plasma esterases to the corresponding alcohol and acid in addition to undergoing oxidative metabolism, such as the formation of *N*-hydroxyurethane from urethane. Thus, the anti-thyroid drug, *Carbimazole*, is rapidly hydrolysed and decarboxylated in blood plasma to *Methimazole* (Williams, 1970).

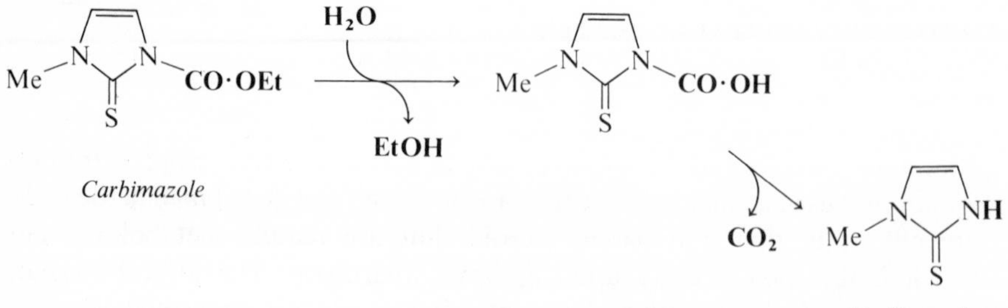

Me — N — N — CO·OEt — S

Carbimazole

H_2O → EtOH

Me — N — N — CO·OH — S

CO_2

Me — N — NH — S

Methimazole

Nitriles

Nitriles of aromatic acids are normally metabolised by ring hydroxylation, but hydrolysis to the corresponding carboxylic acid also occurs to a limited extent. Aliphatic nitriles, such as acetonitrile, however, undergo metabolism with release of the cyanide ion. The overall process is an oxidative one and it is not clear whether oxidation precedes hydrolysis or *vice versa*, though the former appears to be likely from the metabolites identified from phenylacetonitrile. Cyanide (CN^-), which is extremely toxic, undergoes rapid and marked detoxification by conversion to thiocyanate (CNS^-) in a reaction with thiosulphate catalysed by the enzyme, rhodanase (thiosulphate sulphur transferase), which is present in liver and other tissues.

$$Ph \cdot CH_2 \cdot CN \xrightarrow{\text{Oxidation}} Ph \cdot CH(OH) \cdot CN \longrightarrow Ph \cdot CHO + CN^-$$

$$CN^- \xrightarrow[S_2O_3^{2-} \;\rightarrow\; SO_3^-]{\text{Rhodanase}} CNS^-$$

Amides

Enzyme-catalysed hydrolysis of amides, hydrazides and similar carboxylic acid derivatives by proteolytic enzymes, like acidic and basic hydrolysis, is generally much slower than that of esters. Thus, *Procainamide* is far more stable to hydrolysis than *Procaine* (Mark, Kayden, Steele, Cooper, Berlin, Rovenstine and Brodie, 1951), whilst the stability of the amide link in *Isoniazid* is such that it is metabolised to *N*-acetylisoniazid in preference to hydrolysis. *N*-Deacetylation in contrast to *O*-deacetylation usually represents only a minor pathway for the metabolism of *N*-acetylamides such as phenacetin (Brodie and Axelrod, 1949).

$H_2N-C_6H_4-CO \cdot NH \cdot CH_2 \cdot CH_2 \cdot NEt_2$

Procainamide

$CO \cdot NH \cdot NH_2$ (on pyridine ring, 4-position)

Isoniazid

The stability of amides to hydrolysis is markedly influenced by steric and electronic effects. The steric effects of ring substituents in the 2- and 6-positions adjacent to the amide link of *Lignocaine* are particularly important (**1**, 16; **2**, 2). Cyclic amides and imides, also, show marked differences in stability *in vivo*. Thus, imide ring systems in barbiturates and *Glutethimide* are stable metabolically, but both imide rings of thalidomide undergo hydrolysis. The principal urinary excretion products are α(*o*-carboxybenzamido)glutarimide (16% of administered dose), 2- and 4-phthalimidoglutaramic acids (11%), 2- and 4-(*o*-carboxybenzamido)glutaramic acids, and 2-(*o*-carboxybenzamido)glutaric acid (29 %). None of these metabolites is teratogenic (Fabro, Smith and Williams, 1967).

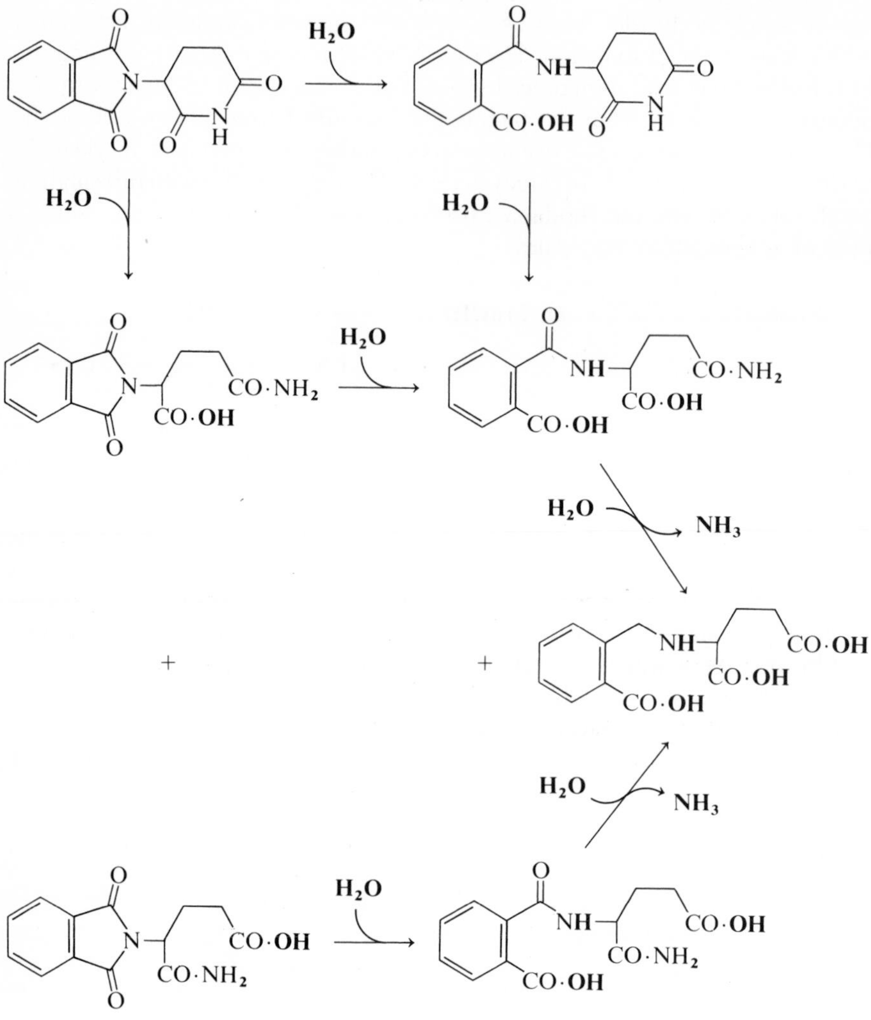

CONJUGATION

Introduction

Conjugation reactions usually represent the final step in the metabolism of lipid-soluble compounds to water-soluble derivatives capable of being readily excreted. The term **conjugation** is applied to reactions in which the drug, or more usually one of its metabolites, combines with an endogenous compound. It includes the formation of glucuronides, sulphates, amino acid and mercapturic acid conjugates, acylation and *O*-, *N*- and *S*-methylation. Conjugates are usually more water-soluble than the products from which they are formed, and hence more readily excreted.

Conjugation reactions are effected by transferase enzymes, which transfer the conjugating moiety from its donor coenzyme to the metabolite substrate. These enzymes are named systematically to include, in order, the donor coenzyme, the substrate (unless the enzyme is highly non-specific) and the group transferred. The more important enzymes involved in conjugation reactions are listed in Table 41.

Table 41. Some Transferases Involved in Conjugation of Drugs and Metabolites

E. C. Number	Systematic Name
2.1.1.a	*S*-Adenosylmethyl: catechol *O*-methyltransferase (Catechol *O*-methyltransferase)
2.3.1.5	Acetyl-CoA: arylamine *N*-acetyltransferase (Arylamine acetyltransferase)
2.3.1.13	Acyl-CoA: glycine *N*-acyltransferase (Glycine acyltransferase)
2.4.1.17	UDP-Glucuronate: glucuronlytransferase (UDP-glucuronyltransferase)[1]
2.8.2.1	3′-Phosphoadenylylsulphate: phenol sulphotransferase (Aryl sulphotransferase)
2.2.2.2	3′-Phosphoadenylylsulphate: 3β-hydroxysteroid sulphotransferase) (3β-hydroxysteroid sulphotransferase)

[1] Substrate unspecified

Methylation

Methylation represents a fairly minor pathway for the metabolism of drugs, despite its importance in the normal intermediary metabolism of catecholamine (Axelrod, Senoch and Witkop, 1958), and it is only compounds which are structurally related to catechol, i.e. *o*-dihydric phenols, which are subject to methylation by catechol *O*-methyltransferase. This enzyme is widely distributed in the soluble fraction of liver, kidney, skin, blood cells and adrenergic nerves, and requires Mg^{2+} ions and *S*-methyladenosine as coenzyme and the source of methyl groups.

O-Methylation may be depicted as shown on p. 270. Catecholamines, and related derivatives, are methylated in either the *meta* or *para* position to the alkyl substituent (Masri, Booth and DeEds, 1962), but gallic acid and other pyrogallol derivatives are methylated exclusively on the central hydroxyl (Masri, Robbins, Emerson and DeEds, 1964).

Monohydric phenols are not usually methylated, but there is good evidence that butyl 3,5-di-iodo-4-hydroxybenzoate is converted to butyl 3,5-di-iodo-4-methoxybenzoate by an iodophenol *O*-methyltransferase present in the soluble

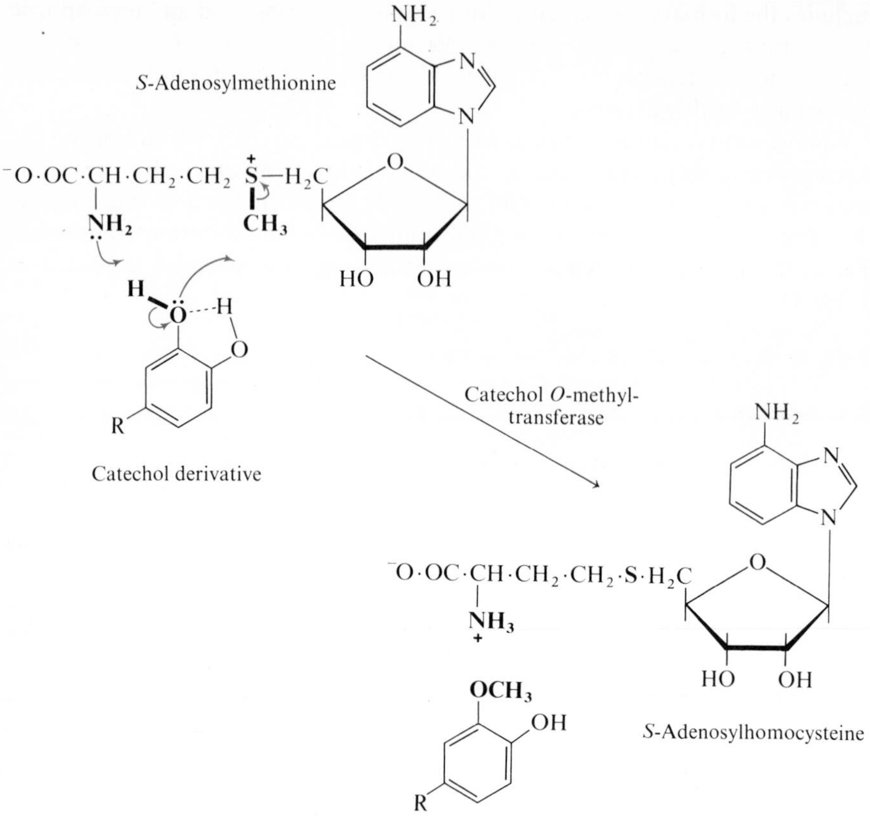

S-Adenosylmethionine

$^{-}O{\cdot}OC{\cdot}CH(NH_2){\cdot}CH_2{\cdot}CH_2\,\overset{+}{S}(CH_3){-}H_2C$

Catechol derivative

Catechol *O*-methyl-transferase

$^{-}O{\cdot}OC{\cdot}CH(\overset{+}{N}H_3){\cdot}CH_2{\cdot}CH_2{\cdot}S{\cdot}H_2C$

S-Adenosylhomocysteine

fraction of human liver (McLagen and Wilkinson, 1954). This enzyme is distinguishable from catechol-*O*-methyltransferase by its greater heat stability and non-requirement of Mg^{2+}, but also utilises *S*-methyladenosine as methyl donor. It does not, however, methylate either *Thyroxine* or *Liothyronine* to any appreciable extent (Tomita, Cha and Lardy, 1964).

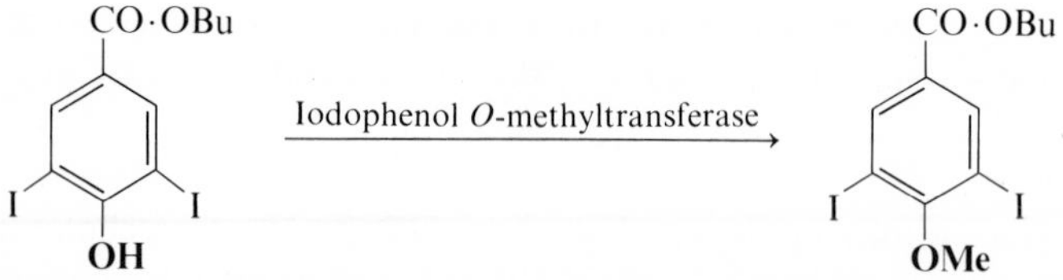

Iodophenol *O*-methyltransferase

Hydroxyindole *O*-methyltransferase which *O*-methylates *N*-acetylserotonin and other related hydroxyindoles has only been identified in the pineal gland and retina of certain vertebrates (Quay, 1965).

S-Adenosylmethionine also acts as methyl donor in a number of *N*-methylations catalysed by *N*-methyltransferases which form part of normal intermediary metabolism. Some, like imidazole *N*-methyltransferase, are specific for the methylation of 1-methylhistamine (Brown, Tomchick and Axelrod, 1959).

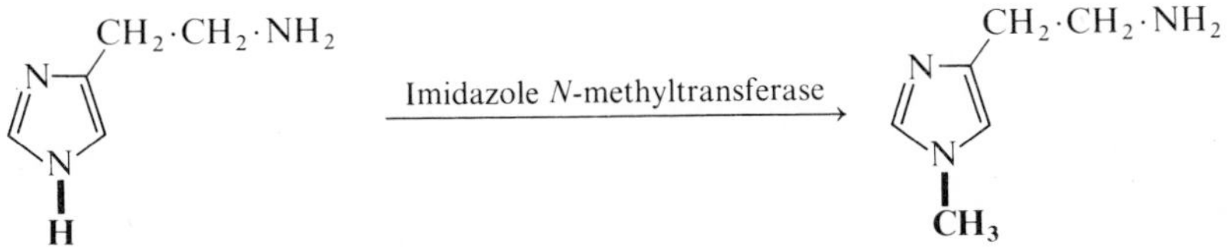

Imidazole *N*-methyltransferase →

Other *N*-methyltransferases are far less specific. Thus, the *N*-methyltransferase present primarily in rabbit lung, but also to a lesser extent in the adrenals and kidneys, which is readily capable of *N*-methylating serotonin, will also *N*-methylate aniline, norpethidine, normorphine, *Norephedrine* (Phenylpropanolamine) and *Ephedrine*, though rather less effectively (Axelrod, 1962).

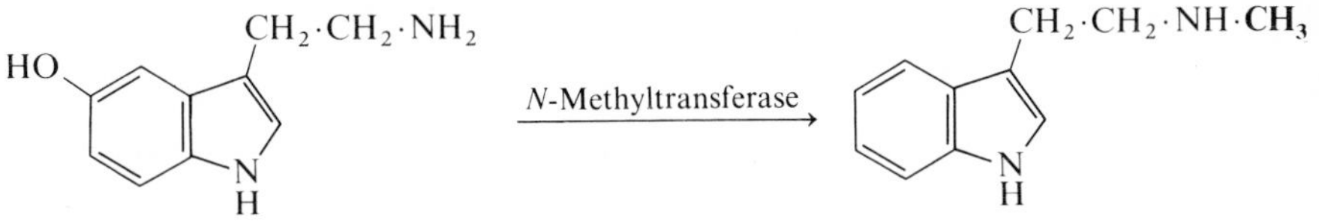

N-Methyltransferase →

Serotonin

N-Methylserotonin

Aniline

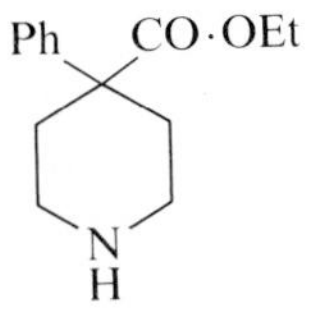

Norpethidine

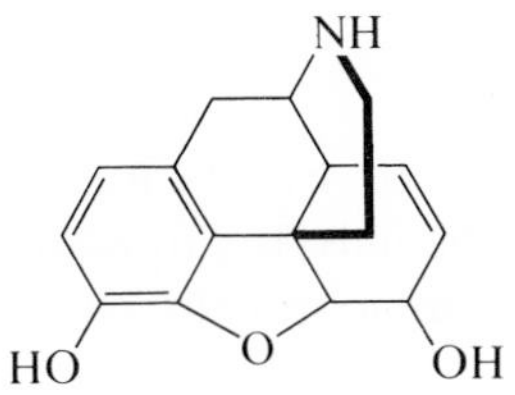

Normorphine

$Ph \cdot CH(OH) \cdot CH(NH_2) \cdot CH_3$

Norephedrine

Acetylation

Acetylation of aromatic (but not aliphatic) amines by acetyl-$\overline{\text{CoA}}$: arylamine *N*-acetyltransferase is the main pathway for the metabolism of aromatic amines, such as aniline, *p*-aminobenzoic acid and sulphonamides. The latter are acetylated to give N^1- and N^4-, as well as the N^1N^4-diacetyl derivatives.

SO_2NH_2 (Sulphanilamide, NH_2) → $SO_2 \cdot NH_2$ / $NH \cdot CO \cdot CH_3$ (N^4-Acetylsulphanilamide) → $SO_2 \cdot NH \cdot CO \cdot CH_3$ / $NH \cdot CO \cdot CH_3$ (N^1N^4-Diacetylsulphanilamide)

Sulphanilamide → $SO_2 \cdot NH \cdot CO \cdot CH_3$ / NH_2 (N^1-Acetylsulphanilamide) → N^1N^4-Diacetylsulphanilamide

The enzyme, which is present in the reticulo-endothelial tissue of liver, lung and spleen (Govier, 1965) is probably also the enzyme responsible for the acetylation of *Isoniazid*, since purification does not alter its relative activity towards the two types of substrate (Weber and Cohen, 1967). Kinetic studies indicate that acetylation takes place in two stages, both bimolecular, the first step being acetylation of the enzyme by acetylcoenzyme A, followed by a nucleophilic attack by the amino compound on the acetyl enzyme.

$$\text{Enz—H} \xrightarrow[\text{CH}_3\cdot\text{CO}\cdot\text{SCoA} \;\to\; \text{HSCoA}]{} \text{Enz—CO}\cdot\text{CH}_3 \xrightarrow[\text{R}\cdot\ddot{\text{N}}\text{H}_2 \;\to\; \text{R}\cdot\text{NH}\cdot\text{CO}\cdot\text{CH}_3]{} \text{Enz—H}$$

A number of other *N*-acetyltransferases have been identified, which catalyse the transfer of acetyl from *N*-hydroxyacetamido compounds to aromatic amines (Booth, 1966), as for example the acetylation of 4-aminoazobenzene.

N-Hydroxyacetamidobiphenyl (N(OH)CO·CH$_3$) → Biphenylhydroxylamine (N(OH)H)

4-Aminobenzene (—N=N—, —NH$_2$) —Acetyltransferase→ 4-Acetylaminoazobenzene (—N=N—, —NH·CO·CH$_3$)

The resulting hydroxylamines are themselves potent carcinogens and are known to be capable of undergoing metabolic transformation to nitroso-compounds, which are even more potent in this respect.

Amino Acid Conjugation

Most aromatic carboxylic acids and certain aliphatic acids are metabolised and excreted as amino acid conjugates. The acyltransferases, which catalyse these conjugations, are present in liver mitochondria, and are usually specific for a particular amino acid characteristic of intermediary nitrogen metabolism in that species (Parke, 1968). In man, benzoic and other aromatic acids are conjugated with glycine. Adenyl benzoate and benzoyl-S$\overline{\text{CoA}}$ are intermediates in the process (Moldave and Meister, 1957a) in the same way as acyl adenylates and acyl-S$\overline{\text{CoA}}$ act as intermediates in the metabolism of fatty acids generally.

$$\text{Ph·}\mathbf{CO·OH} \xrightarrow[\text{ATP + HS·}\overline{\text{CoA}} \;\to\; \text{AMP + PP}]{} \text{Ph·}\mathbf{CO·S}\overline{\mathbf{CoA}} \xrightarrow[\text{H}_2\text{N·CH}_2\text{·CO·OH (glycine)} \;\to\; \text{HS·}\overline{\text{CoA}}]{} \text{Ph·}\mathbf{CO·NH}\text{·CH}_2\text{·CO·OH}$$

Substituted aliphatic acids incapable of conversion to either acetyl or propionyl-S$\overline{\text{CoA}}$, including phenylacetic acid, indolylacetic acid, cinnamic acid and certain cholic acids, on the other hand, are excreted as glutamine conjugates (Moldave and Meister, 1957b). Synthesis of phenylacetylglutaminic acid in human liver and kidney is increased by coenzyme A and ATP in accord with the formation of phenylacetylcoenzyme A as intermediate, and a similar conjugation mechanism to that which gives rise to benzoylglycine. Unlike benzoic acid, which is probably mainly exogenous, arising from the diet, phenylacetic acid is almost certainly derived from the breakdown of phenylalanine.

Glucuronide Conjugation

Glucuronide formation is probably the most important pathway for conjugation of hydroxylated endogenous compounds and drugs in man. Amino, sulphydryl and some acids also form glucuronide conjugates. UDP-Glucuronyl transferase (EC 2.4.1.17, uridinediphosphosphate glucuronate glucuronyltransferase), the enzyme which catalyses glucuronide formation, is present mainly in the microsomal fraction of the liver, but also to a lesser extent in kidney, skin, and the gastro-intestinal tract (Dutton, 1966). The coenzyme, uridinediphosphoglucuronic acid (UDPGA), which acts as glucuronyl donor, is also formed in the liver, though in the soluble fraction, and is synthesised from glucose-1-phosphate by the following pathway (Strominger, Kalckar, Axelrod and Maxwell, 1954; Maxwell, Kalckar and Strominger, 1956; Strominger, Maxwell, Axelrod and Kalckar, 1957).

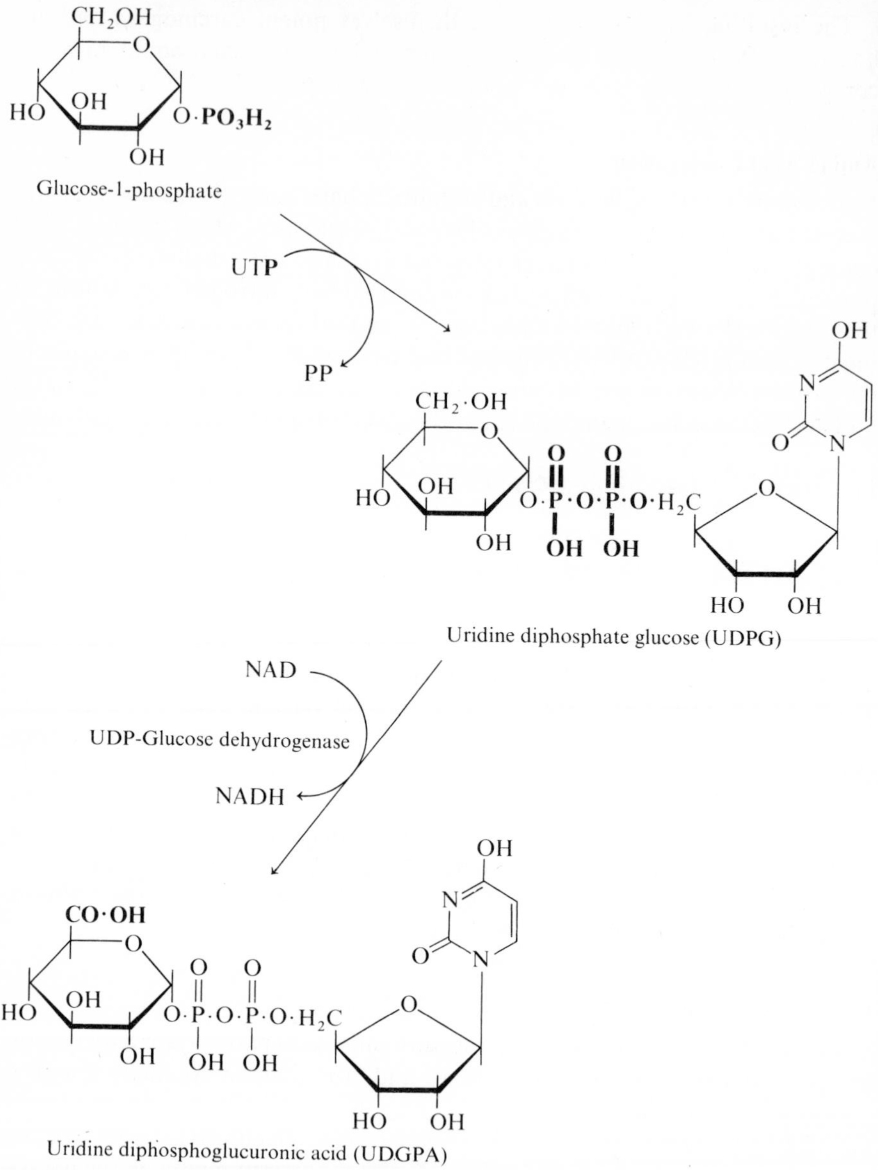

CH_2OH

Glucose-1-phosphate

UTP

PP

$CH_2 \cdot OH$

Uridine diphosphate glucose (UDPG)

NAD

UDP-Glucose dehydrogenase

NADH

CO·OH

Uridine diphosphoglucuronic acid (UDGPA)

Ether *O*-Glucuronides

Ether *O*-glucuronide formation from UDPGA and appropriate substrates in the presence of glucuronyl transferase is a bimolecular displacement reaction with inversion at C-1 of the glucuronide moiety giving rise to a β-glucuronide (Storey and Dutton, 1955).

UDPGA + H–O–Ph ⇌ (Glucuronyltransferase / β-Glucuronidase) β-Phenylglucoside + UDP

The enzyme catalysing glucuronyl transfer is membrane-dependent (Dutton and Storey, 1954). It is inactivated by degradation of the microsomal phospholipid membrane by phospholipase A or C in the presence of Ca^{2+} (Graham and Wood, 1969), but activity is restored by phosphatidylcholine or by the addition of mixed microsomal phospholipid micelles, which consist mainly of phosphatidylcholine (Attwood, Graham and Wood, 1971). Reactivation also occurs with low concentrations of cationic detergents, such as cetylpyridinium bromide. Higher concentrations of cationic detergents, however, and anionic detergents are potent inactivators, but non-ionic detergents have little effect. Similarly, low concentrations (*ca* 0.5%) of detergents, notably Triton X-100 at 0°, enhance the activity of the untreated enzyme (Leuders and Kuff, 1967; Wood and Graham, 1972) in an irreversible process which is not accompanied by solubilisation. In contrast, higher concentrations of detergent inactivate.

O-Glucuronide formation is typical of phenols, e.g. *o*-aminophenol (Storey and Dutton, 1955) and hydroxyguanoxan (Jack, Stenlake and Templeton, 1972). Glucuronidation is the normal pathway for metabolism of endogenous phenols, including *Thyroxine* (Taurog, Briggs and Chaikoff, 1951), *Liothyronine* (Roche, Michel and Tata, 1953) and *Oestriol* (Estriol; Felger and Katzman, 1961). Glucuronidation is also characteristic of steroid metabolism (Jayle and Pasqualini, 1966) and of other secondary and tertiary alcohols resistant to oxidation. The proportion of alcohol excreted as glucuronide generally increases with molecular complexity (Table 42). Steric factors are also important. 3α-Hydroxysteroids, but not 3β-hydroxysteroids, are excreted as glucuronide conjugates. In contrast, 3β-hydroxysteroids undergo sulphate conjugation exclusively. 3α-Hydroxysteroids, however, are usually metabolised to glucuronides, though some appear partly as sulphates. The progesterone metabolites, pregnane-3α,20α-diol, allopregnane-3α,20α-diol and pregnan-3α-ol-20-one, however, are all excreted entirely as glucuronides (Crépy, Judas, Rulleau-Meslin and Jayle, 1962).

O-Glucuronides are generally highly water-soluble and strongly acidic, with pK_a between 3.0 and 4.0 (Crépy, Jayle and Rulleau-Meslin, 1957). With few exceptions, they form water-soluble salts with alkalis. They are, however, only poorly soluble in organic solvents, and are best extracted from acidic aqueous media, such as urine, by butanol or ethanol–ether (1:3). They are stable in alkaline solution and in mild acid (down to pH 2), but slowly hydrolysed

Table 42. Glucuronidation of Alcohols

Primary alcohols	Percent Excreted as Glucuronide
$CH_3 \cdot CH_2 \cdot OH$	0.5
$CH_3 \cdot CH_2 \cdot CH_2 \cdot OH$	0.9
$CH_3(CH_2)_3 \cdot OH$	1.8
$CH_3(CH_2)_4 \cdot OH$	6.7
Secondary alcohols	
$(CH_3)_2 \cdot CH \cdot OH$	10
$CH_3 \cdot CH_2 \cdot CH(CH_3) \cdot OH$	14
$CH_3 \cdot (CH_2)_2 \cdot CH(CH_3) \cdot OH$	45
$CH_3 \cdot (CH_2)_3 \cdot CH(CH_3) \cdot OH$	54
Tertiary alcohols	
$(CH_3)_3C \cdot OH$	24
$CH_3 \cdot CH_2 \cdot C(CH_3)_2 \cdot OH$	58

by prolonged exposure to strong acid at elevated temperatures. *O*-Glucuronides do not reduce Benedict's solution. Enol glucuronides, such as those of 4-hydroxycoumarin and androstanedione, are hydrolysed much more readily at pH 3.5 in 1 hr at 37° (Wakabayashi, Wotiz and Fishman, 1961). Enol glucuronides are also unstable to alkali.

N-Hydroxyamines, whether administered as such or formed metabolically as products of *N*-hydroxylation are sometimes, though by no means always, excreted as glucuronides. Thus, *N*-acetyl-*N*-phenylhydroxylamine (Kato, Ide, Hirohata and Fishman, 1967) and *N*-hydroxy-2-acetylaminofluorene, the

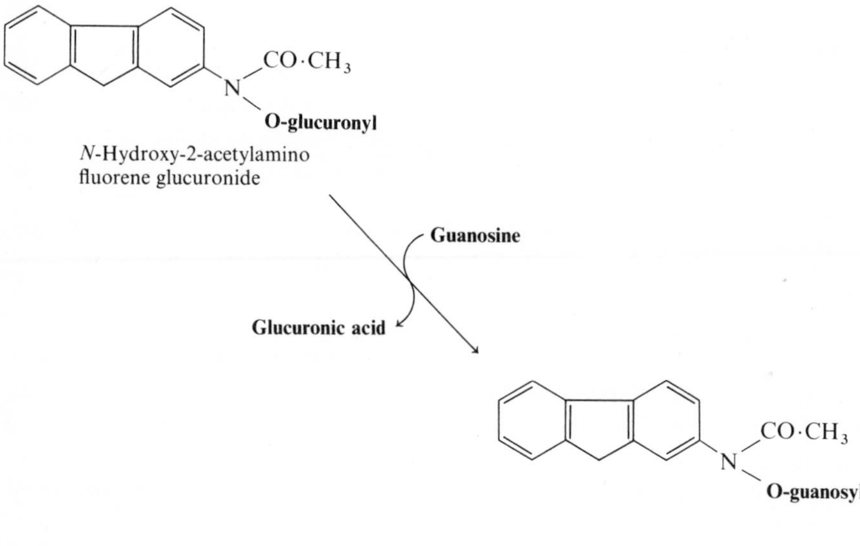

$CO \cdot CH_3$
N
O-glucuronyl

N-Hydroxy-2-acetylamino fluorene glucuronide

Guanosine

Glucuronic acid

$CO \cdot CH_3$
N
O-guanosyl

metabolite of 2-acetylaminofluorene (Irving, 1965), are excreted as *O*-glucuronides. Recently, Irving (1971) has shown that both glucuronide and sulphate conjugates of *N*-hydroxy-2-acetylaminofluorene are reactive in non-enzymic reactions leading to covalent attachment of 2-acetylaminofluorene to RNA and DNA. Thus, the glucuronide of *N*-hydroxy-2-acetylaminofluorene reacts *in vitro* with guanosine to yield *N*-(guanosin-8-yl)acetylaminofluorene, and similarly, with yeast RNA, rat liver RNA and calf thymus DNA. The rates of these reactions are enhanced at alkaline pH, due to hydrolysis to the glucuronide of *N*-2-fluorenylhydroxylamine which is even more reactive with nucleophiles.

β-Glucuronidase

The enzyme, β-glucuronidase, which is capable of hydrolysing β-glucuronides, is present in almost all mammalian tissues (Levvy and Conchie, 1966). High levels of activity are found in man, particularly in liver, kidney and spleen; high levels are often found in cancerous tissue. There is evidence that β-glucuronidase is present in the intestinal wall of the newborn child (Brodersen and Hermann, 1963). β-Glucuronidase in the large intestine of humans, which plays an important part in enterohepatic circulation (**2**, 4) is probably of microbiological origin (Williams, Millburn and Smith, 1965; **2**, 4). Although β-glucuronidase is present in liver tissue, it is probably normally subject to suppression by glucurono(1 → 4)-lactone, the lactone of glucuronic acid, a powerful inhibitor of the enzyme which has been identified in human bile (Matsushiro, 1965).

β-Glucuronidase catalyses the hydrolysis of alkyl, aryl and acyl β-glucuronides (Levvy and Marsh, 1959). The rate of hydrolysis of steroid glucuronides varies substantially with the structure of the steroid, decreasing in the order oestrogenic C-3 (phenolic)glucuronides > C-3β > C-17β > C-3α- (Kellie, 1966). The optimum pH of the enzyme is about 4.5. This shifts to more acidic pH on inhibition by Ag^{+}, Hg^{2+} and Cu^{2+} ions (Fernley, 1962). The enzyme is also inhibited competitively by *p*-chloromercuribenzoate with reversal by cysteine in a manner typical of sulphydryl compounds. Not all glucuronides undergo hydrolysis by β-glucuronidase in the gut, as for example the ester glucuronide of *Iopanoic Acid.*

Ester Glucuronides

The formation of ester glucuronides is typical of bilirubin (**1**, 23) and particularly of carboxylic acids sterically hindered by substituents on the α-carbon, such as phenylacetic and diphenylacetic, *Carbenoxolone* (Iveson, Lindup, Parke and Williams, 1971), *Indomethacin* (Harman, Meisinger, Davis and Kuehl, 1964) and *Iopanoic Acid* (McChesney and Hoppe, 1954, 1956). *Indomethacin* conjugation in man appears to take place mainly in the kidney. In contrast to *O*-glucuronides, ester glucuronides are unstable in dilute alkali, and, as a result of the release of the sugar moiety from substitution at C-1, also reduce Benedict's reagent.

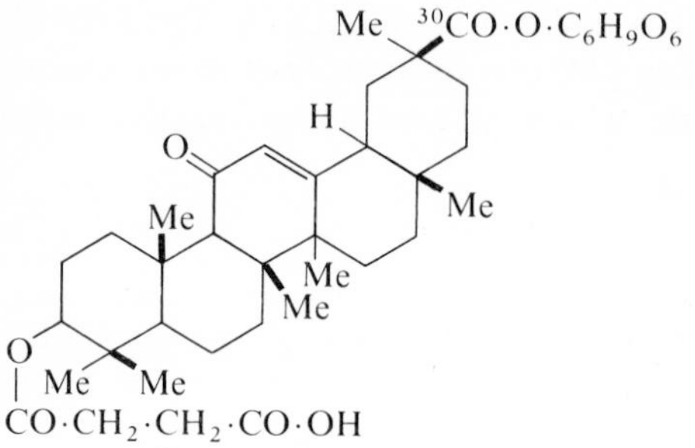

Carbenoxolone glucuronide

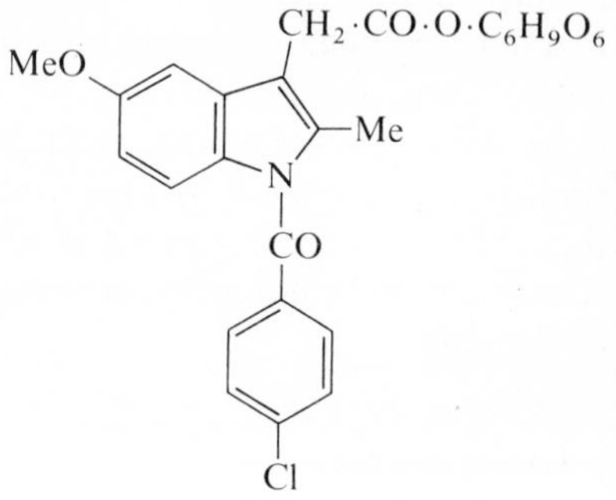

Indomethacin glucuronide

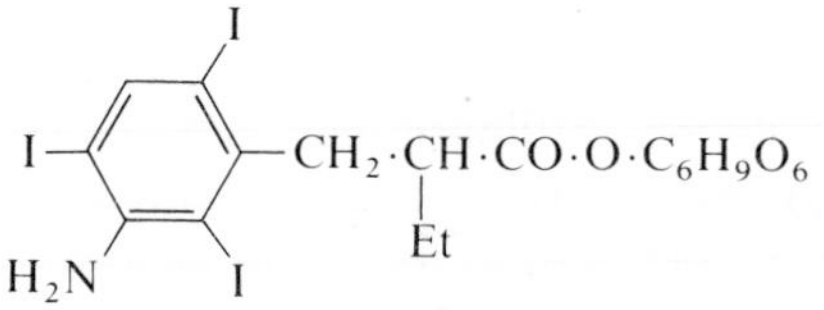

Iopanoic Acid glucuronide

The glucuronyl transferase responsible for the formation of the bilirubin ester conjugate is distinguishable from that involved in the formation of ether *O*-glucuronides (Tomlinson and Yaffé, 1966). Thus, Mg^{2+} increases bilirubin conjugation, but inhibits *p*-nitrophenylglucuronide conjugation; *p*-nitrophenol acts as a potent inhibitor of bilirubin conjugation, but bilirubin does not affect *p*-nitrophenylglucuronidation; and bilirubin conjugation is inhibited by snake venom at pH 9.0, whereas *p*-nitrophenylglucuronidation is enhanced.

N-Glucuronides

A number of primary aromatic amines, such as β-naphthylamine (Boyland, Manson and Orr, 1957) and *Dapsone* (Bushby and Woiwod, 1956) are known to form *N*-glucuronides.

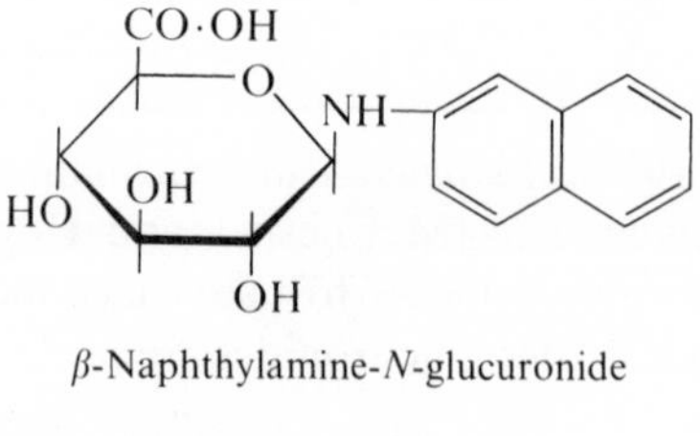

β-Naphthylamine-*N*-glucuronide

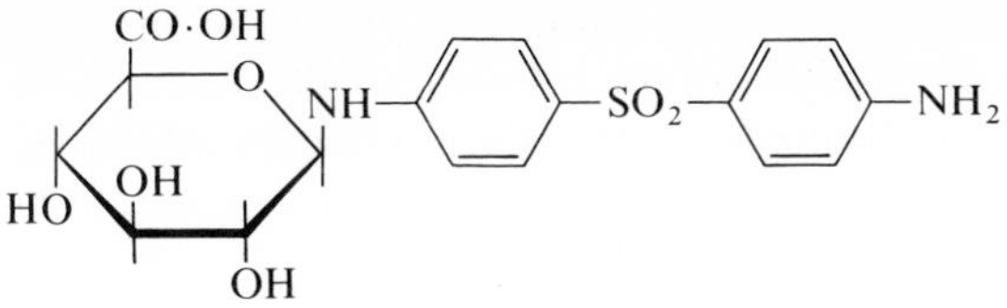

Dapsone-N-glucuronide

These amino N-glucuronides are probably mainly artefacts formed from the amine and glucuronic acid in the mildly acidic conditions of normal urine, though there is some evidence for the existence of a specific N-glucuronyl-transferase (Leventer, Buchanan, Ross and Tapley, 1965). N-Glucuronides are decomposed readily in strong acid, and in alkaline media, so that like ester glucuronides, they reduce Benedict's reagent.

Meprobamate (Tsukamoto, Yoshimura and Tatsumi, 1963) and certain sulphonamides, such as *Sulphadimethoxine* (Bridges, Kibby and Williams, 1964) also form amido N-glucuronides, which are somewhat more stable to both acid and alkali, and do not for example reduce Benedict's solution.

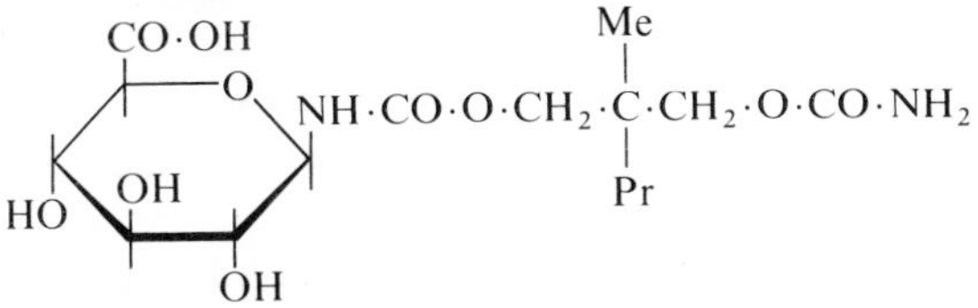

Meprobamate-N-glucuronide

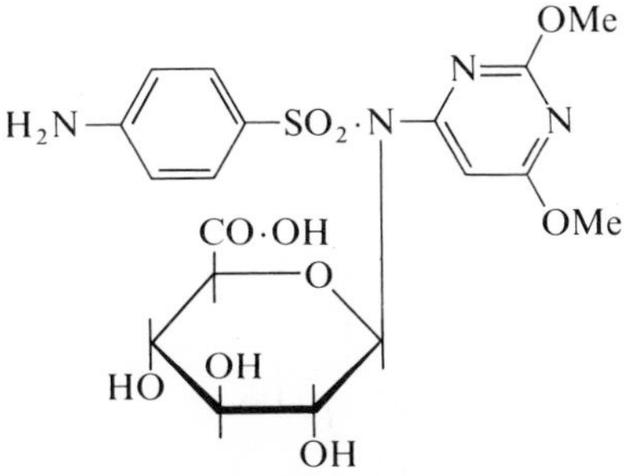

Sulphadimethoxine-N-glucuronide

S-Glucuronides

Certain sulphydryl (-SH) compounds form glucuronide conjugates as, for example, diethyldithiocarbamic acid, which is a metabolite of *Disulfiram*, the drug which is used in the treatment of alcoholism (Kaslander, 1963).

Sulphate Conjugation

Conjugation as sulphate esters and sulphamates provides an important alternative pathway to glucuronidation for the metabolism of phenols, alcohols and certain amines (Robbins and Lipmann, 1957; Böstrom and Vestermark, 1960) as well as for the biosynthesis of the polysaccharides, chondroitin and *Heparin*. Thus, sulphate ester formation forms the normal metabolic pathway for the metabolism of phenols, *Oestrone* (Estrone) and 3β-hydroxysteroids (Crépy, Judas, Rulleau-Meslin and Jayle, 1962).

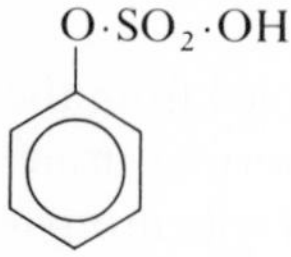

$O \cdot SO_2 \cdot OH$

Phenol sulphate

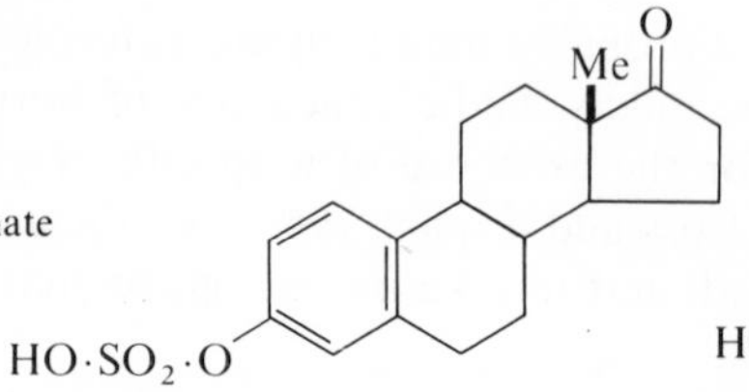

$HO \cdot SO_2 \cdot O$

Oestrone sulphate

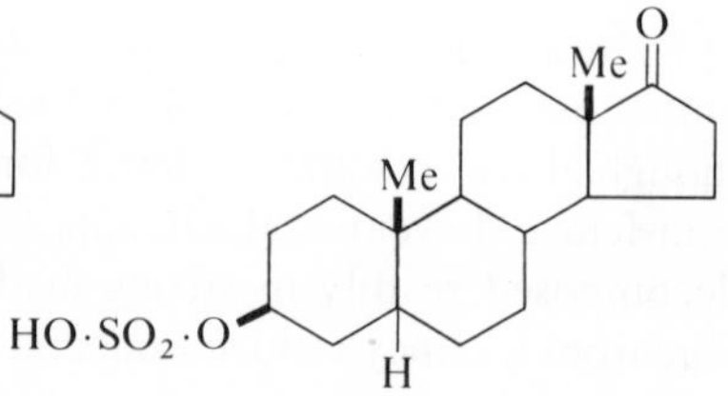

$HO \cdot SO_2 \cdot O$

3β-Hydroxy-5α-androstane-17-one-sulphate

In all cases, the sulphate group is transferred from 3′-phosphoadenosine-5′-phosphosulphate (PAPS) by specific sulphotransferases (sulphokinases) present in liver, kidney and the intestinal mucosa. These sulphotransferases show a degree of substrate specificity as, for example, phenol sulphotransferase (Gregory and Lipmann, 1957) which does not catalyse the formation of either steroidal alcohol sulphates (Nose and Lipmann, 1958) or aromatic amine sulphamates (Roy, 1960). This particular enzyme, which is relatively non-specific with regard to phenol structure, is found in the soluble fraction of liver, kidney and intestinal mucosa. *Oestrone* sulphotransferase and the sterol sulphotransferases are much more specific and for the the most part concentrated in liver tissue.

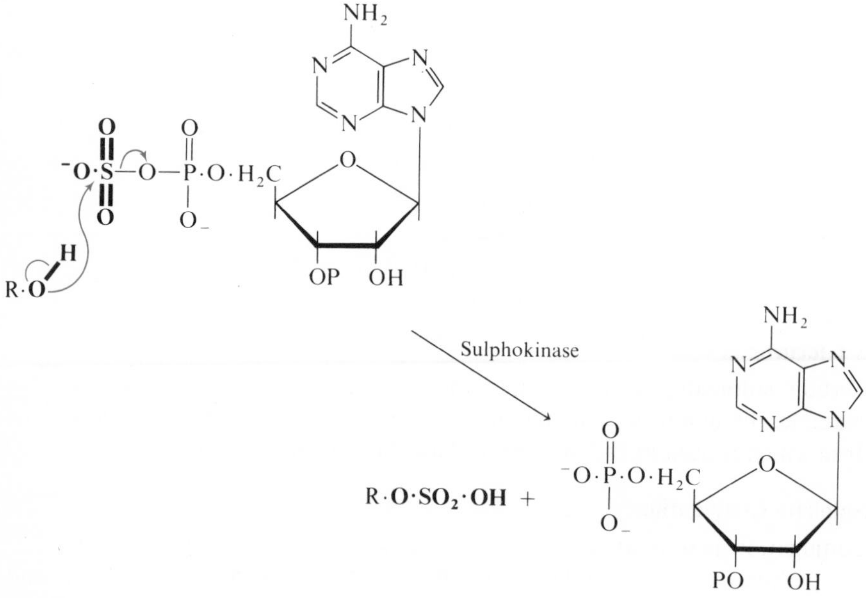

NH_2

$^{-}O \cdot S - O - P \cdot O \cdot H_2C$

$R \cdot O - H$

OP OH

Sulphokinase

$R \cdot O \cdot SO_2 \cdot OH$ + $^{-}O \cdot P \cdot O \cdot H_2C$

PO OH

The formation of PAPS from adenosine triphosphate (ATP) and sulphate occurs in two steps catalysed as follows by ATP sulphurylase (ATP sulphate adenylyltransferase) and APS phosphotransferase (ATP-adenylyl sulphate 3′-phosphotransferase).

NH$_2$

P·P·P·O·H$_2$C

HO OH

Adenosine Triphosphate (ATP)

ATP sulphate adenylyltransferase

SO_4^{2-}

PP

NH$_2$

$^-$O·S·O·P·O·H$_2$C

O$_-$

OH OH

Adenosine-5′-phosphosulphate (APS)

ATP

ATP-adenylylsulphate 3′-phosphotransferase

ADP

NH$_2$

$^-$O·S—O·P·O·H$_2$C

O$_-$

O OH

O=P—O$^-$

O$^-$

PAPS

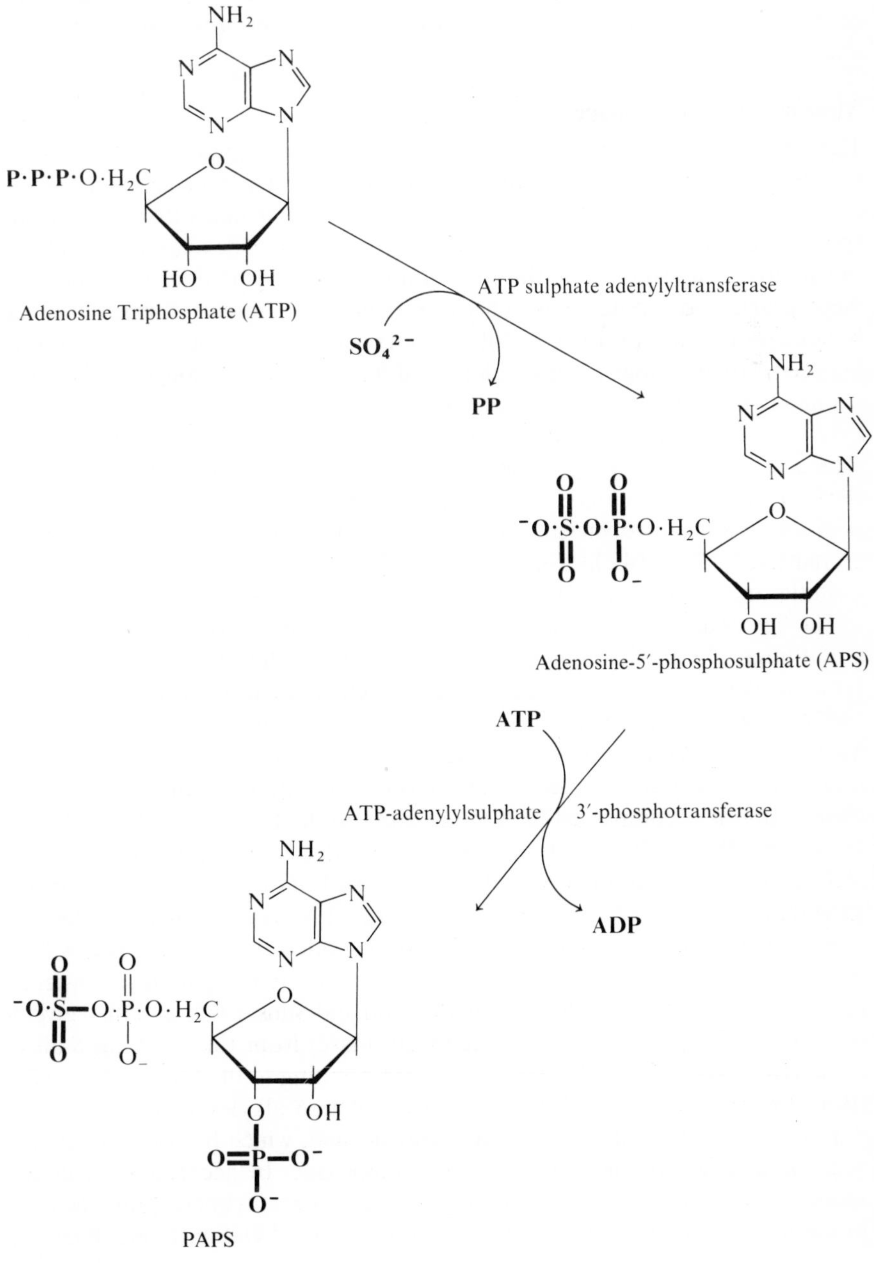

Capacity for sulphate conjugation in man is limited (Levy and Matsuzawa, 1967), and hence vulnerable to saturation and competition. Levy and Yamada (1971) have shown that competitive inhibition of sulphate conjugation occurs between *Paracetamol* (Acetaminophen) and *Salicylamide*. Co-administration of *Ascorbic Acid* (Vitamin C), which is also in part metabolised to its sulphate ester, with *Salicylamide* similarly decreases metabolism of the latter (Houston and Levy, 1975).

Mercapturic Acid Conjugates

The formation of mercapturic acid conjugates forms an important pathway for the metabolism of halogen compounds in mammals. The much greater toxicity of simple alkyl halides in man compared with the rat and other mammalian species provides a good indication of the relative extent of development, and also the relative importance of this degradative pathway (Johnson, 1966). The mercapturic acid conjugates which are excreted are themselves formed by metabolic degradation of products resulting from nucleophilic displacement reactions between labile halogen compounds and glutathione. Halogen compounds conjugated by this process include not only simple alkyl halides (Johnson, 1963, 1966) and aralkyl halides, such as benzyl chloride (Stekol, 1939), but also aromatic compounds (Bray, James and Thorpe, 1957; Booth, Boyland and Sims, 1961), in which the halogen is activated by electron-withdrawing substituents, as in 3,4-dichloronitrobenzene. Suitable halogen-substituted π-deficient hetero-aromatics, e.g. 2- and 4-chloropyridines and 2-chlorotriazines (Hutson, Hoadley, Griffiths and Donniger, 1970) are also conjugated by the same mechanism.

The key reaction is a nucleophilic displacement of halogen by the tripeptide glutathione (glutamylcysteinylglycine) mediated by a glutathione *S*-alkyl- or glutathione-*S*-aryl-transferase. These enzymes, which are quite distinct (Johnson, 1963) are widely distributed in a variety of tissues, but mainly in the soluble fraction of liver, kidney and heart muscle. Evidence that glutathione conjugation represents the key step comes both from the observation that mercapturic acid conjugation is accompanied by a fall in glutathione levels in rat liver (Barnes, James and Wood, 1959; Johnson, 1963), and also from reports that *S*-alkyl-mercapturic acids and their *S*-oxides are excreted in the urine of rats following subcutaneous injections of *S*-alkylglutathiones (Foxwell and Young, 1964).

Both *S*-alkyl and *S*-arylglutathiones are degraded by the same pathway to mercapturic acids. The glutamyl residue is first cleaved by γ-glutamyltransferase (glutathionase; Binkley, 1961; Booth, Boyland and Sims, 1960), and then glycine is displaced by peptidases (i.e. cysteinylglycinase) from the resulting *S*-alkyl- (or aryl)- cysteinylglycine to form the corresponding *S*-alkyl (or aryl)-cysteine (Bray, Franklin and James, 1959). Acetylation of the *S*-alkylcysteine by coenzyme A results in the formation of the mercapturic acid, which is either excreted as such or oxidised to the corresponding sulphoxide. In accordance with this sequence of reactions, both *S*-methylcysteine and *S*-methylmercapturic acid are found in the urine following administration of *S*-methylglutathione (Barnsley,

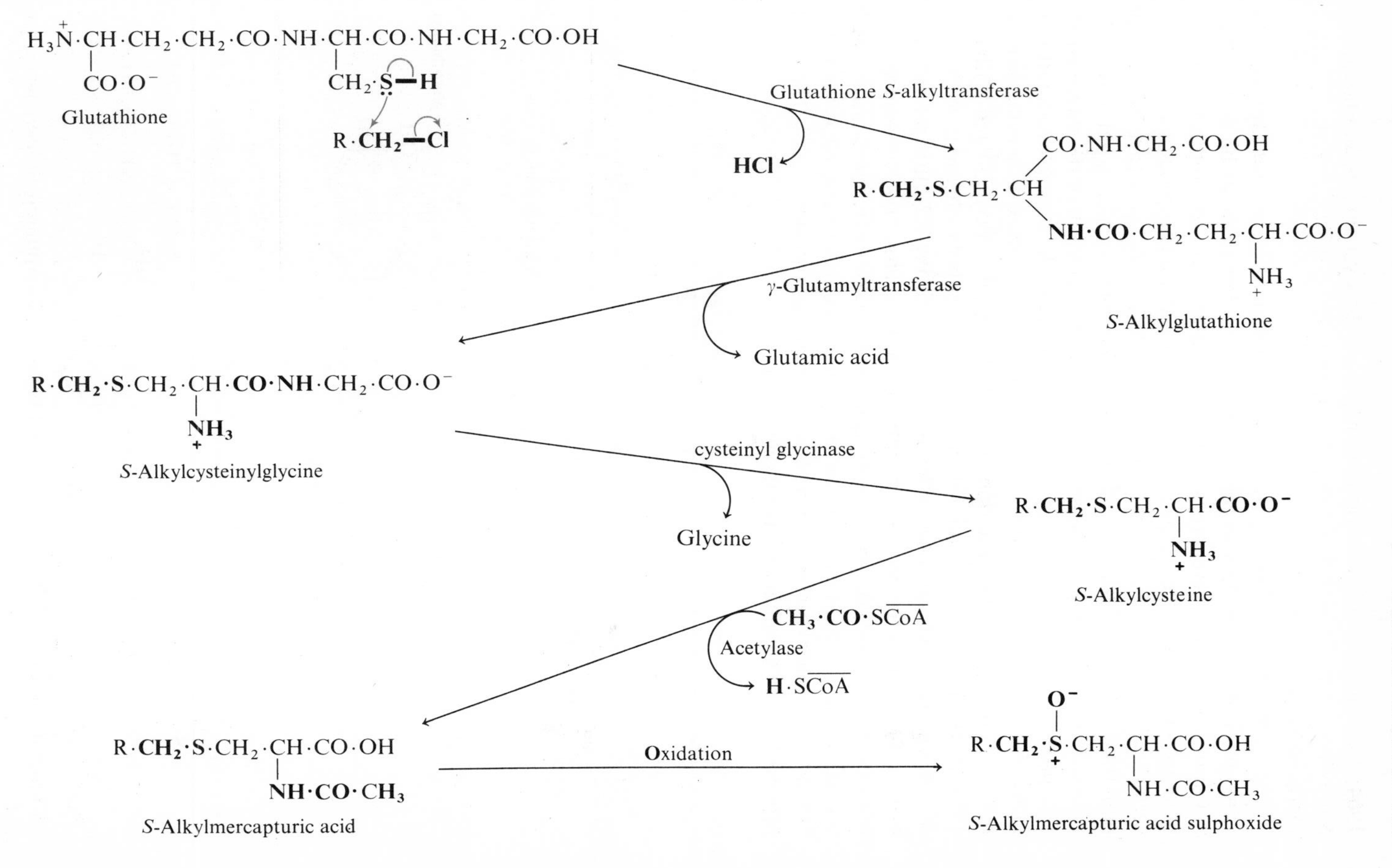
Glutathione
Glutathione S-alkyltransferase
HCl
S-Alkylglutathione
γ-Glutamyltransferase
Glutamic acid
S-Alkylcysteinylglycine
cysteinyl glycinase
Glycine
S-Alkylcysteine
Acetylase
S-Alkylmercapturic acid
Oxidation
S-Alkylmercapturic acid sulphoxide

1964). Similarly, administration of either *S*-ethyl-L-cysteine, its sulphoxide, or *S*-ethylmercapturic acid, has been shown to lead to the excretion of *S*-ethylmercapturic acid sulphoxide.

The complete metabolic pathway for the formation of mercapturic acid conjugates from alkyl halides can, therefore, be summarised as on p. 283.

Glutathione, and mercapturic acid conjugates, are also implicated in the metabolism of a number of fused polycyclic aromatic compounds and also of certain $\alpha\beta$-unsaturated carbonyl compounds. Polycyclic hydrocarbons, such as naphthalene, forms 1,2-dihydro-2-hydroxy-1-mercapturic acids, but only in the presence of NADPH and oxygen, which leads to formation of the intermediate 1,2-epoxy-compound. The latter undergoes nucleophilic attack by glutathione in the presence of glutathione-*S*-epoxytransferase, an epoxide specific enzyme present in the supernatant fraction of liver homogenates (Boyland and Williams, 1965; Boyland and Chasseaud, 1969). The usual metabolic degradative cleavage of the glutathionyl substituent leads to *S*-(2-hydroxy-1,2-dihydronaphthyl)-mercapturic acid as the principal metabolic product. This, however, undergoes a non-enzymic acid-catalysed dehydration under the acidic conditions which prevail in the normal urine, so that the product actually excreted is the metabonate *S*(1-naphthyl)-mercapturic acid (Gillham and Young, 1968). The entire process may be summarised as follows.

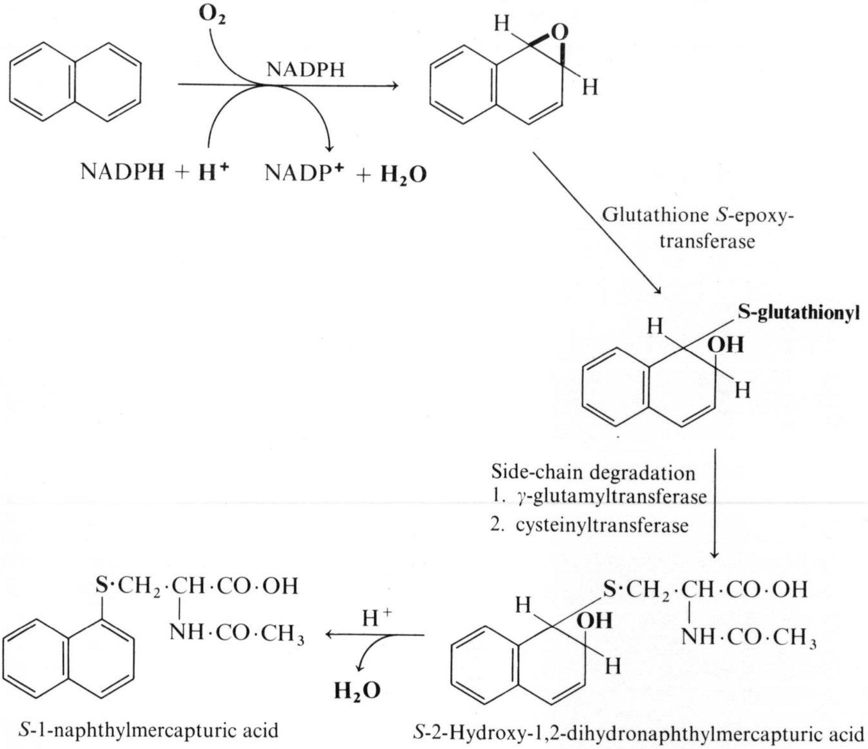

O_2

NADPH

NADPH + H^+ NADP$^+$ + H_2O

Glutathione *S*-epoxy-transferase

S-glutathionyl

OH

Side-chain degradation
1. γ-glutamyltransferase
2. cysteinyltransferase

S·CH$_2$·CH·CO·OH
NH·CO·CH$_3$

H^+

H_2O

S-1-naphthylmercapturic acid

S-2-Hydroxy-1,2-dihydronaphthylmercapturic acid

It is of interest that *S*-arylcysteines derived from carcinogenic hydrocarbons have been detected in protein tissues which bind these hydrocarbons (Smith and Wood, 1959; Bucovaz and Wood, 1964).

Unsaturated compounds, in which the double bond is activated by electron-withdrawing substituents, including $\alpha\beta$-unsaturated carbonyl compounds, also undergo enzyme-catalysed nucleophilic attack by glutathione (Boyland and Chasseaud, 1967). There is evidence of more than one glutathione-*S*-alkylene-transferase (Boyland and Chasseaud, 1969), indicative of a high degree of both glutathione and substrate specificity. One such enzyme appears to be involved in the metabolism of *Ethacrynic Acid* (Baer and Beyer, 1966).

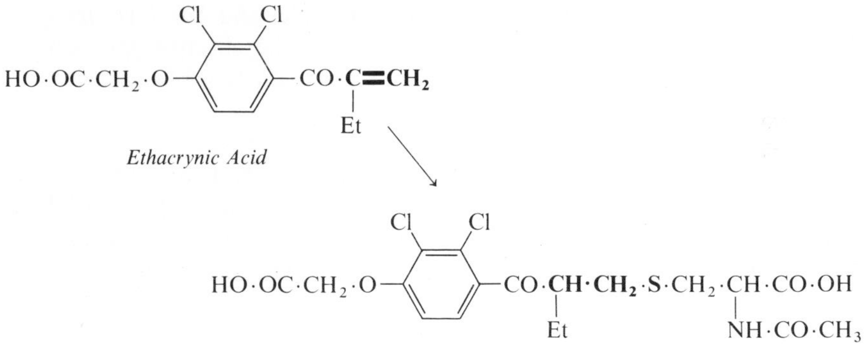

Cl Cl

$HO \cdot OC \cdot CH_2 \cdot O$—[ring]—$CO \cdot C{=}CH_2$ (Et on C)

Ethacrynic Acid

↓

Cl Cl

$HO \cdot OC \cdot CH_2 \cdot O$—[ring]—$CO \cdot CH \cdot CH_2 \cdot S \cdot CH_2 \cdot CH \cdot CO \cdot OH$ (Et on CH; $NH \cdot CO \cdot CH_3$ on CH)

MICROBIAL METABOLISM

The Rôle of Intestinal Micro-flora in Drug Metabolism

Microbial metabolism is important both in relation to the action and design of antimicrobial agents, and also to an understanding of the rôle of gut microflora in the metabolism and action of drugs which reach the intestines whatever the route of administration. The importance of microbial action in drug metabolism can be judged from the fact that more than 60 bacterial species have been isolated from the intestinal tracts of healthy animals, including man. It is not surprising, therefore, that a wide variety of chemical transformations due to the action of gut microflora have been encountered, or that certain inter-species and within-species variations in metabolism are attributable to the gut microflora as, for example, with *Sodium Cyclamate* (Williams, 1971; Drasar, Renwick and Williams, 1971). The extent and importance of gut microbial metabolism has frequently only been revealed in the treatment of intestinal infections with antibiotics, such as *Neomycin* and *Streptomycin*, which are not appreciably absorbed from the gastro-intestinal tract. Such treatment can be disruptive of microbiological metabolism by such normal bowel inhabitants as *E. coli*, which plays a major rôle in enterohepatic circulation (**2**, 4), and other organisms responsible for the synthesis and supply of vitamins such as *Cyanocobalamin*.

The lower bowel always contains large numbers of organisms, such as *E. coli*, which are particularly rich in β-glucuronidase, and similar enzymes capable of hydrolysing glycosides, esters and amides. This facility for the cleavage of glucuronides and other drug conjugates accounts largely for the phenomenon of enterohepatic circulation (**2**, 4), since only non-conjugated drugs, such as *Diazepam* (Marcucci, Mussini, Fanelli and Garattini, 1970), are appreciably absorbed from the intestines. Microbiological glucuronide hydrolysis is also essential for the release of bile acids from their conjugates in the bowel, enabling them to play their essential rôle in fat absorption (**1**, 22). Other aspects of drug metabolism in which the gut microflora has been shown to play a prominent part include the hydrolysis of laxative anthraquinone glycosides, amidic hydrolysis of *Phthalylsulphathiazole* (Phthalylsulfathiazole) and *Succinylsulphathiazole* (Succinylsulfathiazole; **1**, 14), with release of the effective antibacterial, *Sulphathiazole* (Sulfathiazole; Scheline, 1968), and hydrolysis of *Carbenoxolone* (Iveson, Parke and Williams, 1966).

Decarboxylation of *p*-hydroxybenzoic acids (Dacre and Williams, 1962, 1968; Scheline, 1966; Dacre, Scheline and Williams, 1968), nitro-reduction of *Chloramphenicol* (Glazko, Dill and Wolf, 1952), azo reduction of *Prontosil* (**1**, 14; Gingill, Bridges and Williams, 1971) and *Tartrazine* (Ryan, Welling and Roxon, 1969) are all representative of important pathways of gut microbial metabolism.

Drug Metabolism in Transferable Drug Resistance

It is now well established that transferable drug resistance can arise in one of two ways (Shaw, 1971), either through chromosomal gene mutation or through transmission of extra-chromosomal **resistance transfer factor** (RTF) bodies, known as **episomes** or **plasmids**, which contain DNA-transcribed genetic information for resistance transfer. Resistance to individual antibiotics and groups of antibiotics may, therefore, be either chromosomal or extra-chromosomal in origin or both.

Resistance manifests itself in a number of different ways, but mainly either in decreased cell accumulation of the antibiotic or in the inhibition or enhancement of particular cellular enzymes, or some combination of the two which renders the antibiotic ineffective against the micro-organism. Reduction in the ability of the cell to accumulate antibiotics is characteristic of resistance to tetracyclines (**1**, 22), *Chloramphenicol* (**1**, 15), sulphonamides (**1**, 14), and the macrolide antibiotics (**1**, 13). Thus, resistant strains of *Staphylococcus aureus* show something like a one-hundredfold reduction in the uptake of tetracyclines compared with non-resistance strains.

Sulphonamide resistance is not only due to decreased cellular uptake, but also to modification of the dihydropteroic acid synthesis which is blocked by sulphonamides in non-resistant organisms. Similarly, transferable resistance to *Chloramphenicol* in *E. coli* and Salmonella is mediated both by decreased cell accumulation and by development of an inactivating enzyme, chloramphenicol acetyltransferase, which converts the active antibiotic into its antibiotically inactive 1,3-diacetate.

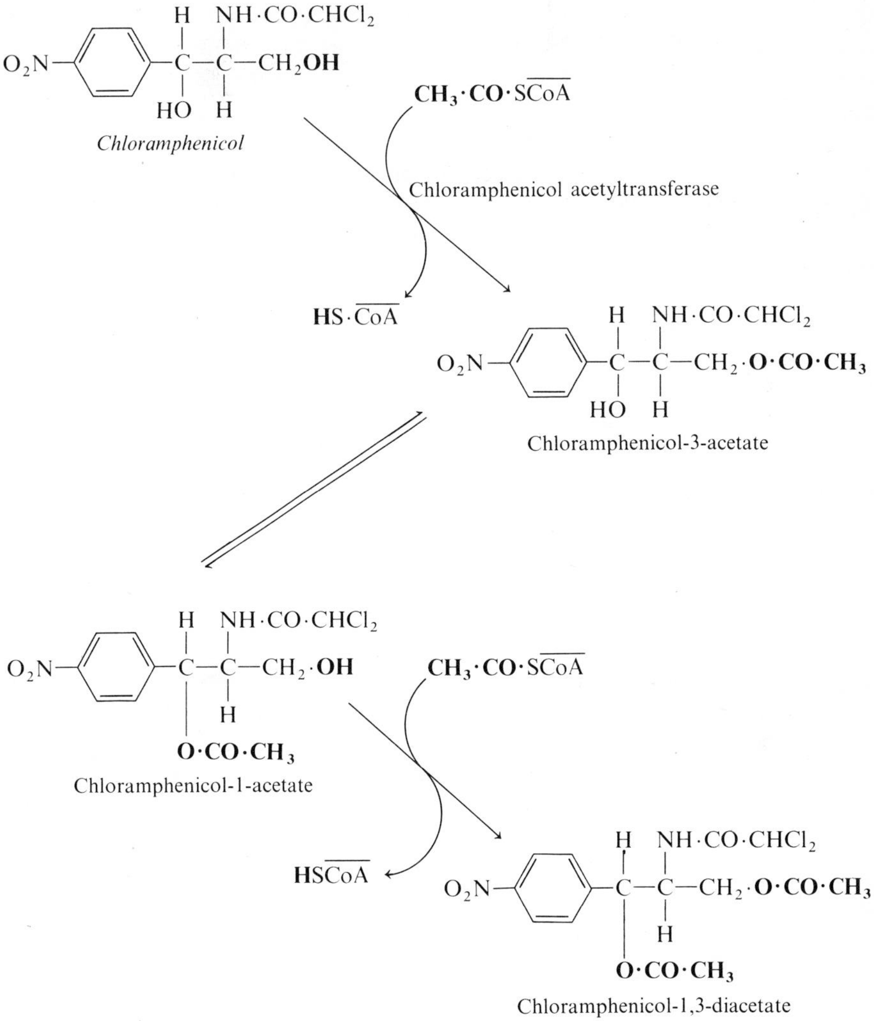

$O_2N-C_6H_4-CH(OH)-CH(NH \cdot CO \cdot CHCl_2)-CH_2\mathbf{OH}$

Chloramphenicol

$\mathbf{CH_3 \cdot CO \cdot}S\overline{CoA}$

Chloramphenicol acetyltransferase

$HS \cdot \overline{CoA}$

$O_2N-C_6H_4-CH(OH)-CH(NH \cdot CO \cdot CHCl_2)-CH_2 \cdot \mathbf{O \cdot CO \cdot CH_3}$

Chloramphenicol-3-acetate

$O_2N-C_6H_4-CH(\mathbf{O \cdot CO \cdot CH_3})-CH(NH \cdot CO \cdot CHCl_2)-CH_2 \cdot \mathbf{OH}$

Chloramphenicol-1-acetate

$\mathbf{CH_3 \cdot CO \cdot}S\overline{CoA}$

$HS\overline{CoA}$

$O_2N-C_6H_4-CH(\mathbf{O \cdot CO \cdot CH_3})-CH(NH \cdot CO \cdot CHCl_2)-CH_2 \cdot \mathbf{O \cdot CO \cdot CH_3}$

Chloramphenicol-1,3-diacetate

3-Deoxychloramphenicol, although ineffective as an antibiotic and not reactive to enzymatic acetylation, is nonetheless an inducer of TRF. 3-Deoxy-1-oxochloramphenicol, however, is an effective antibiotic against resistant *Staph. aureus* (Kono, O'Hara, Honda and Mitsuhashi, 1969). *Staph. aureus* TRF-induced chloramphenicol acetyltransferase, although almost identical to that induced in *E. coli* in catalytic properties, pH optimum and molecular weight (*ca* 800 000), differs in heat stability, substrate specificity and affinity, electrophoretic mobility and immunological reactivity.

Cell accumulation is unaffected in resistance to penicillins and cephalosporins, which is characterised simply by greatly enhanced β-lactamase activity (**1**, 18). Cell accumulation is similarly unaffected in aminoglycoside resistance, though there is a marked decrease in ribosomal activity and enzymic inactivation of the antibiotic. This is achieved by *N*-acetylation of the 6-aminodeoxyglucose of *Kanamycin A* (Umezawa, Okanishi, Utahara, Maeda and Kondo, 1967b), and by C-3 phosphorylation in *Kanamycins A* and *C*, *Neomycin*, and *Paromomycin* (Okanishi, Kondo, Utahara and Umezawa, 1968; Kondo, Okanishi, Utahara, Maeda and Umezawa, 1968). A second phosphorylase (Ozanne, Benveniste, Tipper and Davis, 1969) and an adenylate synthetase (Yamada, Tipper and Davies, 1968) is specific for the C-3 hydroxyl of the *N*-methyl-L-glucosamine fragment of *Streptomycin*.

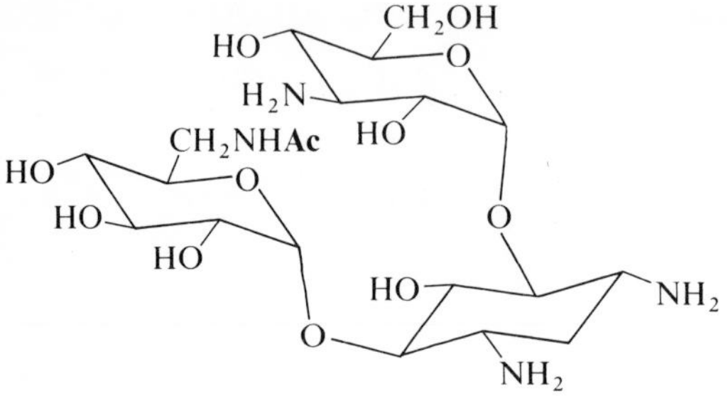

N-Acetylkanamycin A

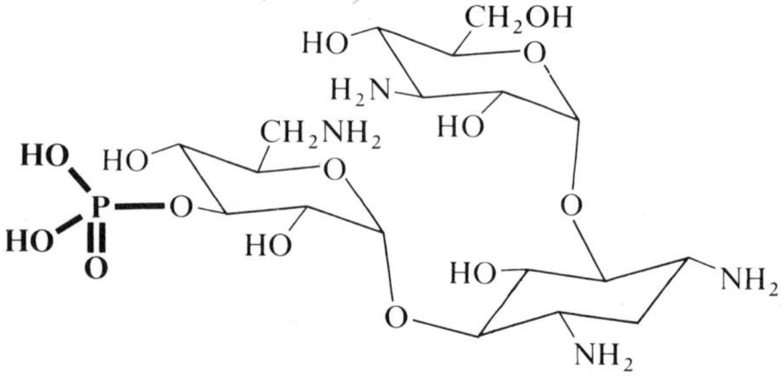

O-Phosphorylkanamycin A

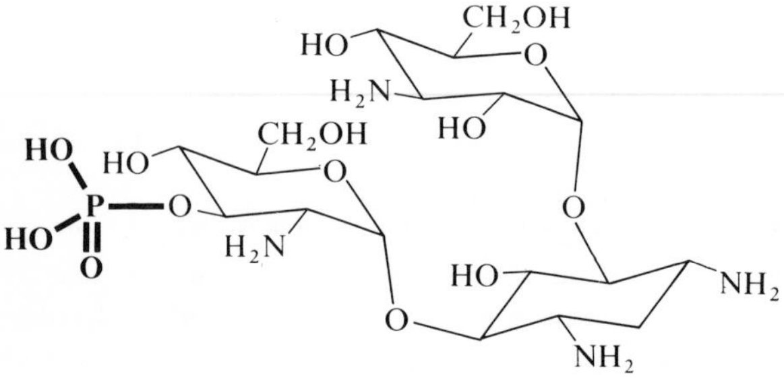

O-Phosphorylkanamycin C

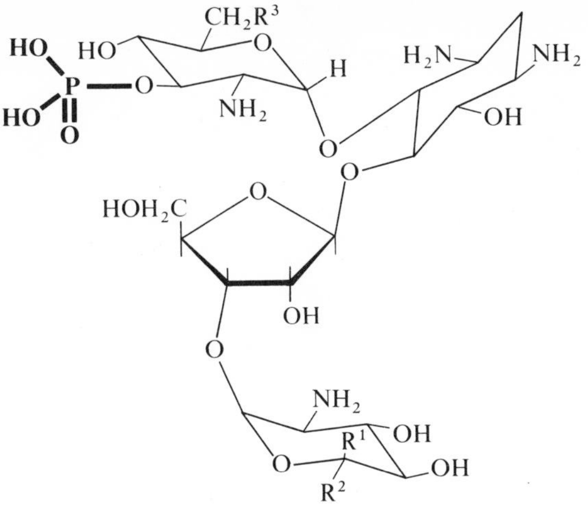

	R^1	R^2	R^3
O-Phosphorylneomycin B	CH_2NH_2	H	NH_2
O-Phosphorylneomycin C	H	CH_2NH_2	NH_2
O-Phosphorylparomomycin	CH_2NH_2	H	OH
O-Phosphorylparomomycin II	H	CH_2NH_2	OH

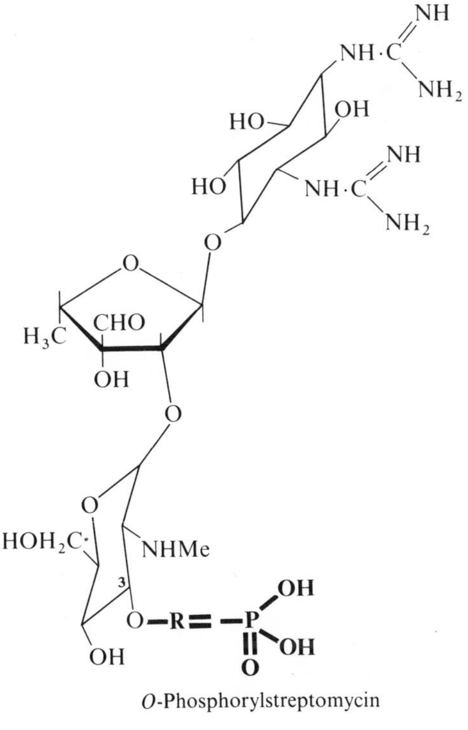

O-Phosphorylstreptomycin

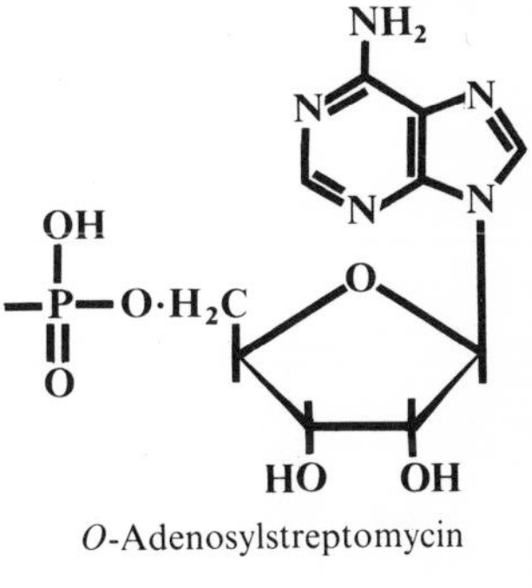

O-Adenosylstreptomycin

Practical Pharmaceutical Chemistry

Third Edition

A. H. Beckett *and* J. B. Stenlake

Part 1 : General Pharmaceutical Chemistry

The third edition of this successful textbook takes full account of the effect on the practice of pharmaceutical analysis and quality control of the passing of t Medicines Act 1968 and of the publication of the European and British Pharmacopoeias. Treatment of general analytical procedures applicable to the control of pharmaceutical dosage forms now includes all the more important products in common use. A broader interpretation of pharmaceutical analysis to include the analysis of drugs and their distribution characteristics in the human body is a further feature.

'. . . the best book on quantitative analysis ever written for students of pharmacy or analytical chemistry . . . no school of pharmacy can afford to omit "Beckett & Stenlake" from its list of compulsory books'.
Pharmaceutical Journal

0 485 11156 x

Part 2 : Physical Methods of Analysis

This standard work provides a broad coverage of physical and instrumental methods used in the analysis of drugs and pharmaceutical products. In this revised edition particle size analysis now includes thermal analysis under the single new heading of solid state analysis. Other new topics include high pressure liquid chromatography, circular dichroism, the use of non-selective electrodes, coulometry, atomic absorption spectroscopy and radio-immunoassay techniques. The extended interpretation of pharmaceutical analysis, begun in Part 1, includes the analysis of drugs and their distribution characteristics in the human body.

'. . . this is a most (probably the most) valuable book for all concerned with analytical aspects of pharmaceutical chemistry, students, teachers, industrial and hospital workers alike.' *Pharmaceutical Journal*

0 485 11159 4

References

Adams, D. H. and Whittaker, V. P. (1950). *Biochim. biophys. Acta*, **4**, 543.
Adamson, R. H., Dixon, R. L., Francis, F. L. and Rall, D. P. (1965). *Proc. natn. Acad. Sci. U.S.A.*, **54**, 1386.
Adler, T. K. (1963). *J. Pharmac. exp. Ther.*, **140**, 155.
Aguiar, A. J., Krc, J., Kinkel, A. W. and Samyn, J. C. (1967). *J. Pharm. Sci.*, **56**, 847.
Aguiar, A. J. and Zelmer, J. E. (1969). Ibid., **58**, 983.
Ahlquist, R. P. (1948). *Am. J. Physiol.*, **153**, 586.
— (1968). *A. Rev. Pharmac.*, **8**, 259.
Ahmed, F. R., Barnes, W. H. and Masironi, L. D. M. (1963). *Acta cryst.*, **16**, 237.
Ahmed, R. (1972). *M.Sc. Thesis*, Strathclyde University.
Akers, M. J., Lach, J. L. and Fischer, L. J. (1973). *J. Pharm. Sci.*, **62**, 391.
Akhtar, M., Munday, K. A., Rahimhula, A. D., Watkinson, I. A. and Wilton, D.C. (1969). *Chem. Commun.*, 1287.
Albert, A. (1965). *Selective Toxicity*, 3rd edn, Methuen, London.
— (1968). Ibid., 4th edn, Methuen, London.
Albert, A. and Goldacre, R. J. (1948). *Nature, Lond.*, **161**, 95.
Albert, A., Rubbo, S. D. and Goldacre, R. J. (1941). Ibid., **147**, 332.
Albert, A., Rubbo, S. D., Goldacre, R. J. and Balfour, B. G. (1947). *Br. J. exp. Path.*, **28**, 69.
Albert, A., Rubbo, S. D., Goldacre, R. J., Davey, M. E. and Stone, J. D. (1945). Ibid., **26**, 160.
Allison, J. L., O'Brien, R. L. and Hahn, F. E. (1965). *Science, N.Y.*, **149**, 1111.
Allsop, I. L., Cole, A. R. H., White, D. E. and Willix, R. L. S. (1956). *J. chem. Soc.*, 4868.
Alpar, O., Deer, J. J., Hersey, J. A. and Shotton, E. (1969). *J. Pharm. Pharmac.*, **21** (*suppl.*), 6S.
Antonucci, R., Bernstein, S., Heller, M., Lenhard, R., Littell, R. and Williams, J. H. (1953). *J. org. Chem.*, **18**, 70.
Ariëns, E. J. (1954). *Archs int. Pharmocodyn. Thér.*, **99**, 32.
Ariëns, E. J., Simonis, A. M. and van Rossum, J. M. (1964). *Molecular Pharmacology*, Vol. 1, ed. Ariens, E. J., Academic Press, New York.
Armett, C. J. and Cooper, J. R. (1965). *Experientia*, **21**, 605.
Armstrong, P. D., Cannon, J. G. and Long, J. P. (1968). *Nature, Lond.*, **220**, 65.
Arrhenius, E. (1970). *Chem.-Biol. Interactions*, **1**, 361.
Arthur, H. R., Cole, A. R. H., Thieberg, K. J. L. and White, D. E. (1956). *Chemy Ind.*, 926.
Atherden, L. M. (1959). *Biochem. J.*, **71**, 411.
Atkinson, R. M., Bedford, C., Child, K. J. and Tomich, E. G. (1962). *Nature, Lond.*, **193**, 588.
Attwood, D., Graham, A. B. and Wood, G. C. (1971). *Biochem. J.*, **123**, 875.
Axelrod, J. (1956). Ibid., **63**, 634.
— (1962). *J. Pharmac. exp. Ther.*, **138**, 28.
Axelrod, J., Senoh, S. and Witkop, B. (1958). *J. biol. Chem.*, **233**, 697.
Babior, B. M. and Bloch, K. (1966). Ibid., **241**, 3643.
Baer, J. E. and Beyer, K. H. (1966). *A. Rev. Pharmac.*, **6**, 261.
Baeyer, von, A. (1885). *Ber. dt. chem. Ges.*, **18**, 2269.
Baker, B. R. (1967). *Design of Active-Site-Directed Irreversible Enzyme Inhibitors*, John Wiley & Sons Inc., New York.
Ballantyne, F. C., Fleck, A. and Carson Dick, W. (1971). *Ann. rheum. Dis.*, **30**, 265.
Barbour, A. K. (1969). *Chemy. Brit.*, **5**, 250.
Barker, H. A. (1967). *Biochem. J.*, **105**, 1.
Barnes, M. M., James, S. P. and Wood, P. B. (1959). Ibid., **71**, 680.
Barnsley, E. A. (1964). Ibid., **90**, 9P.
— (1964). Ibid., **93**, 15P.
Bayne, W. F., Rodgers, G. and Crisologo, N. (1975). *J. Pharm. Sci.*, **64**, 402.

Beck, K. and Richter, E. (1962). *Klin. Wschr.*, **40**, 75.
Beckett, A. H. (1959). *Progress in Drug Research*, Vol. 1, p. 455, ed. Jucker, E., Birkhäuser, Basel.
— (1962). *Enzymes and Drug Action*, pp. 15 and 238, ed. Mongar, J. L., Churchill, London.
— (1971). *The Biological Oxidation of Nitrogen in Organic Molecules, Xenobiotica*, p. 53, eds. Bridges, J. W., Gorrod, J. W. and Parke, O. V., Taylor and Francis, London.
Beckett, A. H., Boyes, R. N. and Appleton, P. J. (1966). *J. Pharm. Pharmac.*, **18**, 76S.
Beckett, A. H., Boyes, R. N. and Triggs, E. J. (1968a). Ibid., **20**, 92.
Beckett, A. H. and Hewick, D. S. (1967). Ibid., **19**, 134.
Beckett, A. H. and Moffat, A. C. (1968b). Ibid., **20** (*suppl.*), 239S.
— (1970). Ibid., **22**, 15.
Beckett, A. H. and Rowland, M. (1964). *Nature, Lond.*, **204**, 1203.
— (1965). *J. Pharm. Pharmac.*, **17**, 628; *suppl.*, 109S.
Beckett, A. H. and Triggs, E. J. (1967). Ibid., **19** (*suppl.*), 31S.
Beckett, A. H. and Walker, J. (1955). Ibid., **7**, 1039.
Behnke, A. R. and Yarbrough, O. D. (1938). *U.S. nav. med. Bull.*, **36**, 542.
— (1939). *Am. J. Physiol.*, **126**, 409.
Belleau, B. (1958). *Can. J. Biochem. Physiol.*, **36**, 731.
— (1960). *Adrenergic Mechanisms*, ed. Wolstenholme, G. E. W., Vane, J. R. and O'Connor, M., Churchill, London.
— (1963). *Proceedings of the First International Pharmacological Meeting*, Vol. 7, ed. Uvnäs, B., Pergamon, New York.
— (1966). *Pharmac. Rev.*, **18**, 131.
— (1967). *Ann. N.Y. Acad. Sci.*, **139**, 580.
Belleau, B. and Pauling, P. (1970). *J. med. Chem.*, **13**, 737.
Belleau, B. and Puranen, J. (1963). Ibid., **6**, 325.
Bender, P., Flowers, D. L. and Goering, H. L. (1955). *J. Am. chem. Soc.*, **77**, 3463.
Benoiton, L. and Rydon, H. N. (1960). *J. chem. Soc.*, 3328.
Bente, D., Hippius, H., Poeldinger, W. and Stach, K. (1964). *Arzneimittel-Forsch.*, **14**, 486.
Bergel, F. (1958). *Ann. N.Y. Acad. Sci.*, **68**, 1238.
Bergström, B., Dahlqvist, A., Lundh, G. and Sjövall, V. (1957). *J. clin. Invest.*, **36**, 1521.
Bergström, S., Lindstedt, S. and Samuelsson, B. (1958). *J. Am. chem. Soc.*, **80**, 2338.
Bessman, S. P., Rubin, M. and Leikin, S. (1954). *Pediatrics, Springfield*, **14**, 201.
Bickel, M. H. (1971). *The Biological Oxidation of Nitrogen in Organic Molecules, Xenobiotica*, p. 1, eds. Bridges, J. W., Gorrod, J. W. and Parke, D. V., Taylor and Francis, London.
Bickerton, D., Coutts, R. T. and Johnson, W. J. (1965). *Proc. Can. Fed. biol. Soc.*, **8**, 44.
Binkley, F. (1961). *J. biol. Chem.*, **236**, 1075.
Biot, J. B. (1815). *Bull. soc. philomath., Paris*, 190.
Bird, A. E. and Marshall, A. C. (1967). *Biochem. Pharmac.*, **16**, 2275.
Birtley, R. D. N., Burton, J. S., Kellett, D. N., Oswald, B. J. and Pennington, J. C. (1973). *J. Pharm. Pharmac.*, **25**, 859.
Blackwood, J. E., Gladys, C. L., Loening, K. L., Petrarca. A. E. and Rush, J. E. (1968). *J. Am. chem. Soc.*, **90**, 509.
Blasberg, R. (1968). *Progress in Brain Research*, **29**, 245.
Blaschko, H. (1950). *Proc. R. Soc.*, **B**, **137**, 307.
— (1952). *Pharmac. Rev.*, **4**, 415.
Blaschko, H., Himms, J. M. and Strömblad, B. C. R. (1955). *Br. J. Pharmac. Chemother.*, **10**, 442.
Bledsoe, T., Island, D. P., Ney, R. L. and Liddle, G. W. (1964). *J. clin. Endocr. Metab.*, **24**, 1303.
Bliss, A. F. (1951). *Archs Biochem. Biophys.*, **31**, 197.
Bloom, B. and Topper, Y. J. (1958). *Nature, Lond.*, **181**, 1128.
Bloom, B. M., Goldman, I. M. and Belleau, B. (1969). *Pharmac. Rev.*, **21**, 131.
Blout, E. R., de Lozé, C. and Asadourian, A. (1961). *J. Am. chem. Soc.*, **83**, 1895.
Bock, H.-G. and Fleischer, S. (1975). *J. biol. Chem.*, **250**, 5774.
Bogaert, M. G. and Rosseel, M. T. (1972). *J. Pharm. Pharmac.*, **24**, 737.
Bonting, S. L. (1969). *Curr. Topics Bioenerg.*, **3**, 351.
Bonting, S. L. and Bangham, A. D. (1967). *Expl. Eye Res.*, **6**, 400.
Booth, J. (1966). *Biochem. J.*, **100**, 745.
Booth, J., Boyland, E. and Sims, P. (1960). Ibid., **74**, 117.
— (1961). Ibid., **79**, 516.
Borzelleca, J. F. and Cherrick, H. M. (1965). *J. oral Therap. Pharmac.*, **2**, 180.

Boström, H. and Vestermark, A. (1960). *Acta physiol. scand.*, **48**, 88.
Boyland, E. and Chasseaud, L. F. (1967). *Biochem. J.*, **104**, 95.
— (1968). Ibid., **109**, 651.
— (1969). *Adv. Enzymol.*, **32**, 173.
Boyland, E., Manson, D. and Orr, S. F. D. (1957). *Biochem. J.*, **65**, 417.
Boyland, E. and Nery, R. (1965). Ibid., **94**, 198.
Boyland, E. and Williams, K. (1965). Ibid., **94**, 190.
Bradbury, A. F., Smyth, D. G. and Snell, C. R. (1976). *Nature, Lond.*, **260**, 165.
Bradley, P. B., Briggs, I., Gayton, R. J. and Lambert, L. A. (1976). Ibid., **261**, 425.
Bray, H. G., Franklin, T. J. and James, S. P. (1959). *Biochem. J.*, **73**, 465.
Bray, H. G., Hybs, Z., Clowes, R. C., Lake, H. J. and Thorpe, W. V. (1951). Ibid., **49**, LXV.
Bray, H. G., James, S. P. and Thorpe, W. V. (1957). Ibid., **65**, 483.
Breimer, D. D. and van Rossum, J. M. (1973). *J. Pharm. Pharmac.*, **25**, 762.
Brennan, T. F., Ross, F. K., Hamilton, W. C. and Shefter, E. (1970). Ibid., **22**, 724.
Bridges, J. W., Kibby, M. R. and Williams, R. T. (1964). *Biochem. J.*, **91**, 12P.
Brink, C., Hodgkin, D. C., Lindsey, J., Pickworth, J., Robertson, J. H. and White, J. G. (1954). *Nature, Lond.*, **174**, 1169.
Brockman, R. W. (1961). *Clin. Pharmac. Therap.* **2**, 237.
Brodersen, R. and Hermann, L. S. (1963). *Lancet*, **i**, 1242.
Brodie, B. B. (1952). *Fedn Proc. Fedn Am. Socs exp. Biol.*, **11**, 632.
— (1956). *J. Pharm. Pharmac.*, **8**, 1.
Brodie, B. B. and Axelrod, J. (1949). *J. Pharmac. exp. Ther.*, **97**, 58.
Brodie, B. B., Gillette, J. R. and La Du, B. N. (1958). *A. Rev. Biochem.*, **27**, 427.
Brodie, B. B. and Hogben, C. A. M. (1957). *J. Pharm. Pharmac.*, **9**, 345.
Brodie, B. B., Maichel, R. P. and Jondorf, W. R. (1958). *Fedn Proc. Fedn Am. Socs exp. Biol.*, **17**, 1163.
Brodie, B. B., Mark, L. C., Papper, E. M., Lief, P. A., Bernstein, E. and Rovenstine, E. A. (1950). *J. Pharmac. exp. Ther.*, **98**, 85.
Broome, J. D. (1961). *Nature, Lond.*, **191**, 1114.
Brown, D. D., Tomchick, R. and Axelrod, J. (1959). *J. biol. Chem.*, **234**, 2948.
Brown, P. K. and Wald, G. (1956). Ibid., **222**, 865.
Brown, R. R., Miller, J. A. and Miller, E. C. (1954). Ibid., **209**, 211.
Bruce, I. C. L., Giles, C. H. and Jain, S. K. (1958). *J. chem. Soc.*, 1610.
Bucovaz, E. T. and Wood, J. L. (1964). *J. biol. Chem.*, **239**, 1151.
Büechi, J. and Perlia, X. (1971). *International Encyclopoedia of Pharmacology and Therapeutics*, Section 8, Vol. 1. *Local Anaesthetics*, Pergamon, Oxford, 39.
Bueding, E. (1962). *Fedn Proc. Fedn Am. Socs exp. Biol.*, **21**, 1039.
Bueding, E. and Mansour, J. M. (1957). *Br. J. Pharmac. Chemother.*, **12**, 159.
Buffoni, F., Della Corte, L. and Knowles, P. F. (1968). *Biochem. J.*, **106**, 575.
Buist, G. J., Burton, J. S. and Elvidge, J. A. (1973). *J. Pharm. Pharmac.*, **25**, 854.
Burdick, K. H. (1972). *Acta Derm.-Vener* (*suppl.*), **52**, 19; *Chem-Abstr.*, 1972, **77**, 66165d.
Büscher, H. H., Hill, R. C., Römer, D., Cardinaux, F., Closse, A., Hauser, D. and Pless, J. (1976). *Nature, Lond.*, **261**, 423.
Bushby, S. R. M. and Woiwod, A. J. (1956). *Biochem. J.*, **63**, 406.
Bustard, T. M. and Egan, R. S. (1971). *Tetrahedron*, **27**, 4457.
Cahn, R. S. (1964). *J. chem. Educ.*, **41**, 116.
Cahn, R. S., Ingold, C. K. and Prelog, V. (1956). *Experientia*, **12**, 81.
Caldwell, J. H., Martin, J. F., Dutta, S. and Greenberger, N. J. (1969). *Am. J. Physiol.*, **217**, 1747.
Canepa, F. G., Pauling, P. and Sörum, H. (1966). *Nature, Lond.*, **210**, 907.
Cannon, J. G., Smith, R. V., Modiri, A., Sood, S. P., Borgman, R. J., Aleem, M. A. and Long, J. P. (1972). *J. med. Chem.*, **15**, 273.
Cano, J. M. and Ranédo, J. (1920). *An. Soc. Esp. Fis. Quim.*, **18**, 184 (*Chem. Abs.*, 1921, **15**, 2672).
Carless, J. E., Moustafa, M. A. and Rapson, H. D. C. (1968). *J. Pharm. Pharmac.*, **20**, 630, 639.
Carrigan, P. J. and Bates, T. R. (1973). *J. Pharm. Sci.*, **62**, 1476.
Casy, A. F. (1970). *Progress in Medicinal Chemistry*, Vol. 7, p. 229, eds. Ellis, G. P. and West, G. B., Butterworths, London.
— (1975). Ibid., **11**, 1.
Casy, A. F. and Myers, J. L. (1964). *J. Pharm. Pharmac.*, **16**, 455.
Casy, A. F., Simmonds, A. B. and Staniforth, D. (1968). Ibid., **20**, 768.

Cavalieri, L. F., Angelos, A. and Balis, M. E. (1951). *J. Am. chem. Soc.*, **73**, 4902.

Cella, J. A., Eggenberger, D. N., Noel, D. R., Harriman, L. A. and Harwood, H. J. (1952). Ibid., **74**, 2061.

Cerami, A., Reich, E., Ward, D. C. and Goldberg, I. H. (1967). *Proc. natn. Acad. Sci. U.S.A.*, **57**, 1036.

Chader, G. J. and Westphal, U. (1968). *J. biol. Chem.*, **243**, 928.

Champion, R. H. and Goldin, D. (1975). *Pharm. J.*, **215**, 328.

Chapman, J. H., Page, J. E., Parker, A. C., Rogers, D., Sharp, C. J. and Staniforth, S. E. (1968). *J. Pharm. Pharmac.*, **20**, 418.

Chen, C. and Lin, C. C. (1968). *Biochim. biophys. Acta*, **170**, 366.

— (1969). Ibid., **184**, 634.

Chen, W., Vrindten, P. A., Dayton, P. G. and Burns, J. J. (1962). *Life Sci.*, **1**, 35.

Cheo, K. L., Elliott, T. H. and Tao, R. C. C. (1967). *Biochem. J.*, **104**, 198.

Childs, A. F., Davies, D. R., Green, A. L. and Rutland, J. P. (1955). *Br. J. Pharmac. Chemother.*, **10**, 462.

Chiou, C. Y., Long, J. P., Cannon, J. G. and Armstrong, P. D. (1969). *J. Pharmac. exp. Ther.*, **166**, 243.

Chothia, C. and Pauling, P. (1969a). *Chem. Commun.*, 626.

— (1969b). *Nature, Lond.*, **223**, 919.

Clark, A. G., Hitchcock, M. and Smith, J. N. (1966). Ibid., **209**, 103.

Clauder, O., Radics, L., Szabo, L. and Varga, J. (1968). *Acta pharm. hung.*, **38**, 260; *Chem. Abstr.*, **69**, 57236K.

Cohen, E. N., Brewer, H. W. and Smith, D. (1967). *Anaesthesiology*, **28**, 309.

Cohen, J. A., Oosterbaan, R. A., Jansz, H. S. and Berends, F. (1959). *J. cell. comp. Physiol.*, **54**, 231.

Cohen, L. A. (1968). *Biochem. Rev.*, **37**, 695.

Cohen, S. N. and Yielding, K. L. (1965). *J. biol. Chem.*, **240**, 3123.

Coleman, J. (1967). Ibid., **242**, 5212.

Collander, R. (1954). *Plant Physiol.*, **7**, 420.

Coltart, J., Howard, M. and Chamberlain, D. (1972). *Br. med. J.*, **2**, 318.

Commission on Enzymes of the International Union of Biochemistry Report (1961), Pergamon Press, Oxford.

Conney, A. H. (1969). *29ᵉ Congrès International Dès Sciences Pharmaceutiques*, London, 1969.

Conney, A. H. and Gilman, A. G. (1963). *J. biol. Chem.*, **238**, 3682.

Connolly, C. K. (1970). *Prescribers Journal*, **10**, 87.

Cooper, J. R. and Brodie, B. B. (1955a). *J. Pharmac. exp. Ther.*, **120**, 75.

— (1955b). Ibid., **114**, 409.

Cooper, J. R., Roth, R. H. and Kini, M. M. (1963), *Nature, Lond.*, **199**, 609.

Cornforth, J. W., Ryback, G. J., Popják, G., Donninger, C. and Schroepfer, G. J. (1962). *Biochem. biophys. Res. Commun.*, **9**, 371.

Cox, B. M. and Weinstock, M. (1964). *Br. J. Pharmac. Chemother.*, **22**, 289.

Cox, J. S. G., Beach, J. E., Blair, A. M. J. N., Clarke, A. J., King, J., Lee, T. B., Loveday, D. E. E., Moss, G. F., Orr, T. S. C., Ritchie, J. T. and Sheard, P. (1970). *Advances in Drug Research*, Vol. 5, p. 115, eds. Harper, N. J. and Simmonds, A. B., Butterworths, London.

Cramer, J. W., Miller, J. A. and Miller, E. C. (1960). *J. biol. Chem.*, **235**, 885.

Crawford, J. S. and Gardiner, J. E. (1956). *Br. J. Anaesth.*, **28**, 154.

Crépy, O., Jayle, M. F. and Rulleau-Meslin, F. (1957). *Acta endocr. Copnh.*, **24**, 233.

Crépy, O., Judas, O., Rulleau-Meslin, F. and Jayle, M. F. (1962). *Bull. Soc. Chim. biol.*, **44**, 327.

Cristol, S. J., Hause, N. L. and Meek, J. S. (1951). *J. Am. chem. Soc.*, **73**, 674.

Cucinell, S. A., Conney, A. H., Sansur, M. and Burns, J. J. (1965). *Clin. Pharmac. Ther.*, **6**, 420.

Cucinell, S. A., O'Dessky, L., Weiss, M. and Dayton, P. G. (1966). *J. Am. med. Ass.*, **197**, 366.

Culp, H. W. and McMahon, R. E. (1968). *J. biol. Chem.*, **243**, 848.

Culvenor, C. C. J. and Ham, N. S. (1966). *Chem. Commun.*, **15**, 537.

Curry, S. H. (1970). *J. Pharm. Pharmac.*, **22**, 193.

Cushny, A. R. (1926). *Biological Relations of Optically Isomeric Substances*, Ballière, Tindall and Cox, London.

Dacre, J. C., Scheline, R. R. and Williams, R. T. (1968). *J. Pharm. Pharmac.*, **20**, 619.

Dacre, J. C. and Williams, R. T. (1962). *Biochem. J.*, **84**, 81P.

— (1968). *J. Pharm. Pharmac.*, **20**, 610.

Daemen, F. J. M. and Bonting, S. L. (1969). *Nature, Lond.*, **222**, 879.

Daly, J. W., Jerina, D. M. and Witkop, B. (1968). *Archs Biochem. Biophys.*, **128**, 517.

Dalziel, K. and Dickinson, F. M. (1966). *Biochem. J.*, **100**, 34.
— (1967). Ibid., **104**, 165.
Danielli, J. F. (1954). *Symp. Soc. exp. Biol.*, **8**, 502.
Danielli, J. F. and Davson, H. (1935). *J. cell. comp. Physiol.*, **5**, 495.
Das, M. L., Orrenius, S. and Ernster, L. (1968). *Eur. J. Biochem.*, **4**, 519.
Datta, P. R., Laug, E. P. and Klein, A. K. (1964). *Science, N.Y.*, **145**, 1052.
Daughaday, W. H. (1956). *J. Lab. clin. Med.*, **48**, 799.
— (1956). *J. clin. Invest.*, **35**, 1434.
De Moor, P., Heirwegh, K., Heremans, J. F. and Declerck-Raskin, M. (1962). Ibid., **41**, 816.
Dent, C. E., Richens, A., Rowe, D. J. F. and Stamp, T. C. B. (1970). *Br. med. J.*, **4**, 69.
Dingell, J. V., Sulser, F. and Gillette, J. R. (1964). *J. Pharmac. exp. Ther.*, **143**, 14.
Dixon, R. L., Shultice, R. W. and Fouts, J. R. (1960). *Proc. Soc. exp. Biol. Med.*, **103**, 333.
Dodds, E. C., Goldberg, L., Grünfeld, E. I., Lawson, W. and Robinson, R. (1944). *Proc. R. Soc.*, **B**, **132**, 83.
Dodds, E. C., Goldberg, L., Lawson, W. and Robinson, R. (1939). Ibid., **127**, 140.
Doherty, J. E. and Perkins, W. H. (1962). *Am. Heart J.*, **63**, 528.
Doherty, J. E., Perkins, W. H. and Flanigan, W. J. (1967). *Ann. intern. Med.*, **66**, 116.
Dolphin, D. H., Johnson, A. W. and Rodrigo, R. (1964). *J. chem. Soc.*, 3186.
Donninger, C. and Ryback, G. (1964). *Biochem. J.*, **91**, 11P.
Douglas, J. F., Ludwig, B. J. and Smith, N. (1963). *Proc. Soc. exp. Biol. Med.*, **112**, 436.
Doyle, T. D., Stewart, J. M., Filipescu, N. and Benson, W. R. (1975). *J. Pharm. Sci.*, **64**, 1525.
Drasar, B. S., Renwick, A. G. and Williams, R. T. (1971). *Biochem. J.*, **123**, 26P.
Duncan, G. W., Lyster, S. C., Clark, J. J. and Lednicer, D. (1963). *Proc. Soc. exp. Biol. Med.*, **112**, 439.
Duncan, W. A. M., MacDonald, G. and Thornton, M. J. (1962). *J. Pharm. Pharmac.*, **14**, 217.
Dunitz, J. D., Eser, H. and Strickler, P. (1964). *Helv. chim. Acta*, **47**, 1897.
Dutton, G. J. (1959). *Biochem. J.*, **71**, 141.
— (1966). *Glucuronic Acid: Free and Combined*, ed. Dutton, G. J., Academic Press, London.
Dutton, G. J. and Greig, C. G. (1957). *Biochem. J.*, **66**, 52P.
Dutton, G. J. and Storey, I. D. E. (1954). Ibid., **57**, 275.
Easson, L. H. and Stedman, E. (1933). Ibid., **27**, 1257.
Ecanow, B. and Siegel, F. P. (1963). *J. Pharm. Sci.*, **52**, 812.
Eckert, T. (1962). *Naturwissenschaften*, **49**, 18.
Eddy, N. B. (1959). *Chemy Ind.*, 1462.
Edgren, R. A. and Johns, W. F. (1960). *Proc. Soc. exp. Biol. Med.*, **105**, 286.
Edwards, D., Stenlake, J. B., Carey, F. M. and Lewis, J. J. (1959). *J. Pharm. Pharmac.*, **21**, 70T.
Edwards, S. W. and Knox, W. E. (1956). *J. biol. Chem.*, **220**, 79.
Eger, E. I., Brandstater, B., Saidman, L. J., Regan, M. J., Severinghaus, J. W. and Munson, E. S. (1965). *Anaesthesiology*, **26**, 771.
Eger, E. I., Lundgren, C., Miller, S. L. and Stevens, W. C. (1969). Ibid., **30**, 129.
Eger, E. I., Saidman, L. J. and Brandstater, B. (1965). Ibid., **26**, 764.
Eger, E. I. and Shargel, R. O. (1969). Ibid., **30**, 136.
Ehrlich, P. (1909). *Ber. dt. chem. Ges.*, **42**, 17.
Eik-nes, K., Schellman, J. A., Lumry, R. and Samuels, L. T. (1954). *J. biol. Chem.*, **206**, 411.
Eliel, E. L. (1962). *Stereochemistry of Carbon Compounds*, McGraw-Hill, New York.
Ellard, G. A., Garrod, J. M. B., Scales, B. and Snow, G. A. (1965). *Biochem. Pharmac.*, **14**, 129.
Elliott, J. S., Sharp, R. F. and Lewis, L. (1959). *J. Urol.*, **81**, 339.
Elliott, T. H. and Hanam, J. (1968). *Biochem, J.*, **108**, 551.
Elliott, T. H., Hanam, J., Parke, D. V. and Williams, R. T. (1964). *Ibid.*, **92**, 52P.
Elliott, T. H., Jacob, E. and Tao, R. C. C. (1969). *J. Pharm. Pharmac.*, **21**, 561.
Elliott, T. H., Robertson, J. S. and Williams, R. T. (1966). *Biochem, J.*, **100**, 403.
Elliott, T. H., Tao, R. C. C. and Williams, R. T. (1965). *Ibid.*, **95**, 70.
Elworthy, P. H. (1963a). *J. chem. Soc.*, 388.
— (1963b). *J. Pharm. Pharmac.*, **15**, 137T.
Ernster, L. and Orrenius, S. (1965). *Fedn Proc. Fedn Am. Socs exp. Biol.*, **24**, 1190
Estabrook, R. W., Cooper, D. Y. and Rosenthal, O. (1963). *Biochem. Z.*, **338**, 741.
Evans, J. G. and Jarvis, E. H. (1972). *Br. med. J.*, **4**, 487.
Fabro, S., Smith, R. L. and Williams, R. T. (1967). *Biochem J.*, **104**, 565, 570.
Felger, C. B. and Katzman, P. A. (1961). *Fedn Proc. Fedn Am. Socs exp. Biol.*, **20**, 199.
Fenner, H. (1970). *Pharmakopsychiatrie-Neuro-Psychopharmakologie*, **3**, 332.

Ferguson, J. (1939). *Proc. R. Soc.*, **B**, **127**, 387.
Fernley, H. N. (1962). *Biochem. J.*, **82**, 500.
Ferone, R., Burchall, J. J. and Hitchings, G. H. (1969). *Molec. Pharmac.*, **5**, 49.
Ferone, R. and Hitchings, G. H. (1966). *J. Protozool.*, **13**, 504.
Fiese, G. and Perrin, J. H. (1968). *J. Pharm. Pharmac.*, **20**, 98.
Fieser, L. (1951). *Ann. intern. Med.*, **15**, 648.
Fildes, P. (1940). *Lancet*, **i**, 955.
Fischer, E. (1894). *Ber. dt. chem. Ges.*, **27**, 2985.
Florence, A. T., Salole, E. G. and Stenlake, J. B. (1974). *J. Pharm. Pharmac.*, **26**, 479.
Foster, R. A. C. (1948). *J. Bact.*, **56**, 795.
Fouts, J. R. and Adamson, R. H. (1959). *Science, N. Y.*, **129**, 897.
Fouts, J. R. and Brodie, B. B. (1957). *J. Pharmac. exp. Ther.*, **119**, 197.
Fouts, J. R., Kamm, J. J. and Brodie, B. B. (1957). Ibid., **120**, 291.
Foxwell, C. J. and Young, L. (1964). *Biochem, J.*, **92**, 50P.
Franks, F. (1969). *Society for Drug Research Medicinal Chemistry Symposium*, London.
Fraser, I. M., Peters, M. A. and Hardinge, M. G. (1967). *Molec. Pharmac.*, **3**, 233.
Free, S. M. and Wilson, J. M. (1964). *J. med. Chem.* **7**, 395.
French, T. C., Dawid, I. B. and Buchanan, J. M. (1963). *J. biol. Chem.*, **238**, 2178, 2186.
French, T. C., Dawid, I. B., Day, R. A. and Buchanan, J. M. (1963). Ibid., **238**, 2171.
Friedman, P. J. and Cooper, J. R. (1960). *J. Pharmac. exp. Ther.*, **129**, 373.
Furesz, S. and Scotti, R. (1961). *Farmaco, Ed. sci.*, **16**, 262.
Gage, J. C. (1953). *Biochem. J.*, **54**, 426.
Galinsky, A. M., Gearien, J. E., Perkins, A. J. and Susina, S. V. (1963). *J. med. Chem.*, **6**, 320.
Gantt, C. L., Gochman, N. and Dyniewicz, J. M. (1961). *Lancet*, **i**, 486.
Garrod, A. E. (1909). *Inborn Errors of Metabolism*, Oxford University Press, London.
Gazzotti, P., Bock, H-G. and Fleischer, S. (1975). *J. biol. Chem.*, **250**, 5782.
Gent, J. P. and Wolstencroft, J. H. (1976). *Nature, Lond.*, **261**, 426.
Gibaldi, M. and Kanig, J. L. (1965). *J. oral. Therap. Pharmac.*, **1**, 440.
Gibaldi, M., Levy, G. and Hayton, W. (1972). *Anaesthesiology*, **36**, 213.
Gibaldi, M. and Perrier, D. (1972). *J. clin. Pharmac.*, **12**, 201; *J. Pharm. Sci.*, **61**, 952.
Gibaldi, M. and Schwartz, M. A. (1968). *Clin. Pharmac. Ther.*, **9**, 345.
Gigon, P. L. and Bickel, M. H. (1971). *Biochem, Pharmac.*, **20**, 1921.
Giles, C. H. and McKay, R. B. (1962). *J. biol. Chem.*, **237**, 3388.
Gill, E. W. (1959). *Proc. R. Soc.*, **B150**, 381.
— (1965). *Progress in Medicinal Chemistry*, Vol. 4, p. 39, eds. Ellis, G. P. and West, G. B., Butterworths, London.
Gillette, J. (1969). *Biochemical Aspects of Antimetabolites and Drug Hydroxylation; Fed. European Biochem. Soc.*, Ed. Shuger, D., **16**, 109, Academic Press, London.
Gillette, J. R. (1959). *J. biol. Chem.*, **234**, 139.
Gillette, J. R., Kamm, J. J. and Sasame, H. A. (1968). *Molec. Pharmac.*, **4**, 541.
Gillham, B. and Young, L. (1968). *Biochem. J.*, **109**, 143.
Gilman, A. and Philips, F. S. (1946). *Science, N. Y.*, **103**, 409.
Gingill, R., Bridges, J. W. and Williams, R. T. (1971). *Xenobiotica*, **1**, 143.
Giri, S. N. and Combs, A. B. (1972). *Chem.-Biol. Interactions*, **5**, 97.
Glazko, A. J. (1966). *Antimicrobial Agents and Chemotherapy*, p. 655, E. & S. Livingstone, Edinburgh.
Glazko, A. J., Dill, W. A. and Wolf, L. M. (1952). *J. Pharmac. exp. Therap.*, **104**, 452.
Godeaux, J. and Tonnesen, M. (1949). *Acta pharmac. tox.*, **5**, 95.
Goldbaum, L. R. and Smith, P. K. (1954). *J. Pharmac. exp. Ther.*, **111**, 197.
Goldberg, L. (1966). *Proc. Eur. Soc. Study Drug Toxicity*, **7**, 171.
Goldstein, A. (1949). *Pharmac, Rev.*, **1**, 102.
Goodman, D. S. (1958). *J. Am. chem. Soc.*, **80**, 3892.
Gorrod, J. W. (1971). *The Biological Oxidation of Nitrogen in Organic Molecules, Xenobiotica*, p. 37, eds. Bridges, J. W., Gorrod, J. W. and Parke, D. V., Taylor and Francis, London.
Gorrod, J. W., Jenner, P., Keysell, G. and Beckett, A. H. (1971). *Chem.-Biol. Interactions*, **3**, 269.
Gourevitch, A., Hunt, G. A. and Lein, J. (1960). *Antibiot. Chemother.*, **10**, 121.
Govier, W. C. (1965). *J. Pharmac. exp. Ther.*, **150**, 305.
Graf, L., Szekely, J. I., Ronai, A. Z., Dunai-Kovaks, Z. and Bajusz, S. (1976). *Nature, Lond.*, **261**, 240.

Graham, A. B. and Wood, G. C. (1969). *Biochem. biophys. Res. Commun.*, **37**, 567.
Graham, A. B., Wood, G. C. and Woodcock, B. G. (1972). *Biochem. J.*, **129**, 22P.
Granchelli, F. E., Neumeyer, J. L., Fuxe, K., Ungerstedt, U. and Corrodi, H. (1971). *Pharmacologist*, **13**, 252.
Gregory, J. D. and Lipmann, F. (1957). *J. biol. Chem.*, **229**, 1081.
Grover, P. L. and Sims, P. (1965). *Biochem. J.*, **96**, 521.
Guroff, G., Daly, J. W., Jerina, D, M., Renson, J., Witkop, B. and Udenfried, S. (1967). *Science, N. Y.*, **157**, 1524.
Guroff, G. and Udenfried, S. (1962). *J. biol. Chem.*, **237**, 803.
Gutman, A. B. and Yü, T. F. (1963). *Am. J. Med.*, **35**, 820.
Gutman, A. B., Yü, T. F. and Sirota, J. H. (1955). *J. clin. Invest.*, **34**, 711.
Haddow, A. and Sexton, W. A. (1946). *Nature, Lond.*, **157**, 500.
Haddow, A. and Timmis, G. M. (1951). *Acta Un. int. Cancr.*, **7**, 469.
Hadzija, B. W. (1969). *J. Pharm. Pharmac.*, **21**, 196.
Hakala, M. T. and Suolinna, E. M. (1966). *Molec. Pharmac.*, **2**, 465.
Halkerston, I. D. K., Eichhorn, J. and Hechter, O. (1961). *J. biol. Chem.*, **236**, 374.
Hall, L. H., Kier, L. B. and Murray, W. J. (1975). *J. Pharm. Sci.*, **64**, 1974.
Hammer, W. and Sjöqvist, F. (1967). *Life Sci.*, **6**, 1895.
Hansch, C. and Clayton, J. M. (1973). *J. Pharm. Sci.*, **62**, 1.
Hansch, C. and Dunn, W. J. (1972). Ibid., **61**, 1.
Hansch, C. and Fujita, T. (1964). *J. Am. chem. Soc.*, **86**, 1616.
Hansch, C., Kutter, E. and Leo, A. (1969). *J. med. Chem.*, **12**, 746.
Hansch, C., Muir, R. M., Fujita, T., Maloney, P. P., Geiger, F. and Streich, M. (1963). *J. Am. chem. Soc.*, **85**, 2817.
Hansch, C., Quinlan, J. E. and Lawrence, G. L. (1968). *J. org. Chem.*, **33**, 347.
Hansch, C. and Steward, A. R. (1964). *J. med. Chem.*, **7**, 691.
Hansch, C., Steward, A. R., Anderson, S. M. and Bentley, D. L. (1967). Ibid., **11**, 1.
Hansch, C., Steward, A. R. and Iwasa, J. (1965a). *Molec. Pharmac.*, **1**, 87.
— (1965b). *J. med. Chem.*, **8**, 868.
Hansen, L. (1940). *J. Lab. clin. Med.*, **25**, 669.
Harbers, E. and Müller, W. (1962). *Biochim. biophys. Res. Commun.*, **7**, 107.
Hare, M. L. C. (1928). *Biochem. J.*, **22**, 968.
Harman, R. E., Meisinger, M. A. P., Davis, G. E. and Kuehl, F. A. (1964). *J. Pharmac, exp. Ther.*, **143**, 215.
Harris, M. (1971). *Triangle*, **10**, 41.
Hartiala, K. J. V., Nänto, V. and Rinne, U. K. (1961). *Acta physiol. scand.*, **53**, 376.
Hartley, B. S. and Kilby, B. A. (1952). *Biochem, J.*, **50**, 672.
Hawkins, D., Pinckard, R. N. and Farr, R. S. (1968). *Science, N. Y.*, **160**, 780.
Hayano, M., Gut, M., Dorfman, R. I., Sebek, O. K. and Peterson, D. H. (1958). *J. Am. chem. Soc.*, **80**, 2336.
Hechter, O. and Halkerston, I. D. K. (1965). *A. Rev. Physiol.*, **27**, 133.
Hewick, D. S. (1967). *Ph.D. Thesis*, University of London.
Heymann, H. and Fieser, L. F. (1948). *J. Pharmac. exp. Ther.*, **94**, 97.
Higuchi, T. and Davis, S. S. (1970). *J. Pharm. Sci.*, **59**, 1376.
Higuchi, W. I., Lau, P. K., Higuchi, T. and Shell, J. W. (1963). Ibid., **52**, 150.
Hill, H. A. O., Pratt, J. M. and Williams, R. J. P. (1969). *Chemy Bul.*, **5**, 156.
Hill, R. K. and Barcza, S. (1966). *Tetrahedron*, **22**, 2889.
Hinderling, P. H., Bres, J. and Garrett, E. R. (1974). *J. Pharm. Sci.*, **63**, 1684.
Hirsch, P. (1918). *Einwirkung der Micro-organisms auf Eiweiskörper*, Berlin. (For Review see *Pharm. Weekblad.* (1919), **56**, 77).
Hitchings, G. H. (1952). *Trans. R. Soc. trop. Med. Hyg.*, **46**, 467.
Hoffman, W. S., Fishbein, W. I. and Andelman, M. B. (1964). *Arch. Envir. Hlth.*, **9**, 387.
Hoffman, R. (1963). *J. chem. Phys.*, **39**, 1397.
Hogenkamp, H. P. C. (1966). *Biochemistry, N. Y.*, **5**, 417.
Holck, H. G. O., Kanan, A. M., Mills, L. M. and Smith, E. L. (1937). *J. Pharmac. exp. Ther.*, **60**, 323.
Holder, G. M., Ryan, A. J., Watson, T. R. and Wiebe, L. I. (1970). *J. Pharm. Pharmac.*, **22**, 375.
Holmdahl, K. H. and Lodin, H. (1959). *Acta radiol.*, **51**, 247.
Holmes, R. and Robins, E. L. (1955). *Br. J. Pharmac. Chemother.*, **10**, 490.

Hospital, M., Busetta, B., Bucourt, R., Weintraub, H. and Baulieu, E. E. (1972). *Molec. Pharmac.*, **8**, 438.
Houston, J. B. and Levy, G. (1975). *J. Pharm. Sci.*, **64**, 607.
Hubbard, R. and Wald, G. (1952). *J. gen. Physiol.*, **36**, 269.
Hucker, H. B., Zacchei, A. G., Cox, S. V., Brodie, D. A. and Cantwell, N. H. R. (1966). *J. Pharmac. exp. Ther.*, **153**, 237.
Hughes, J., Smith, T. W., Kosterlitz, H. W., Fothergill, L. A., Morgan, B. A. and Morris, H. R. (1975). *Nature, Lond.*, **257**, 135.
Hurwitz, J., Furth, J. J., Malamy, M. H. and Alexander, M. (1962). *Proc. natn. Acad. Sci. U. S. A.*, **48**, 1222.
Hutchings, B. L. (1957). *Ciba Foundation Symposium on Chemistry and Biology of Purines*, pp. 177–88, Churchill, London.
Hutson, D. H., Hoadley, E. C., Griffiths, M. H. and Donninger, C. (1970). *J. agric. Fd Chem.*, **18**, 507.
Iba, M. M., Soyka, L. F. and Schulman, M. P. (1975). *Biochem. biophys. Res. Commun.*, **65**, 870.
Ichihara, K., Kusunose, E. and Kusunose, M. (1969a). *Biochim. biophys. Acta*, **176**, 713.
— (1969b). ibid., **176**, 704.
Idéo, G., de Franchis, R., Del Ninno, E. and Dioguardi, N. (1971). *Lancet*, **ii**, 825.
Idson, B. (1975). *J. Pharm. Sci.*, **64**, 901.
Iles, D. H. and Ledwith, A. (1969). *Chem. Commun.*, 364.
Irving, C. C. (1965). *J. biol. Chem.*, **240**, 1011.
— (1971). *The Biological Oxidation of Nitrogen in Organic Molecules, Xenobiotica*, p. 75, eds. Bridges, J. W., Gorrod, J. W. and Parke, D. V., Taylor and Francis, London.
Irwin, G. M., Kostenbauder, H. B., Dittert, L. W., Staples, R., Misher, A. and Swintosky, J. V. (1969). *J. Pharm. Sci.*, **58**, 313.
Itokawa, Y. and Cooper, J. R. (1969). *Biochem. Pharmac.*, **18**, 545.
Iveson, P., Lindup, W. E., Parke, D. V. and Williams, R. T. (1971). *Xenobiotica*, **1**, 79.
Iveson, P., Parke, D. V. and Williams, R. T. (1966). *Biochem. J.*, **100**, 28P.
Jack, D. B., Stenlake, J. B. and Templeton, R. (1971). *J. Pharm. Pharmac.*, **23**, 222S.
— (1972). *Xenobiotica*, **2**, 35.
Jacobs, M. H., Glassman, H. N. and Parpart, A. K. (1935). *J. cell. comp. Physiol.*, **7**, 197.
Janicki, C. A. and Almond, H. R. (1974). *J. Pharm. Sci.*, **63**, 41.
Janssen, P. A. J. and Eddy, N. B. (1969). *J. med. Pharm. Chem.*, **2**, 31.
Jayle, M. F. and Pasqualini, J. R. (1966). *Glucuronic Acid: Free and Combined*, ed. Dutton, G. J., Academic Press, New York.
Jenne, J. W. (1965). *J. clin. Invest.*, **44**, 1992.
Jerina, D. M., Daly, J. W., Witkop, B., Zaltman-Nirenberg, P. and Udenfriend, S. (1968). *Arch. Biochem. Biophys.*, **128**, 176.
Joelsson, I. and Adamsons, K. (1966). *Am. J. Obstet, Gynec.*, **96**, 437.
Johns, D. G., Ianotti, A. T., Sartorelli, A. C., Booth, B. A. and Bertino, J. R. (1964). *Life Sci.*, **3**, 1383.
Johnson, A. W., Mervyn, L., Shaw, N. and Smith, E. L. (1963). *J. chem. Soc.*, 4146.
Johnson, M. K. (1963). *Biochem. J.*, **87**, 9P.
— (1966). Ibid., **98**, 38, 44.
Jondorf, W. R., Maickel, R. P. and Brodie, B. B. (1958). *Biochem. Pharmac.*, **1**, 352.
Jones, R., Ryan, A. J. and Wright, S. E. (1964). *Fd Cos. Toxicol.*, **2**, 447.
Jordan, B. J. and Rance, M. J. (1974). *J. Pharm. Pharmac.*, **26**, 360.
Jordan, D. O. (1960). *The Chemistry of Nucleic Acids*, Butterworths, London.
Juchau, M. R., Niswander, K. R. and Yaffe, J. S. (1968). *Am. J. Obstet. Gynec.*, **100**, 348.
Kalow, W. and Staron, W. (1957). *Can. J. Biochem. Physiol.*, **35**, 1305.
Karig, A. W., Peck, G. E. and Sperandio, G. J. (1973). *J. Pharm. Sci.*, **62**, 811.
Karush, F. (1956). *J. Am. chem. Soc.*, **78**, 5519.
Kaslander, J. (1963). *Biochim. biophys. Acta*, **71**, 730.
Kater, R. M. H., Tobon, F. and Iber, F. L. (1969). *J. Am. med. Ass.*, **207**, 363.
Kato, K. (1953). *J. Antibiot. Japan*, **A6**, **130**, 184.
Kato, K., Ide, H., Hirohata, I. and Fishman, W. H. (1967). *Biochem, J.*, **103**, 647.
Kato, R. (1967). *Biochem. Pharmac.*, **16**, 871.
Kato, R., Chiesara, E. and Vassanelli, P. (1962). Ibid., **11**, 211.
Kato, R., Oshima, T. and Takanaka, A. (1969). *Molec. Pharmac.*, **5**, 487.
Kato, R., Oshima, T. and Tomizawa, S. (1968). *Jap. J. Pharmac.*, **18**, 356.

Kato, R. and Takanaka, A. (1967). Ibid., **17**, 208.
Katz, B. (1973). *Glaxo Volume*, **38**, 27.
Katzung, B. G. and Meyers, F. H. (1965). *J. Pharmac. exp. Ther.*, **149**, 257.
Kauzmann, W. (1959). *Adv. Protein Chem.*, **14**, 1.
Keberle, H., Riess, W. and Hoffman, K. (1963). *Archs int. Pharmacodyn. Thér.*, **142**, 117.
Keberle, H., Riess, W., Schmidt, K. and Hofmann, K. (1963) Ibid., **142**, 125.
Kellie, A. E. (1966). *Biochem. J.*, **100**, 631.
Kersten, W., Kersten, H. and Rauen, H. M. (1960). *Nature, Lond.*, **187**, 60.
Khalil, S. A., Moustafa, M. A., Ebian, A. R. and Motawi, M. M. (1972). *J. Pharm. Sci.*, **61**, 1615.
Khokhar, A. Q. (1971). *Ph.D. Thesis*, Chelsea College, University of London.
Kidd, J. G. and Todd, J. E. (1954). *Proc. Soc. exp. Biol. Med.*, **86**, 781.
Kier, L. B. (1967). *Molec. Pharmac.*, **3**, 487.
— (1968a). Ibid., **4**, 70.
— (1968b). *J. med. Chem.*, **11**, 441.
— (1969). *J. Pharm. Pharmac.*, **21**, 93.
Kier, L. B., Hall, L. H., Murray, W. J. and Randic, M. (1975). *J. Pharm. Sci.*, **64**, 1971.
Kier, L. B., Murray, W. J. and Hall, L. H. (1975). *J. med. Chem.*, **18**, 1272.
Kindler, K. (1928). *Arch. Pharm.*, **266**, 19.
— (1929). Ibid., **267**, 541.
Kini, M. M. and Cooper, J. R. (1961). *Biochem. Pharmac.*, **8**, 207.
Kirk, W. F. (1972). *J. Pharm. Sci.*, **61**, 262.
Kito, Y. and Takezaki, M. (1966). *Nature, Lond*, **211**, 197.
Kitz, R. J. and Kremzner, T. L. (1968). *Molec. Pharmac.*, **4**, 104.
Knowles, J. A. (1965). *J. Pediat.*, **66**, 1068.
Knox, W. E. and Edwards, S. W. (1955). *J. biol. Chem.*, **216**, 489.
Kondo, S., Okanishi, M., Utahara, R., Maeda, K. and Umezawa, H. (1968). *J. Antibiot., Japan*, **Ser. A 21**, 22.
Kono, M., O'Hara, K. Honda, M. and Mitsuhashi, S. (1969). Ibid., **22**, 603.
Koshland, D. E. (1958). *Proc. natn. Acad. Sci. U.S.A.*, **44**, 98.
— (1963). *Science, N. Y.*, **142**, 1533.
Kováts, E. (1961). *Z. Anal. Chem.*, **181**, 351.
Kraemer, R. J. and Dietrich, R. A. (1968). *J. biol. Chem.*, **243**, 6402.
Kraut, J. and Reed, H. J. (1962). *Acta cryst.*, **15**, 747.
Krinsky, N. I. (1958). *Arch. Ophthal., N. Y.*, **60**, 688.
Krueger, H. R. and O'Brien, R. D. (1959). *J. econ. Ent.*, **52**, 1063.
Krupa, R. M. (1966a). *Biochemistry, N. Y.*, **5**, 1983.
— (1966b). Ibid., **5**, 1988.
— (1967). Ibid., **6**, 1183.
Kuntzman, R., Jacobson, M., Levin, W. and Conney, A. H. (1968). *Biochem. Pharmac.*, **17**, 565.
Kusunose, M., Ichihara, K. and Kusunose, E. (1969). *Biochim. biophys. Acta*, **176**, 679.
Lackner, H. (1970). *Tetrahedron Lett.*, 3189.
Ladomery, L. G., Ryan, A. J. and Wright, S. E. (1967). *J. Pharm. Pharmac.*, **19**, 383, 388.
La Du, B. N., Gaudette, L., Trousof, N. and Brodie, B. B. (1955). *J. biol. Chem.*, **214**, 741.
Lajtha, A. and Cohen, S. R. (1972). *Handbook of Neurochemistry*, Vol. VII, p. 534, ed. Lajtha, A., Plenum Press, New York and London.
Lands, A. M., Luduena, F. P. and Tullar, B. F. (1954). *J. Pharmac. exp. Ther.*, **111**, 469.
Langecker, H. and Schulz, E. (1954). *Naunyn-Schmiedeberg's Arch. exp. Path. Pharmak.*, **221**, 160.
Langley, P. F., Lewis, J. D., Mansford, K. R. L. and Smith, D. (1966). *Biochem. Pharmac.*, **15**, 1821.
Lawley, P. D. and Brookes, P. (1959). *Rep. Br. Emp. Cancer Campn.*, **37**, 68.
Laycock, H. H. and Mulley, B. A. (1970). *J. Pharm. Pharmac.*, **22**, 157S.
Lee, G. E., Wragg, W. R., Corne, S. J., Edge, N. D. and Reading, H. W. (1958). *Nature, Lond.*, **181**, 1717.
Le Fevre, P. G. (1961). *Pharmac. Rev.*, **13**, 39.
Lehman, H. and Ryan, E. (1956). *Lancet*, **ii**, 124.
Lenhert, P. G. and Hodgkin, D. C. (1961). *Nature, Lond.*, **192**, 937.
Leo, A., Hansch, C. and Elkins, D. (1971). *Chem. Rev.*, **71**, 525.
Leonard, N. J. and Hauck, F. P. (1957). *J. Am. chem. Soc.*, **79**, 5279.
Lerman, L. S. (1961). *J. molec. Biol.*, **3**, 18.
— (1963). *Proc. natn. Acad. Sci. U.S.A.*, **49**, 94.

Leuders, K. K. and Kuff, E. L. (1967). *Archs. Biochem. Biophys.*, **120**, 198.
Levenberg, B., Melnick, I. and Buchanan, J. M. (1957). *J. biol. Chem.*, **225**, 163.
Leventer, L. L., Buchanan, J. L., Ross, J. E. and Tapley, D. F. (1965). *Biochim. biophys. Acta*, **110**, 428.
Levine, R. M. and Clark, B. B. (1955). *J. Pharmac. exp. Ther.*, **113**, 272.
Levvy, G. A. and Conchie, J. (1966). *Gluruconic Acid: Free and Combined*, ed. Dutton, G. J., Academic Press, London.
Levvy, G. A. and Marsh, C. A. (1959). *Adv. Carbohydrate Chem.*, **14**, 381.
Levy, G. (1967). *J. Pharm. Sci.*, **56**, 1687.
— (1973). *Proceedings of Symposium British Pharmacological Society on Biological Effects of Drugs in Relation to their Plasma Concentration*, eds. Davies, D. S. and Pritchard, B. N. C., Macmillan, London.
Levy, G. and Matsuzawa, T. (1967). *J. Pharmac. exp. Ther.*, **156**, 285.
Levy, G. and Yamada, H. (1971). *J. Pharm. Sci.*, **60**, 215.
Levy, H. R., Loewus, F. A. and Vennesland, B. (1957). *J. Am. chem. Soc.*, **79**, 2949.
Levy, H. R., Talalay, P. and Vennesland, B. (1962). *Progress in Stereochemistry*, Vol. 3, pp. 299–349. Butterworths, London.
Lewis, R. J. and Trager, W. F. (1970). *J. clin. Invest.*, **49**, 907.
Lewkowitsch, J. (1883). *Ber. dt. chem. Ges.*, **16**, 1565.
Liebermann, C. and Limpach, L. (1892). Ibid., **25**, 927.
Lindenbaum, J., Mellow, M. H., Blackstone, M. O. and Butler, V. P. (1971). *New Engl. J. Med.*, **285**, 1344.
Lineweaver, H. and Burk, D. (1934). *J. Am. chem. Soc.*, **56**, 658.
London, F. (1930). *Z. Phys.*, **63**, 245.
Long, W. P. and Gallagher, T. F. (1946). *J. biol. Chem.*, **162**, 511.
Longchampt, J. E., Gual, C., Ehrenstein, M. R. and Dorfman, R. I. (1960). *Endocrinology*, **66**, 416.
Luduena, F. P. (1962). *Archs int. Pharmacodyn. Thér.*, **137**, 155.
Luduena, F. P., von Euler, L., Tullar, B. F. and Lands, A. M. (1957). Ibid., **111**, 392.
Lukus, D. S. and DeMartino, A. G. (1969). *J. clin. Invest.*, **48**, 1041.
Lundquist, F., Fugmann, U., Rasmussen, H. and Svendsen, I. (1962). *Biochem. J.*, **84**, 281.
McCarthy, R. D. (1964). *Biochim. biophysics Acta*, **84**, 74.
McChesney, E. W. and Hoppe, J. O. (1954). *Archs. int. Pharmacodyn. Ther.*, **99**, 127.
— (1956). Ibid., **105**, 306.
MacDonald, M. G., Robinson, D. S., Sylwester, D. and Jaffé, J. J. (1969). *Clin. pharmac. Ther.*, **10**, 80.
McElvain, S. M. and Carney, T. P. (1946). *J. Am. chem. Soc.*, **68**, 2592.
McGilveray, I. G., Matlok, G. L. and Hossie, R. D. (1971). *J. Pharm. Pharmac.*, **23**, 246S.
McKenzie, A. (1902). *J. chem. Soc.*, **81**, 1402.
McLagan, N. F. and Wilkinson, J. H. (1954). *Biochem. J.*, **56**, 211.
McLean, A. E. M. and Verschuuren, H. G. (1969). *Br. J. exp. Path.*, **50**, 22.
McMahon, R. E., Culp, H. W., Mills, J. and Marshall, F. J. (1963). *J. med. Chem.*, **6**, 343.
McMahon, R. E., Culp, H. W. and Occolowitz, J. C. (1969). *J. Am. chem. Soc.*, **91**, 3389.
McMahon, R. E. and Easton, N. R. (1961). *J. med. Pharm. Chem.*, **4**, 67, 437.
McMahon, R. E. and Sullivan, H. R. (1964). *Life Sci.*, **3**, 1167.
McMenamy, R. H. and Oncley, J. L. (1958). *J. biol. Chem.*, **233**, 1436.
Mahler, H. R. and Mehrotra, B. D. (1963). *Biochim, biophys. Acta*, **68**, 211.
Main, A. R. (1964). *Science, N. Y.*, **144**, 992.
Malkinson, F. D. and Ferguson, E. H. (1955). *J. invest. Derm.*, **25**, 281.
March, C. H. and Elliott, H. W. (1954). *Proc. Soc. exp. Biol. Med.*, **86**, 494.
Marchant, B. and Alexander, W. D. (1972). *Endocrinology*, **91**, 747.
Marcucci, F., Mussini, E., Fanelli, R. and Garattini, S. (1970). *Biochem. Pharmac.*, **19**, 1847.
Maren, T. H., Mayer, E. and Wadsworth, B. C. (1954). *Bull. Johns Hopkins Hosp.*, **95**, 199.
Mark, L. C., Kayden, H. J., Steele, J. M., Cooper, J. R., Berlin, I., Rovenstine, E. A. and Brodie, B. B. (1951). *J. Pharmac. exp. Ther.*, **102**, 5.
Marshall, E. K., Cutting, W. C. and Emerson, K. (1937). *Science, N. Y.*, **85**, 202.
Marshall, I. G., Murray, J. B., Smail, G. A. and Stenlake, J. B. (1967). *J. Pharm. Pharmac.*, **19**, 53.
Martin, C. M., Rubin, M., O'Malley, W. E., Garagusi, W. E. and McCauley, C. F. (1968). *J. Am. med. Ass.*, **205**, 23.

Martin, P. J., Martin, J. V. and Goldberg, D. M. (1975). *Br. med. J.*, **i**, 17.
Martin, W. R. (1967). *Pharmac. Rev.*, **19**, 443.
Martin, Y. C. and Hackbarth, J. J. (1976). *J. Pharm. Sci.*, **19**, 1033.
Marvel, J. R., Schlichting, D. A., Denton, C., Levy, E. J. and Cahn, M. M. (1964). *J. invest. Derm.*, **42**, 197.
— (1964). Ibid., **42**, 203.
Mason, H. S. (1957). *Adv. Enzymol.*, **19**, 79.
Masri, M. S., Booth, A. N. and DeEds, F. (1962). *Biochim. biophys. Acta*, **65**, 495.
Masri, M. S., Robbins, D. J., Emerson, O. H. and DeEds, F. (1964). *Nature, Lond.*, **202**, 878.
Masters, B. S. S. and Ziegler, D. M. (1971). *Archs Biochem. Biophys.*, **145**, 358.
Matsushiro, T. (1965). *Tohoku J. exptl. Med.*, **85**, 330.
Mayer, S. E., Maickel, R. P. and Brodie, B. B. (1957). *J. Pharmac. exp. Ther.*, **119**, 35.
Maynert, E. W. and Dawson, J. M. (1952). *J. biol. Chem.*, **195**, 389.
Maxwell, E. S., Kalckar, H. M. and Strominger, J. L. (1956). *Archs Biochem. Biophys.*, **65**, 2.
Mazel, P., Henderson, J. F. and Axelrod, J. (1964). *J. Pharmac. exp. Ther.*, **143**, 1.
Meadows, D. H. and Jardetzky, O. (1968). *Proc. natn. Acad. Sci. U.S.A.*, **61**, 406.
Meadows, D. H., Roberts, G. C. K. and Jardetzky, O. (1969). *J. molec. Biol.*, **45**, 491.
Merril, C. R., Snyder, S. H. and Bradley, D. F. (1966). *Biochim. biophys Acta*, **118**, 316.
Merritt, A. D. and Tomkins, G. M. (1959). *J. biol. Chem.*, **234**, 2778.
Mesley, R. J. (1968). *J. Pharm. Pharmac.*, **20**, 877.
— (1971). Ibid., **23**, 687.
Mezei, M. and Ryan, K. J. (1972). *J. Pharm. Sci.*, **61**, 1329.
Michaelis, L. and Menten, M. L. (1913). *Biochem. Z.*, **49**, 333.
Milborrow, B. V. and Williams, D. A. (1968). *Pl. Physiol.*, **21**, 902.
Miller, C. S., Gordon, J. T. and Engelhardt, E. L. (1953). *J. Am. chem. Soc.*, **75**, 6086.
Miller, J. H. McB. (1972). *Ph.D. Thesis*, Strathclyde University.
Miller, S. L. (1961). *Proc. natn. Acad. Sci. U.S.A.*, **47**, 1515.
Mitoma, C., Yasuda, D. M., Tagg, J. and Tanabe, M. (1967). *Biochim. biophys. Acta*, **136**, 566.
Mizushima, S., Simanouti, T., Nagakura, S., Kuratani, K., Tsuboi, M., Baba, H. and Fujioka, O. (1950). *J. Am. chem. Soc.*, **72**, 3490.
Moldave, K. and Meister, A. (1957a). *Biochim. biophys-Acta*, **25**, 434.
— (1957b). Ibid., **24**, 654.
Monod, J., Changeux, J-P. and Jacob, F. (1963). *J. molec. Biol.*, **6**, 306.
Monod, J., Wyman, J. and Changeux, J. P. (1965). Ibid., **12**, 88.
Moody, J. P. and Williams, R. T. (1962). *Biochem. J.*, **85**, 4P.
Moore, B. and Roaf, H. E. (1904). *Proc. R. Soc.*, **B.73**, 382.
Morato, T., Hayano, M., Dorfman, R. I. and Axelrod, L. R. (1961). *Biochem. biophys. Res. Commun.*, **6**, 334.
Morris, C. R., Andrew, L. V., Whichard, L. P. and Holbrook, D. J. (1970). *Molec. Pharmac.*, **6**, 240.
Morton, R. A. and Pitt, G. A. J. (1955). *Biochem. J.*, **59**, 128.
Moustafa, M. A., Ebian, A. R., Khalil, S. A. and Motawi, M. M. (1971). *J. Pharm. Pharmac.*, **23**, 868.
Moya, F. and Kvisselgaard, N. (1961). *Anaesthesiology*, **22**, 1.
Mueller, G. C. and Miller, J. A. (1950). *J. biol. Chem.*, **185**, 145.
— (1953). Ibid., **202**, 579.
Müller, W. and Crothers, D. M. (1968). *J. molec. Biol.*, **35**, 251.
Muldoon, T. G. and Westphal, U. (1967). *J. biol. Chem.*, **242**, 5636.
Mulliken, R. S. (1952). *J. Am. chem. Soc.*, **74**, 811; *J. phys. Chem.*, **56**, 801.
Mullins, J. D. and Macek, T. J. (1960). *J. Am. pharm. Ass. Sci. Edn.*, **49**, 245.
Murray, W. J., Hall, L. H. and Kier, L. B. (1975). *J. Pharm. Sci.*, **64**, 1978.
Murray, W. J., Kier, L. B. and Hall, L. H. (1976). *J. med. Chem.*, **19**, 573.
Nachmansohn, D. (1952). *Bull. Soc. Chim. biol.*, **34**, 447.
Nagashima, R., Levy, G. and O'Reilly, R. A. (1968). *J. Pharm. Sci.*, **57**, 1888.
Nagwekar, J. B. and Muangnoicharoen, N. (1973). Ibid., **62**, 1439.
Nakanishi, K., Kasai, H., Cho, H., Harvey, R. G., Jeffrey, A. M., Jennette, K. W. and Weinstein, I. B. (1977). *J. Am. chem. Soc.*, **99**, 258.
Narahashi, T., Yamada, M. and Frazier, D. T. (1969). *Nature, Lond.*, **223**, 748.
Nathans, D. (1964). *Proc. natn. Acad. Sci. U.S.A.*, **51**, 585.

Neher, R., Riniker, B., Rittel, W. and Zuber, H. (1968). *Helv. chim. Acta*, **51**, 1900.
Nery, R. (1971a). *The Biological Oxidation of Nitrogen in Organic Molecules, Xenobiotica*, p. 27, eds. Bridges, J. W., Gorrod, J. W. and Parke, D. V., Taylor and Francis, London.
— (1971b). *Biochem. J.*, **122**, 311.
Neuberg, C. and Wahlgemuth, J. (1902). *Hoppe-Seyler's Z. physiol. Chem.*, **35**, 41.
Neunhoeffer, O. (1970). *Z. Naturf.* **B.**, **25**, 299.
Nicholls, P. (1968). *Structure and Function of Cytochromes*, p. 76, eds. Okuniski, K., Kamen, M. D. and Sekuzu, I., University of Tokyo Press, Tokyo.
Nose, Y. and Lipmann, F. (1958). *J. biol. Chem.*, **233**, 1348.
Nunes, M. A. and Brochmann-Hanssen, E. C. (1974). *J. Pharm. Sci.*, **63**, 716.
O'Brien, R. L. and Hahn, F. E. (1965). *Antimicrobial Agents and Chemotherapy*, p. 315.
O'Gorman, J. M., Shand, W. and Schomaker, V. (1950). *J. Am. chem. Soc.*, **72**, 4222.
Ogston, A. G. (1948). *Nature, Lond.*, **162**, 963.
Okanishi, M., Kondo, S., Utahara, R. and Umezewa, H. (1968). *J. Antibiot., Japan, Ser. A.*, **21**, 13.
Oliverio, V. T. and Davidson, J. D. (1962). *J. Pharmac. exp. Ther.*, **137**, 76.
O'Malley, K., Crooks, J., Duke, E. and Stevenson, I. H. (1971). *Br. med. J.*, **3**, 607.
O'Malley, K., Stevenson, I. H. and Crooks, J. (1972). *Clin. Pharmac. Ther.*, **13**, 552.
Orcutt, F. S. and Seevers, M. H. (1937). *J. Pharmac. exp. Ther.*, **59**, 206.
Orloff, J. and Berliner, R. W. (1956). *J. clin. Invest.*, **35**, 223.
Orrenius, S. (1965). *J. Cell Biol.*, **26**, 725.
Orrenius, S. and Ernster, L. (1964). *Biochem. biophys. Res. Commun.*, **16**, 60.
Orrenius, S. and Thor. H. (1969). *Eur. J. Biochem.*, **9**, 415.
Orth, H. D. and Brümmer, W. (1972). *Angew. Chem. Int. Edn.* **11**, 249.
Ozanne, B., Benveniste, R., Tipper, D. and Davies, J. (1969). *J. Bact.*, **100**, 1144.
Park, J. T. (1952). *J. biol. Chem.*, **194**, 877, 885, 897.
Park, J. T. and Strominger, J. L. (1957). *Science, N. Y.*, **125**, 99.
Parke, D. V. (1961). *Biochem. J.*, **78**, 262.
— (1968). *The Biochemistry of Foreign Compounds*, Pergamon, Oxford.
Parli, C. J., Wang, N. and McMahon, R. E. (1971a). *Biochem. biophys. Res. Commun.*, **43**, 1204.
— (1971b). *J. biol. Chem.*, **246**, 6953.
Partington, P., Feeney, J. and Bergen, A. S. V. (1972). *Molec. Pharmac.*, **8**, 269.
Pasteur, L. (1860). *Researches on the Molecular Asymmetry of Natural Organic Products*, Alembic Club Reprints No. 14, Edinburgh, 1897. (See also Radot, 1922.)
Paton, W. D. M. (1961). *Proc. R. Soc.*, **B**, **154**, 21.
Patrick, M. J., Tilstone, W. J. and Reavey, P. (1972). *Lancet*, **i**, 542.
Pauling, L. (1961). *Science, N. Y.*, **134**, 15.
Pauling, L. and Corey, R. B. (1951). *Proc. natn. Acad. Sci. U.S.A.*, **37**, 251, 729.
Pauling, L. and Pressman, D. (1945). *J. Am. chem. Soc.*, **67**, 1003.
Pauling, L. C. (1956). *Enzymes. Units of Biological Structure and Function*, ed. Goebler, O. H., Academic Press, New York.
Peacocke, A. R. and Skerrett, J. N. H. (1956). *Trans. Faraday Soc.*, **52**, 261.
Pearson, R. G. (1963). *J. Am. chem. Soc.*, **85**, 3533.
— (1966). *Science, N. Y.*, **151**, 172.
— (1968). *J. chem. Ed.*, **45**, 581, 643.
Pedersen, L., Hoskins, R. E. and Cable, H. (1971). *J. Pharm. Pharmac.*, **23**, 216.
Penniston, J. T., Beckett, L., Bentley, D. L. and Hansch, C. (1969). *Molec. Pharmac.*, **5**, 333.
Perrin, J. H. and Vallner, J. J. (1970). *J. Pharm. Pharmac.*, **22**, 758.
Peters, J. H., Gordon, G. R. and Brown, P. A. (1965). *Life Sci.*, **4**, 99.
Peters, J. H., Miller, K. S. and Brown, P. A. (1965). *J. Pharmac. exp. Ther.*, **150**, 298.
Peters, R. A. (1936). *Nature, Lond.* **138**, 327.
— (1948). *Br. med. Bull.*, **5**, 313.
Peters, R. A., Wakelin, R. W., Rivett, D. E. A. and Thomas, L. C. (1953). *Nature, Lond.*, **171**, 1111.
Peterson, R. E., Wyngaarden, J. B., Guerra, S. L., Brodie, B. B. and Bunin, J. J. (1955). *J. clin. Invest.*, **34**, 1779.
Petten, G. R. V., Hirsch, G. R. and Cherrington, A. D. (1968). *Can. J. Biochem. Physiol.*, **46**, 1057.
Pettit, F. H., Orme-Johnson, W. H. and Ziegler, D. M. (1964). *Biochem. biophys. Res. Commun.*, **16**, 444.
Pfeiffer, R. R., Yang, K. S. and Tucker, M. A. (1970). *J. Pharm. Sci.*, **59**, 1809.
Piller, M. and Bernstein, A. (1955). *Schweiz. med. Wschr.*, **85**, 104.

Pinckard, R. N., Hawkins, D. and Farr, R. S. (1968). *Nature, Lond.*, **219**, 68.
Pinder, R. M., Buxton, D. A. and Green, D. M. (1971). *J. Pharm. Pharmac.*, **23**, 995.
Poate, H. R. G. (1944). *Lancet*, **ii**, 238.
Poincelot, R. P., Millar, P. M., Kimbel, R. L. and Abrahamson, E. W. (1969). *Nature, Lond.*, **221**, 256.
Poland, A., Smith, D., Kuntzman, R., Jacobson, M. and Conney, A. H. (1970). *Clin. Pharmac. Ther.*, **11**, 724.
Pratt, J. M. (1964). *J. chem. Soc.*, 5154.
— (1972). *Inorganic Chemistry of Vitamin B_{12}*, Academic Press, New York.
Pressman, D., Grossberg, A. L., Pence, L. H. and Pauling, L. (1946). *J. Am. chem. Soc.*, **68**, 250.
Puca, G. A. and Bresciani, F. (1969). *Nature, Lond.*, **223**, 745.
Quay, W. B. (1965). *Life Sci.*, **4**, 983.
Quinn, G. P. Axelrod, J. and Brodie, B. B. (1958). *Biochem. Pharmac.*, **1**, 152.
Randall, L. O. and Lehmann, G. (1948). *J. Pharmac. exp. Ther.*, **93**, 314.
— (1950). Ibid., **99**, 163.
Randic, M. (1975). *J. Am. chem. Soc.*, **97**, 6609.
Raphael, R. A. and Stenlake, J. B. (1953). *Chemy Ind.*, 1286.
Redetzki, H. M., Redetzki, J. E. and Elias, A. L. (1966). *Biochem. Pharmac.*, **15**, 425.
Reich, E., Goldberg, I. H. and Rabinowitz, M. (1962). *Nature, Lond.*, **196**, 743.
Reid, R. E. (1972). *Ph.D. Thesis*, Chelsea College, University of London.
Rekker, R. F., Engel, D. C. J. and Nys, G. G. (1972). *J. Pharm. Pharmac.*, **24**, 589.
Remmer, H. (1959). *Arch. exp. Path. Pharmak.*, **235**, 279.
Remmer, H. and Merker, H. J. (1965). *Ann. N. Y. Acad. Sci.*, **123**, 79.
Remmer, H., Schenkman, J. B., Estabrook, R. W., Sasame, H. A., Gillette, J. R., Narasimhulu, S., Cooper, D. Y. and Rosenthal, O. (1966). *Molec. Pharmac.*, **2**, 187.
Renson, J., Weissbach, H. and Udenfriend, S. (1965). Ibid., **1**, 145.
Richter, J. J. and Wainer, A. (1971). *J. Neurochem.*, **18**, 613.
Riegelman, S., Loo, J. C. K. and Rowland, M. (1968). *J. Pharm. Sci.*, **57**, 117.
Riniker, B. von., Neher, R., Maier, R., Kahnt, F. W., Byfield, P. G. H., Gudmundsson, T. V., Galante, L. and MacIntyre, I. (1968). *Helv. chim. Acta*, **51**, 1738.
Robbins, P. W. and Lipmann, F. (1957). *J. biol. Chem.*, **229**, 837.
Robinson, D. and Williams, R. T. (1958). *Biochem. J.*, **68**, 23P.
Roche, J., Michel, R. and Tata, J. (1953). *Biochim. biophys. Acta*, **11**, 543.
Roe, F. J. C. (1957). *Cancer Res.*, **17**, 64.
Rogers, M. J. and Strittmatter, P. (1973). *J. biol. Chem.*, **248**, 800.
— (1975). Ibid., **250**, 5713.
Rollo, I. M. (1955). *Br. J. Pharmac. Chemother.*, **10**, 208.
Rose, I. A. (1958). *J. Am. chem. Soc.*, **80**, 5835.
Rosenthal, S. M. (1926). *J. biol Chem.*, **70**, 129.
Rosner, F., Rubin, H. and Parise, F. (1971). *Cancer Chemother. Rep.*, **55**, 167.
Ross, W. C. J. (1958). *Ann. N. Y. Acad. Sci.*, **68**, 669.
— (1962). *Biological Alkylating Agents*, Butterworths, London.
Roy, A. B. (1960). *Biochem. J.*, **74**, 49.
Roze, U. and Strominger, J. L. (1966). *Molec. Pharmacol.*, **2**, 92.
Rubin, E. and Lieber, C. S. (1968). *Science, N. Y.*, **162**, 690.
Rudofsky, S. and Crawford, J. S. (1966). *Pharmacologist*, **8**, 181.
Ryan, A. J., Roxon, J. J. and Sivayavirojana, A. (1968). *Nature, Lond.*, **219**, 854.
Ryan, A. J., Welling, P. G. and Roxon, J. J. (1969). *Fd Cos. Toxicol.*, **7**, 297.
Ryan, A. J. and Wright, S. E. (1962). *Nature, Lond.*, **195**, 1009.
Ryan, K. J. (1959). *J. biol. Chem.*, **234**, 268.
Salem, L. (1962). *Nature, Lond.*, **193**, 476.
Sandler, M. and Youdim, M. B. H. (1972). *Pharmac. Rev.*, **24**, 331.
Sanger, F. (1963). *Proc. chem. Soc.*, 76.
Sasame, H. A. and Gillette, J. R. (1969). *Molec. Pharmac.*, **5**, 123.
Sato, T., Fukuyama, T., Suzuki, T. and Yoshikawa, H. (1963). *J. Biochem., Tokyo*, **53**, 23.
Saunders, F. J. (1966). *Proc. Soc. exp. Biol. Med.*, **123**, 303.
Scatchard, G. (1949). *Ann. N. Y. Acad. Sci.*, **51**, 660.
Scatchard, G., Coleman, J. S. and Shen, A. L. (1957). *J. Am. chem. Soc.*, **79**, 12.
Scatchard, G., Scheinberg, I. H. and Armstrong, S. H. (1950). ibid., **72**, 535, 540.

Schaffer, N. K., May, S. C. and Summerson, W. H. (1954). *J. biol. Chem.*, **206**, 201.
Schanker, L. S. (1962). *Pharmac. Rev.*, **14**, 501.
Schanker, L. S., Shore, P. A., Brodie, B. B. and Hogben, C. A. M. (1957). *J. Pharmac. exp. Ther.*, **120**, 528.
Scheline, R. R. (1966). *J. Pharm. Pharmac.*, **18**, 664.
— (1968). *Acta Pharmac. tox.*, **26**, 332.
Schellenberg, K. A. and Coatney, G. R. (1961). *Biochem. Pharmac.*, **6**, 143.
Schellman, J. A., Lumry, R. and Samuels, L. T. (1954). *J. Am. chem. Soc.*, **76**, 2808.
Schepartz, B. and Gurin, S. (1949). *J. biol. Chem.*, **180**, 663.
Schmid, K., Cornu, F., Inhof, P. and Keberle, H. (1964). *Schweiz. med. Wschr.*, **94**, 235.
Schmid, R., Buckingham, S., Mendilla, G. A. and Hammaker, L. (1959). *Nature, Lond.*, **183**, 1823.
Schoenwald, R. D. and Ward, R. L. (1976). *J. Pharm. Sci.*, **65**, 777.
Schwarzenbach, G. (1936). *Z. phys. Chem.*, **A 176**, 133.
Searl, R. O. and Pernarowski, M. (1967). *Can. med. Ass. J.*, **96**, 1513.
Sgent-Györgyi, A. E. (1960). *Introduction to a Submolecular Biology*, Academic Press, New York.
Shaw, J. E., Chandrasekaran, S. K. and Taskovich, L. (1975). *Pharm. J.*, **215**, 325.
Shaw, T. R. D., Carless, J. E., Howard, M. R. and Raymond, K. (1973). *Lancet*, **ii**, 209.
Shaw, W. V. (1971). *Advances in Pharmacology and Chemotherapy*, Vol. 9, p. 131, Academic Press, London.
Shore, P. A., Brodie, B. B. and Hogben, C. A. M. (1957). *J. Pharmac. exp. Ther.*, **119**, 361.
Sidbury, J. B., Bynum, J. C. and Fetz, L. L. (1953). *Proc. Soc. exp. Biol. Med.*, **82**, 226.
Sieber, P., Brugger, M., Kamber, B., Riniker, B. and Rittel, W. (1968). *Helv. chim. Acta*, **51**, 2057.
Singh, T., Stein, R. G. and Biel, J. H. (1969). *J. med. Chem.*, **12**, 368.
Sisodia, C. S. and Stowe, C. M. (1964). *Ann. N. Y. Acad. Sci.*, **111**, 650.
Skipski, V. P., Barclay, M., Barclay, R. K., Fetzner, V. A., Good, J. J. and Archibald, F. M. (1967). *Biochem. J.*, **104**, 340.
Slaunwhite, W. R. and Sandberg, A. A. (1959). *J. clin. Invest.*, **38**, 384.
Smith, C. C. (1969). *Ann. N. Y. Acad. Sci.*, **162**, 604.
Smith, H. W. (1951). *The Kidney: Structure and Function in Health and Disease*, p. 53, Oxford University Press, New York.
Smith, J. T. and Wood, J. L. (1959). *J. biol. Chem.*, **234**, 3192.
Soloway, A. H., Whitman, B. and Messer, J. R. (1960). *J. Pharmac, exp. Thér.*, **129**, 310.
Solter, A. W. (1965). *J. Pharm. Sci.*, **54**, 1755.
Spinks, A. and Young, E. H. P. (1958). *Nature, Lond.*, **181**, 1397.
Spinks, A., Young, E. H. P., Farrington, J. A. and Dunlop, D. (1958). *Br. J. Pharmac. Chemother.*, **13**, 501.
Stearns, B., Losee, K. A. and Bernstein, J. (1963). *J. med. Chem.*, **6**, 201.
Steeno, O., Heyns, W., van Baelen, H. and de Moor, P. (1968). *Annls Endocr.*, **29**, 141.
Stekol, J. A. (1939). *J. biol. Chem.*, **128**, 199.
Stenlake, J. B. (1953). *Chemy Ind.*, 1089.
— (1968a). *Ann. Pharm. Franc.*, **26**, 185.
— (1968b). Unpublished information.
— (1974). *The Analyst, Lond.*, **99**, 824.
— (1975). *Harrison Memorial Lecture, Pharm. J.*, **215**, 533.
Stenlake, J. B. and Skellern, G. G. (1972). Unpublished information.
Stenlake, J. B., Templeton, R. and Taylor, D. C. (1968). *J. Pharm. Pharmac.*, **20**, 142S.
Stenlake, J. B., Williams, W. D., Davidson, A. G., Downie, W. and Whaley, K. (1969). Ibid., **21**, 451.
Stenlake, J. B., Williams, W. D. and Skellern, G. G. (1973). *Xenobiotica*, **3**, 121.
Stephenson, R. P. (1956). *Br. J. Pharmac. Chemother.*, **11**, 379.
Sternglanz, H., Yielding, K. L. and Pruitt, K. M. (1969). *Molec. Pharmac.*, **5**, 376.
Stevenson, I. H., Browning, M., Crooks, J. and O'Malley, K. (1972). *Br. med. J.*, **4**, 322.
Stone, J. V. and Townshend, A. (1972). *Chem. Commun.*, 502.
Storey, I. D. E. and Dutton, G. J. (1955). *Biochem. J.*, **59**, 279.
Strange, R. E. (1956). Ibid., **64**, 23P.
Strominger, J. L., Kalckar, H. M., Axelrod, J. and Maxwell, E. S. (1954). *J. Am. chem. Soc.*, **76**, 6411.
Strominger, J. L., Maxwell, E. S., Axelrod, J. and Kalckar, H. M. (1957). *J. biol. Chem.*, **224**, 79.
Suwalsky, M., Traub, W. and Shmueli, U. (1969). *J. molec. Biol.*, **42**, 363.

Symes, A. L., Missala, K. and Sourkes, T. L. (1971). *Science, N. Y.*, **174**, 153.
Szasz, G. J., Sheppard, N. and Rank, D. H. (1948). *J. chem. Phys.*, **16**, 704.
Tabor, H. (1962). *Biochemistry, N. Y.*, **1**, 496.
Takagi, M. (1963). *Biochim. biophys. Acta*, **66**, 328.
Tanaka, C. and Cooper, J. R. (1968). *J. Histochem. Cytochem.*, **16**, 362.
Tanford, C. (1962). *J. Am. chem. Soc.*, **84**, 4240.
Tanford, C., Swanson, S. A. and Shore, W. S. (1955). Ibid., **77**, 6414.
Taurog, A., Briggs, F. N. and Chaikoff, I. L. (1951). *J. biol. Chem.*, **191**, 29.
Taylor, K. M. and Snyder, S. H. (1972). *Molec. Pharmac.*, **8**, 300.
Taylor, W. J. (1948). *J. chem. Phys.*, **16**, 257.
Teorell, T. (1937). *Archs intern. Pharmacodyn. Thér.*, **57**, 205.
Teresi, J. D. and Luck, J. M. (1952). *J. biol. Chem.*, **194**, 823.
Terry, D. H. and Shelanski, H. A. (1952). *Modern Sanitation*, **4**, 61.
Theorell, H. (1938). *Biochem. Z.*, **298**, 242.
Theorell, H. and Bonnichsen, R. (1951). *Acta chem. scand.*, **5**, 1105.
Thomas, R. C. and Ikeda, G. J. (1966). *J. med. Pharm. Chem.*, **9**, 507.
Thorpe, J. M. (1964). *Absorption and Distribution of Drugs*, p. 64. ed. Binns, T. E., E. & S. Livingstone, Edinburgh.
Thyrum, P. T., Luchi, R. J. and Thyrum, E. M. (1969). *Nature, Lond.*, **223**, 747.
Timmis, G. M. (1951). *12th Int. Congr. Pure and Appl. Chem. Abst.*, 334.
— (1960). *Biochem. Pharmac.*, **4**, 49.
Tipper, D. J. and Strominger, J. L. (1968). *J. biol. Chem.*, **243**, 3169.
Tipton, K. F. (1967). *Biochim. biophys. Acta*, **135**, 910.
Tomita, K., Cha, C-J. M. and Lardy, H. A. (1964). *J. biol. Chem.*, **239**, 1202.
Tomlinson, G. A. and Yaffé, S. J. (1966). *Biochem. J.*, **99**, 507.
Tsukamoto, H., Yoshimura, H. and Tatsumi, K. (1963). *Chem. Pharm. Bull. Tokyo*, **11**, 421.
Tute, M. S. (1971). *Advances in Drug Research*, Vol. 6, p. 1, eds. Harper, N. J. and Simmonds, A. B., Academic Press, London.
Uehleke, H. (1971). *The Biological Oxidation of Nitrogen in Organic Molecules, Xenobiotica*, p. 15, eds. Bridges, J. W., Gorrod, J. W. and Parke, D. V., Taylor and Francis, London.
Ullrich, V. (1969). *Z. phys. Chem.*, **350**, 357.
Umar, M. T. and Mitchard, M. (1968). *Biochem. Pharmac.*, **17**, 2057.
Umezawa, H., Okanishi, M., Utahara, R., Maeda, K. and Kondo, S. (1967b). *J. Antibiot., Japan, Ser. A.*, **20**, 136.
Usui, T. and Yamasaki, K. (1960). *J. Biochem., Tokyo*, **48**, 226.
Vallee, B. L. and Riordan, J. F. (1969). *A. Rev. Biochem.*, **38**, 733.
van Abbé, N. J., Spearman, R. I. C. and Jarrett, A. (1969). *Pharmaceutical and Cosmetic Products for Topical Administration*, Heinemann, London.
van Baelen, H., Heyns, W., Schonne, E. and de Moor, P. (1968). *Annls. Endocr.* **29**, 153.
Van den Bossche, H. and Janssen, P. A. J. (1967). *Life Sci.*, **6**, 1781.
van Dyke, R. A. (1966). *J. Pharmac. exp. Ther.*, **154**, 364.
van Dyke, R. A. and Chenoweth, M. B. (1965). *Biochem. Pharmac.*, **14**, 603.
Van Oudtshoorn, M. C. B. and Potgieter, F. J. (1972). *J. Pharm. Pharmac.*, **24**, 357.
Varley, A. B. (1968). *J. Am. med. Ass.*, **206**, 1745.
Vesell, E. S. and Page, J. G. (1968). *J. clin. Invest.*, **47**, 2657.
Vickers, C. F. H. (1963). *Archs. Derm. Syph.*, **88**, 20.
Vitali, R., Gardi, R., Falconi, G. and Ercoli, A. (1966). *Steroids*, **8**, 527.
Vree, T. B., Gorgels, J. P. M. C., Muskens, A. Th. M. J. and van Rossum, J. M. (1971). *Clinica chim. Acta*, **34**, 333.
Vree, T. B., Muskens, A. Th. M. J. and Van Rossum, J. M. (1971). *The Biological Oxidation of Nitrogen in Organic Molecules, Xenobiotica*, p. 73, eds. Bridges, J. W., Gorrod, J. W. and Parke, D. V., Taylor and Francis, London.
Wade, O. L. (1970). *Prescribers Journal*, **10**, No. 4, 79.
Waite, M. (1969). *Biochemistry, N. Y.*, **8**, 2536.
Wakabayashi, M., Wotiz, H. H. and Fishman, W. H. (1961). *Biochim. biophys. Acta*, **48**, 198.
Walshe, J. M. (1975). *Br. med. J.*, **3**, 701.
Waser, P. G. (1961). *Pharmac. Rev.*, **13**, 465.
Way, E. L., Swanson, R. and Gimble, A. I. (1947). *J. Pharmac. exp. Ther.*, **91**, 178.
Weber, W. W. and Cohen, S. N. (1967). *Molec. Pharmac.*, **3**, 266.

Weiner, I. M., Levy, R. I. and Mudge, G. H. (1962). *J. Pharmac. exp. Ther.*, **138**, 96.
Weiner, M., Shapiro, S., Axelrod, J., Cooper, J. R. and Brodie, B. B. (1950). *ibid.*, **99**, 409.
Weiner, N. D., Hart, F. and Zografi, G. (1965). *J. Pharm. Pharmac.*, **17**, 350.
Welles, J. S., Root, M. A. and Anderson, R. C. (1961). *Proc. Soc. exp. Biol. Med.*, **107**, 583.
Werk, E. E., MacGee, J. and Sholiton, L. J. (1964). *J. clin. Invest.*, **43**, 1824.
Wessely, F. and Welleba, H. (1941). *Ber. dt. chem. Ges.*, **74**, 777.
Westphal, U. and Ashley, B. D. (1958). *J. biol. Chem.*, **233**, 57.
White, T. A. and Evans, D. A. P. (1968). *Clin. Pharmac. Therap.*, **9**, 80.
Whitecar, J. P., Bodey, G. P., Harris, J. E. and Freireich, E. J. (1970). *New Engl. J. Med.*, **282**, 732.
Wilbrandt, W. and Rosenberg, T. (1961). *Pharmac. Rev.*, **13**, 109.
Wilhelm, M. and Kuhn, R. (1970). *Pharmacopsychiatric-Neuro-Psychopharmacologie*, **3**, 317.
Wilhelm, M. and Schmidt, P. (1969). *Helv. chim. Acta*, **52**, 1385.
Wilkinson, J. H. (1970). *Topics in Medicinal Chemistry*, **3**, 25.
Williams, R. T. (1971). *Ann. N. Y. Acad. Sci.*, **179**, 141.
Williams, R. T., Millburn, P. and Smith, R. L. (1965). Ibid., **123**, 110.
Williams, W. D. (1970). Unpublished observation.
Wilson, I. B. and Bergmann, F. (1950). *J. biol. Chem.*, **185**, 479.
Wilson, I. B., Harrison, M. A. and Ginsburg, S. (1961). Ibid., **236**, 1498.
Wilson, I. B. and Meislich, E. K. (1953). *J. Am. chem. Soc.*, **75**, 4628.
Witkop, B. (1969). *Microsomes and Drug Oxidation*, p. 235, eds. Gillette, J. R., Conney, A. H., Cosmides, G. J., Estabrook, R. W., Fouts, J. R. and Mannering, G. J., Academic Press, New York.
Wohl, A. J. (1970). *Molec. Pharmac.*, **6**, 189, 195.
Wood, G. C. and Graham, A. B. (1972). *Biochim. biophys. Acta*, **276**, 392.
Wood, G. C. and Woodcock, B. G. (1970). *J. Pharm. Pharmac.*, **22**, 60S.
Wood, J. M., Kennedy, F. S. and Rosen, C. G. (1968). *Nature, Lond.*, **220**, 173.
Woodcock, B. G. and Wood, G. C. (1971). *Biochem. Pharmac.*, **20**, 2703.
Wroblewski, F. and Gregory, K. F. (1960). *Proc. int. Congr. clin. Chem.*, **4**, 62.
Wulf, R. J. and Featherstone, R. M. (1957). *Anaesthesiology*, **18**, 97.
Yamada, T., Tipper, D. and Davies, J. (1968). *Nature, Lond.*, **219**, 288.
Yarbro, C. L. (1956). *J. Urol.*, **75**, 216.
Yoshida, H., Kaniike, K. and Namba, J. (1963). *Nature, Lond.*, **198**, 191.
Yoshimura, H., Tsuji, H. and Tsukamoto, H. (1966). *Chem. Pharm. Bull., Tokyo*, **14**, 939.
Ziegler, D. M., Mitchell, C. H. and Jollon, D. (1969). *Microsomes and Drug Oxidation*, p. 173, eds. Gillette, J. R., Conney, A. H., Cosmides, G. J., Estabrook, R. W., Fouts, J. R. and Mannering, G. J., Academic Press, New York.

Index